Mobile Robots

Mobile Robots
Inspiration to Implementation

Second Edition

Joseph L. Jones
Anita M. Flynn
Bruce A. Seiger

A K Peters
Natick, Massachusetts

Editorial, Sales, and Customer Service Office

A K Peters, Ltd.
63 South Avenue
Natick, MA 01760

Library of Congress Cataloging-in-Publication Data

Jones, Joseph L., 1953-
Mobile robots : inspiration to implementation / Joseph L. Jones,
Anita M. Flynn, Bruce A. Seiger. — 2nd ed.
p. cm.
Includes bibliographical references and index.
ISBN 1-56881-097-0
1. Mobile robots. I. Seiger, Bruce A. II. Flynn, Anita M.
III. Title.
TJ211.415.J65 1998 98-42605
629.8'92—dc21

Printed in the United States of America
03 02 01 00 99 10 9 8 7 6 5 4 3 2

From Joseph L. "Joe Lee" Jones:

To my mother, who read me science books when I was a kid.

From Anita M. Flynn:

To my family and the Mobot Lab, for the big dreams
(And all the little tiny ones, too!)

From Bruce A. Seiger:

To the memory of my grandmother, Catharine Bowers,
who gave me more than she could ever know. I appreciate
the encouragement of my wife, Ruth M. Seiger, and my parents,
Herman and Pearl Seiger.

Contents

Preface to the Second Edition

Welcome to the second edition of *Mobile Robots: Inspiration to Implementation.* In the five years since the original publication of our book we have witnessed a transformation in mobile robotics. Once an arcane craft practiced only in a rarefied set of university research labs, mobile robotics is now gaining an expanding following. Robots are regularly featured in TV shows and magazines, instruction in robotics is available at numerous high schools and colleges, dozens of robot contests are held annually around the world, robot clubs have sprung up everywhere, and thousands of web pages are devoted to robots.

As we had hoped and expected, the pace of robotic development has quickened and the achievements of mobile robots have become ever more impressive. Thus, we felt it was high time to update, revise, and expand *Mobile Robots* to reflect this progress.

To the many readers of our first edition, we offer our heart-felt thanks and appreciation. The success of *Mobile Robots: Inspiration to Implementation* has both surprised and gratified us. The book has been reprinted a dozen times and has been translated into Japanese, German, and French. Enthusiasts new to robotics and vetern roboticists alike have responded to our text more positively than we ever imagined. Our goal was to widely disseminate knowledge of how to design and build robots. The achievement of this goal has outpaced our dreams.

Changes

This new edition has several expanded and updated chapters as well as three completely new chapters.

Chapter 2 has been completely revised using Fischer-Technik parts to build the TuteBot. These parts are easier to acquire in single quantities than LEGO components which were used for the original TuteBot.

Chapter 3 has been expanded. It now includes a full explanation of how to add memory-mapped devices to your robot.

The chapter on sensors, Chapter 5, includes new devices and more driver code. Revisions to other chapters have been made as well.

To answer the oft asked question "But what can you *do* with a robot?" we have added a new chapter, Chapter 10. Here a number of engaging projects suitable for Rug Warrior, the robot presented in the book, are described. Some projects even include crucial code segments.

Examples are key to understanding. And there can be few more pertinent examples in robotics than attempts by commercial and research interests to build "real" robots. A second new chapter, Chapter 11, contains examples of this kind.

In the course of our continuing experience in robotics we have seen many examples of less than optimal robot design. A third new chapter, Chapter 12, contains heuristics and advice we hope will be helpful in this regard. Here we list principles that have guided us toward the successful completion of robot projects.

Appendix B has been completely revised. In the first edition, this appendix contained a single all-encompassing, but perhaps confusing program. The revised appendix contains numerous simpler and more easily understood programs for Rug Warrior.

The World Wide Web offers to the robot builder informational resources vastly greater than the ones with which we originally worked. The Yellow Pages, Appendices C-E, one of the most applauded features of the first edition, have been updated and revised and now include web references wherever possible.

Finally, a new appendix, Appenndix F details the large and growing list of robot contests.

We have included the preface to the first edition of *Mobile Robots* for completeness. The reader should note that the chapter references there reflect the original chapter numbering which has changed for this new edition. Also, the Motorola MC68HC811A0 microprocessor which was indicated at the time to be out of production is in fact still available.

Opening the Black Box

Modern technology sometimes succeeds too well at hiding the details of its functionality. In years past, a child could gain an understanding of mechanics by disassembling a wind-up clock. A child of today will find inside the black box of a digital clock only more black boxes. The function of these boxes (integrated circuit chips) is hidden at a level impenetrable to eyes of the curious.

Robots are at a stage of development where they can bring back some of this lost "discoverability." You can still take off the top of the robot, poke around inside, change some things and see what happens. This is a key goal for us: to open the black box and reveal what is inside. And it is for this reason that we strongly encourage you to work with a robot of your own. A deep understanding of the technology can be attained not by reading, but only by doing.

Rug Warrior and Rug Warrior Pro

For the first edition of *Mobile Robots: Inspiration to Implementation* we developed a robot called Rug Warrior. Rug Warrior was a convenient vehicle for us as it gave us a consistent platform on which to hang our examples. We imagined that readers would acquire the components for Rug Warrior and build their own robots from scratch. Toward that end we included in the first edition detailed information about how specific components could be purchased.

Unfortunately, building a robot from scratch is a daunting task especially if you do not have access to a fully-supplied stock room. Many readers who tried to construct a robot in this way found that they learned much more about ordering from vendors, fulfilling minimum purchase requirements, and tracking down scarce components than they learned about robot building. It was the vocal distress of such readers that led our publisher to offer Rug Warrior in kit form.

Recently, a new version of the robot called Rug Warrior Pro™ was developed. Rug Warrior Pro™ is fully compatible with the original version but has increased functionality. The original Rug Warrior, however, is highly embedded in the text. This led to a dilemma in regard to updating the book. Should we simply ignore the advances of Rug Warrior Pro™? Or should we redo every example to conform to the new robot even when the changes are not material to the point under discussion?

We resolved the problem by adopting a two-robot policy. Rug Warrior continues in its original roll as our primary example robot for describing sensor and actuator attachment. Information about Rug Warrior's circuit board and components is provided for those stalwarts who wish to build from scratch. Significant new features implemented in Rug Warrior Pro™ are described separately. These new features need not be lost to builders of Rug Warrior—an inexpensive upgrade module that effectively converts Rug Warrior into Rug Warrior Pro™ is available from the publisher.

Acknowledgments

Many people have read early drafts of this book and offered helpful comments. We would like to thank Colin Angle, Rodney Brooks, Roger Chen, Jill Crisman, CDR H.R. Everett, Dorothy Flynn, Kathleen Flynn, Richard Flynn, Douglas Gage, Mattew Good, Ken Good, Tina Kapur, Ken Livingston, Fred Martin, James McLurkin, Michael Noakes, Lynne Parker, Alison Reid, John Richardson, Rick Shafer, Wendy Taylor, William Wells, Masaki Yamamoto, and Holly Yanco.

We appreciate the interest of Bruce Seiger and Don McAleer and their students at Wellesley High School, who beta-tested the material in the first edition of this book. We would also like to acknowledge and thank Randy Sargent and Fred Martin of the MIT Media Laboratory, who were instrumental in making this book possible, both through their efforts in creating new robot software development tools and in contributing to the actual manuscript. We were fortunate to have the help and encouragement of our publishers, Alice and Klaus Peters, who pushed this book to aim for as wide an audience as possible. We are grateful for the patience, love, and support of Sue, Kate, and Emily during the many days that Daddy was off playing with robots.

The second edition benefited from the comments and suggestions of Phil Veatch, Jim Maddox, and Mark Chiappetta who offered insights into the development of commercial robots.

Finally, thanks to all those Robot Olympians whose enthusiasm and participation in the Robot Olympics inspired and instigated this book.

Preface

The design and construction of mobile robots is as much an art as a science. The intent of *Mobile Robots: Inspiration to Implementation* is to explain the skills involved in a manner amenable to as broad an audience as possible. Our aim is to teach you, the reader, how to build a robot. With the recent wide availability of home computers and the tremendous reductions in costs for microelectronics, building mobile robots with an assortment of sensors and actuators is within the reach of nearly everyone.

This book is designed to appeal to readers on a variety of levels. First, for novices and those eager to jump in and get their hands dirty, there are basic lessons on the tools of the trade and the craft of building things and long appendices of suppliers and distributors of interesting robot parts. Chapter 2 plunges right in and leads the reader through a tutorial design example of possibly the world's simplest robot, but nevertheless a complete system. This is TuteBot (for Tutorial Robot), an obstacle-avoiding robot comprised solely of two motors, two wheels, two bump switches, and a few discrete electronic components. The TuteBot exercise should conjure up a plethora of questions and incite the imagination for many ways to make the robot better and act more intelligently.

With TuteBot as a warmup, we then introduce a more sophisticated robot based on software control, Rug Warrior. The remaining chapters after TuteBot are designed to convey basic knowledge about the building-block technologies that make up a robot: sensors, actuators, a power supply, and an intelligence system. The progression of Chapters 3 through 8 instructs you on how to put together the hardware subsystems of Rug Warrior: microprocessor-controlled sensors and actuators, the mechanics of a locomotion system, and a capable battery supply. Rug Warrior has enough sensors and actuators to enable a richer class of behaviors than TuteBot (such as chasing people, avoiding obstacles, moving towards noises, hiding in the dark, and playing music).

Our purpose is not to publish a cookbook but rather to put together an exposé on enough basic skills so that a generation of enthusiasts will not only widen their imaginations but also have the requisite tools to implement those dreams. This is, to us, the real excitement of robotics.

Chapter 9 of the book is directed at just that issue: How can we put all the pieces together to build truly intelligent systems? As we add more sensors, more actuators, and more software, how do we manage complexity? How do we coerce interesting behaviors to emerge? And in the end, how can such machines solve useful problems for us? We conclude our book in Chapter 10, with some discussion of new directions in artificial intelligence and arising technologies that may take these ideas to the next step.

Although this book is intended to be an exposition on building mobile robots rather than a literature review of the field, we have included some annotated references at the end of each chapter, pointing to sources of further reading or background of concepts mentioned. There is a full bibliography at the end of the book.

Mobile Robots has grown out of research at the MIT Artificial Intelligence (AI) Laboratory under Rodney Brooks and his mobile robot group. The half dozen years that the "mobot" lab has been in existence have seen the birth of a wide variety of artificial creatures: some avoid obstacles, some collect things, a few wander and build maps, several walk and climb over rough terrain and a tiny one hides in dark corners. While the research has focused on the issue of how to organize the "insect-level" intelligence of these mobile robots, we have found that we have also had to do extensive engineering throughout several generations of newly available technology.

In 1989, we staged a Robot Talent Show, transferring much of this technology to the AI Lab as a whole. Students were given kits of parts and computers and were encouraged to pick their own problems and solve them. Vacuum cleaners, laser tag-playing robots, autonomous blimps, and cross-country skiers were a few of the resulting mechanical participants in the talent show. Photographs from that night are included at the end of this section. We put together a robot builder's manual before the event, outlining the basics of building autonomous creatures, and handed it out to all the students. The idea for this book sprang directly from that first manual.

Our expectations and experiences in building mobile robots over the years have not always matched, but the lessons learned have been invaluable and we hope to share these with you. Our method is to give general background in each chapter on how different robot subsystems work and then to ground the discourse in specific examples with a robot we have designed solely for this book as a teaching aid. In this way, specific circuits and bits of code are sprinkled throughout, and readers who follow along can implement their own robots and see them evolve step by step. The complete system is laid out in one place in the appendices at the end of the book. A gives the schematic for Rug Warrior's brain along with all the interface electronics to drive its sensors and actuators and B lists a program that defines Rug Warrior's behaviors. The entire robot has been reduced to eight chips and six connectors, a very minimalist example of a mobile robot.

Getting started in robotics involves not only learning how to build things but knowing where to get materials. The remaining appendices contain a compendium of parts, suppliers, and information that we have found helpful. Appendix C lists a yellow pages of over 150 suppliers and distributors for robot parts, such as motors, sensors, prototyping equipment, electronic components, and power supplies. Hopefully, this collection will help you overcome the inertia of getting started, whether it be in a basement workshop or in a university laboratory.

Technology changes rapidly, and while a book such as this can provide a general foundation, it cannot be dynamic enough to provide up-to-date information on new product announcements. Staying abreast of technology is crucial in making design decisions. We have discovered that systems we engineered in house one day would often become commercially available the next, or that components we relied on for years would suddenly become discontinued, so in Appendix D, we have listed a number of magazines, trade journals, and electronic bulletin boards that we have found invaluable for staying current. (In fact, just as this book is going to press, the microprocessor that we chose for Rug Warrior has gone out of production. Fortunately, however, the Motorola MC68HC811A0 microprocessor mentioned throughout is upward compatible with the Motorola MC68HC11A1, so simply substitute that part into Rug Warrior.)

Semiconductor manufacturers' data books are another source of current technology; we have annotated our collection in E. G adds a few more tables that are handy to have in one place, such as the resistor color code and the ASCII code for alphanumeric symbols.

As technology marches on, a book that emphasizes specific hardware will quickly become outdated. But the art and the means and the basic concepts survive, and these we hope to share with you.

Cambridge, MA
April, 1993

Anita M. Flynn
Joseph L. Jones

1

Introduction

The rise in popularity of the single-chip microcomputer and the drastic reductions in size and cost of integrated circuits in recent years have opened up huge new arenas for creating intelligent systems. Building a robot, however, requires more expertise than simple programming. A roboticist must be a generalist. The robot designer must own a compendium of basic skills from fields such as mechanical engineering, electrical engineering, computer science, and artificial intelligence (AI). Unfortunately, few people have the opportunity to study so broadly. In this book, we attempt to outline a few basic ideas from each of those areas and, more importantly, to suggest strategies for putting the pieces together. Hopefully, with a little creativity, you will be able to later use this toolbox of techniques to design far more intriguing machines than those outlined in this book.

Robotics is about building systems. Locomotion actuators, manipulators, control systems, sensor suites, efficient power supplies, well-engineered software—all of these subsystems have to be designed to fit together into an appropriate package suitable for carrying out the robot's task. Where do we start?

We think of a robot as an intelligent connection of perception to action. The implementation of that goal might take on a variety

Figure 1.1. TuteBot is a very simple robot, yet it can exhibit two distinct behaviors. It will follow a wall and when it bumps into something, it will backup and turn. TuteBot's brain is an analog computer, which is programmed only by adjusting potentiometers.

of "costumes," from mechanical logic to microprocessor control to networks of neuron-like gates. Our approach is to create abstraction barriers in terms of thinking about the intelligent capabilities our robot might possess and then to gradually break them down by explaining the specific hardware details that we might employ to create those competences. The theme throughout is to build systems early and build systems often—to start with very simple systems that connect perception to action and to gradually move to more sophisticated machines.

We start with a tutorial in the next chapter that describes how to build a robot, TuteBot, that is able to wander around a room and avoid obstacles. This example robot, pictured in Figure 1.1, is implemented without recourse to a microprocessor. TuteBot is merely an agglomeration of switches, relays, motors, and discrete electronic components, all of which can be assembled rather easily. You will be able to adjust TuteBot's reflexes by tweaking two potentiometers.

From this very simple example of a robot, we introduce the microprocessor and the advantages of using software to manage the complexity of large numbers of sensors and actuators. The viewpoint from this moment on is to build systems with the intent of getting to software as soon as possible. To keep parts count, size, and costs down for our readers, we describe minimalist ways to interface sensors, motors, and power supplies in another example robot, Rug Warrior. The microprocessor becomes the heart of Rug Warrior, and the following chapters describe the workings of mechanical and electrical components and the interface circuitry that enables them to be driven from a microprocessor. Software-primitive operations are threaded throughout the book as each new perception or locomotion system is introduced.

Although this book describes the details involved in actually building robots, we hope also to raise some deeper points about models of intelligence. What is *intelligence*? Is it the contemplative thought involved in playing chess? Is it the reflexive action that occurs as you try to keep the gnats out of your eyes while walking down the street on a hot, muggy summer night? Or is it the common-sense reasoning used in deciding what to make for breakfast? We will stick with the notion that *intelligence* is the foundation for how people act most of the time. It will be interesting to keep some of these questions in mind as we investigate the sorts of mechanisms we can use to endow our example robots with low-level behaviors.

Other features of intelligence have to do with the role the environment plays in our view of cleverness. How connected are sensing and actuation to intelligence? How much of what we acknowledge as complex behavior is merely a reflection of simple behaviors off of a complex environment? For instance, if we observe the behavior of ants scurrying around their anthills, we might begin to wonder whether their complex paths result from careful planning and deep contemplation, or perhaps merely from simple rules of behavior acted out in an environment full of uneven terrain, obstacles to climb over and other ants.

TuteBot and Rug Warrior will not answer many of these questions pertaining to the structure of intelligence, but we hope that they can be the platforms for an inexpensive, easily attainable AI input/output device–a collection of sensors and actuators that provide a little bit of input, a little bit of output, and a little bit of computation to readers interested in experimenting with some of these issues.

Many of the modern theories in artificial intelligence grew from work in a number of other fields. Cybernetics, in the 1940s and 1950s, was a field of research that tried to understand intelligence through the study of the control of machines. Cybernetics developed in parallel with classical control theory. Its model of computation was analog, and it tried also to understand intelligence in animals by modeling them as machines. Our example of TuteBot is very much in the same spirit as the early work in cybernetics.

For instance, Figure 1.2 illustrates the extent of TuteBot's talents. The long dashed lines at the bottom of the figure exemplify one initial behavior, where TuteBot moves forward in a straight line until it hits an obstacle. It then backs up, turning left for some period, and then proceeds forward again in a straight-line motion.

A number of mechanisms could be imagined necessary to achieve this behavior. We could suggest contemplative recognition of chair legs and walls and TuteBot making explicit decisions concerning when to back up and how far to turn, but TuteBot has no such model of the world. Instead, TuteBot has a simple analog electrical circuit for a control system, which directs TuteBot's two wheels to move it forward until a bump sensor on the front detects a collision. The signal from the bump sensor directs both motors to reverse direction, and TuteBot then backs up. What makes it turn is an element of state, or timing, in the system that is implemented with a resistor-capacitor (RC) circuit, one for each wheel. If the RC circuit on each wheel is set differently, one wheel will back up for a longer period of time than the other wheel, causing TuteBot to turn. When TuteBot resumes forward motion, it no longer has the same heading and so avoids ramming the obstacle it first bumped.

A second behavior can be added to TuteBot using a similar strategy. If, during the forward motion, one wheel is allowed to turn faster than the other (for instance, by adding a resistor in series with one motor) TuteBot will move in an arc. The short dashed lines at the top of Figure 1.2 illustrate this behavior. As TuteBot moves forward, arcing to the left, it tends to bump into obstacles. When this happens, the initial obstacle-avoiding behavior just described is triggered and TuteBot backs up, turns toward the right, and proceeds forward in a new direction. However, the bias between the wheels causes the robot to veer off to the left again. The result of these two

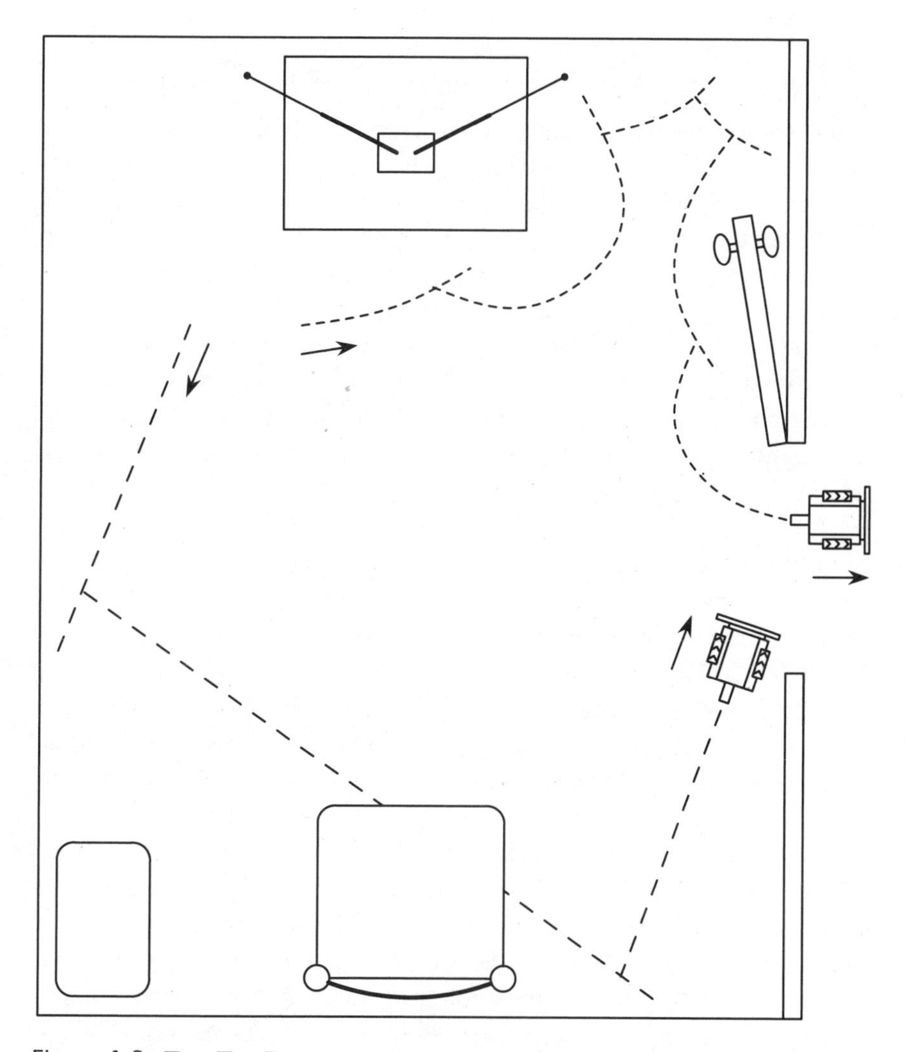

Figure 1.2. Two TuteBots each displaying a different behavior. Dashed lines indicate the paths they have traveled. In one behavior (long dashed lines) TuteBot moves along a straight path until it encounters an obstacle. It then backs up, turns left to change its heading, and proceeds forward again, performing straight-line navigation. In the second behavior (short dashed lines), the robot's forward motion forms an arc to the left. When it bumps into an obstacle, the robot backs up and turns right; then it arcs to the left once again as it moves forward. With this strategy, TuteBot demonstrates a wall-following behavior.

behaviors is that TuteBot tends to follow along the edges of clutter. We call this behavior *wall following.*

In the 1960s after cybernetics, and with the rise of the digital computer, the field of artificial intelligence was born, and with it came computational models of intelligence. The contributions of AI to the understanding of intelligence were the notions of representation, search, and modularity. Information could be explicitly represented in data structures inside a computer, which could then be searched for the desired answer. Representations could be more easily formulated, as the model of computation was no longer time-varying analog signals, but bits and numbers. This capability enabled modularity and led to increasingly sophisticated information-processing systems. Chess-playing programs, expert systems, natural language interpreters, and problem solvers were some of the demonstrations developed in this era of traditional AI.

Unfortunately, some of the ideas involved with representation led to problems when intelligence systems were designed for machines that interacted with the dynamically changing real world. Traditional AI had formulated the problem of robot intelligence as sensing, building a world-model representation from the fusing of sensor data and then planning actions based upon that model. Computational bottlenecks, noisy sensors, and the complexity of reality led some researchers to look for new models of intelligence that would be robust and would work in real time. These new ideas have collectively come to be known as *Nouvelle AI* or *behavior-based robotics.* Rodney Brooks at the MIT Mobile Robot Lab proposed the *subsumption architecture* which is a way of organizing the intelligence system by means of layering task-achieving behaviors without recourse to world models or sensor fusion. This book grew directly from that research, and Rug Warrior is our example robot that illustrates many of the ideas in a subsumption architecture.[1]

The word *subsumption* is used to describe the mechanism of arbitration between the layers of task-achieving behaviors. *Arbitration* is the process of deciding which behavior should take precedence when many conflicting behaviors are triggered. In a subsumption architecture, the designer of the intelligence system lays out the behaviors in

[1]Subsumption architecture is an example of what is now more commonly referred to as behavior-based robotics or behavior control.

such a way that higher-level behaviors subsume lower-level behaviors when the higher-level behaviors are triggered.

For instance, if the lowest-level behavior enables a wandering action and the highest-level behavior initiates following light, then normally, the robot will wander around, moving along randomly chosen headings. However, should someone point a flashlight at Rug Warrior, the highest-level behavior would trigger, suppressing wandering for the duration of time that the flashlight is directed at the robot. Instead of random headings, Rug Warrior's wheels would be commanded to turn toward the point of highest light intensity and move forward in that direction. If the flashlight were turned off, the follow-light behavior would no longer be activated and would cease subsuming the wandering behavior. Random wandering would then resume.

In order to experiment with a richer set of behaviors than mere wandering and following of lights, we have designed Rug Warrior to have as many different kinds of sensors as possible, within the constraints of trying to keep it as simple, and inexpensive as we could. We have built several versions of Rug Warrior, each very different from the other. Two are shown in Figure 1.3. We think of Rug Warriors as a *class* of robots rather than an *instance*. Basically, we will refer to a Rug Warrior as any robot that incorporates our electronics (illustrated in Appendix A) but where vehicle mechanics and software behaviors may vary widely.

The Rug Warrior on the left in Figure 1.3 has two drive wheels, which enable the robot to spin around its center point, and a passive caster for three-point stability. The Plexiglas ring around the robot is a bump skirt, which is mounted on three switches; this feature tells the robot it has bumped into an obstacle. The motors used in this robot came from a surplus dealer, and the chassis was made from Plexiglas that was cut, drilled, and punched in a machine shop.

The Rug Warrior on the right in the figure, running over this book, is a tank-drive robot made from LEGO bricks, gears, axles, and treads. The two motors used in this version of Rug Warrior are model airplane servo motors, ordered through a hobbyist catalog.

While the robots look and act rather differently, their electronics are the same. The board we have designed (which you can prototype yourself using Speedwire or Scotchflex prototyping technology, as

Figure 1.3. Rug Warriors I and II, wandering around their environment, bumping into chairs and driving over books.

discussed in Chapter 4.) is 3.4” × 4.5” in size and contains a Motorola MC68HC11A0 microprocessor, 32K bytes of memory, a serial port, two motor drivers, a piezoelectric buzzer, and a number of sensors. Three bump sensors detect collisions, two near-infrared proximity detectors notice obstacles up to one foot away, two photoresistors sense light level, a microphone listens for noises, and a pyroelectric sensor detects moving sources of heat (such as humans, cats, and, oops, sometimes even fireplaces).

Figure 1.4 illustrates a day in the life of these Rug Warriors. Rug Warrior I, the wheeled version, moves across the room in a straight line until it bumps into the television set. As it turns left to a new heading, the microphone detects a loud noise from the TV, which triggers a behavior to play “Bicycle Built for Two” on the piezo buzzer. As it wanders on, near-infrared proximity detectors see an imminent collision and a wall-following behavior becomes active. Wall following times out after a few moments, and straight-line motion resumes. A low-lying (but very interesting) book on mobile robots is in the path, but the near-infrared proximity detectors are pointed upward and miss it. Rug Warrior I then drives into the book, but the bump skirt detects the collision, causing the robot to back up and turn away to a new heading. It catches a peak of light intensity coming from the doorway, and a follow-light behavior becomes activated. Rug Warrior I then leaves the room.

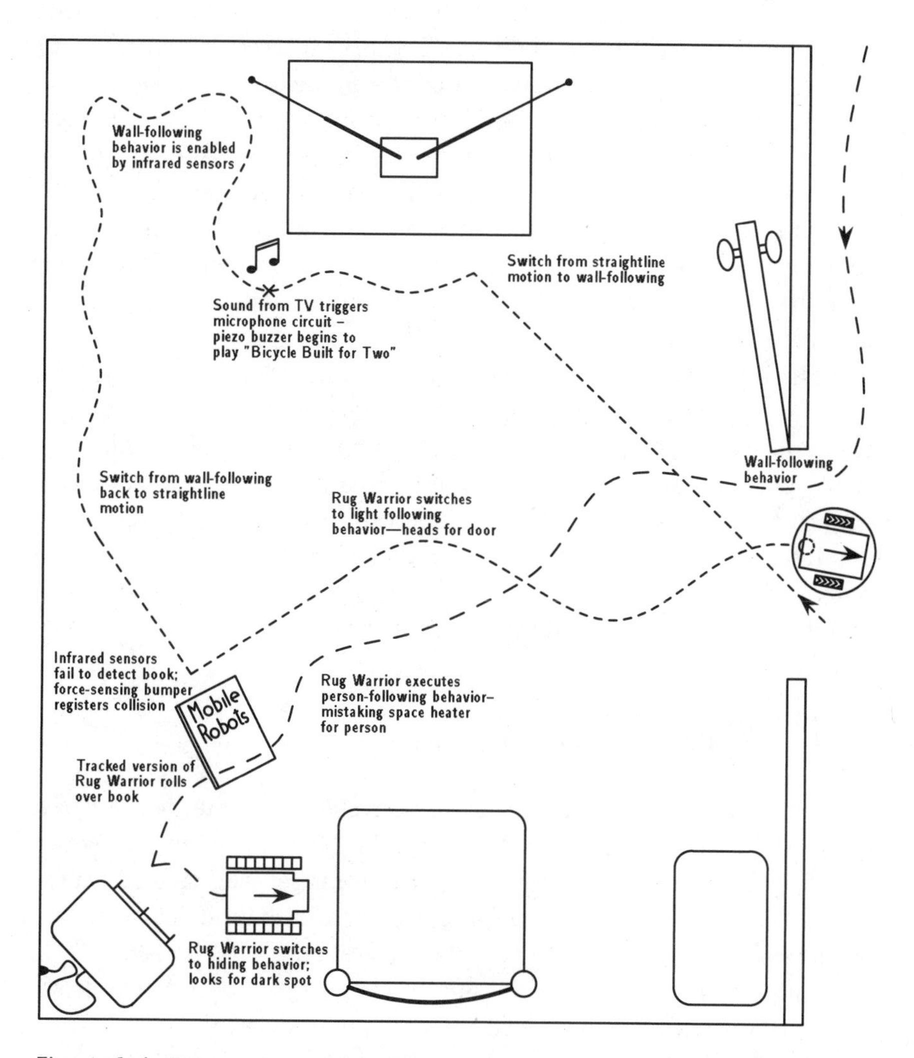

Figure 1.4. Two versions of Rug Warrior, wandering around their environment. Subsumption networks for the intelligence systems prescribe a layering of behaviors that become active upon the proper triggers. Behaviors such as wall following, straight-line motion, obstacle avoidance, noticing sounds, playing music, homing in on light sources, and hiding in the dark are all possible with the sensors available on Rug Warriors.

In the meantime, Rug Warrior II, the tank, has been following walls outside the room and now comes maneuvering down the hallway. As it nears the open door, the wall-following behavior causes the robot to turn to the right, as if the wall were still there. As it does this, the cone of detection of the pyroelectric sensor sweeps past the space heater, mistakingly triggering a people-following behavior. Rug Warrior II does not see its favorite book lying in the way and drives right over it. As it nears the space heater, the people-following behavior happens to time out and a hide-in-dark-corners behavior activates. This directs Rug Warrior II to veer off on a new heading, wandering around until it lands in a shadow, where it sits and hides under a chair.

These illustrations are meant to give a flavor of a subsumption architecture intelligence system. The main idea is that there are no explicit geometric representations of the world from which the robot plans its actions. Instead, there are a number of control loops granting a very tight coupling of perception to action, and from the interaction of many simple behaviors, complex activity seems to emerge. The following chapters will expand on these ideas and reveal the details involved in making things work.

1.1 References

A long history of research predates nouvelle AI. Some of the early ideas from cybernetics can be found in Norbert Wiener's works (1948, 1961). Grey Walter (1950, 1951) built several vacuum tube-based robots that could home in on goals and exhibit learning behaviors. Many years later, Valentino Braitenberg's work (1984) with imaginary vehicles containing simple connections between sensors and actuators nicely illustrated many of these ideas.

Marvin Minsky (1986) proposed the notion of multiagent intelligence systems in which parallel processes interact to produce emergent behavior. The first work on subsumption architectures, incorporating the modularity of layered behaviors was presented in Brooks (1986). One influence during this time was work in the field of ethology, the study of animals in their environments. Rüdiger Wehner (1987) underscored the fact that, in animals, many sensors are specifically matched to their environments. A paper by Brooks (1991b) gives a more thorough exposition on the prior work and contributing ideas that gave birth to behavior-based robotics. Brooks' forthcoming book (Brooks 1998) spells out his approach in great detail. We will return again to this subject toward the end of this book.

But enough of history and philosophy. Let's get started!

TuteBot

2.1 A Tutorial Robot

Building a robot can be a lot of work. All the more so, if the first plan is unnecessarily complex. This chapter is intended to help get you started with building robots while also illustrating some key points about designing a robot's intelligence system without becoming too encumbered in the myriad of details involved in creating a more sophisticated creature. We will show just how simple a robot can be and launch you on your way to building one.

Before proceeding to the more sophisticated Rug Warrior described in the next several chapters, we will begin here by constructing TuteBot—a robot which is simple yet complete. Do not underestimate the elegance of simplicity, however. It is often the simplest solution which takes the longest to comprehend, and yet it is also often the simplest solution which illustrates the main lessons with the most clarity. Experienced designers of robotics and automation systems agree that the first way they design something is usually the most complex way. Difficulty usually arises when trying to simplify the system.

TuteBot will exemplify how a robot as a system, a collection of sensors, actuators and computational elements, can be organized in

Figure 2.1. TuteBot is a robot that can explore its surroundings, escape from collisions with obstacles, and be programmed to follow walls.

such a way that intelligent actions result in response to certain stimuli. TuteBot will consist of a circuit, a chassis, a sensor, a battery, and two motors. It can be programmed by adjusting two potentiometers. The entire robot will be built from Fischer-Technik parts and a few electronic components which are readily available from Radio Shack and other electronic stores.

What will the TuteBot be able to do? Its repertoire of behaviors will endow it with the capabilities to explore its world, escape from objects with which it collides, and follow along walls that it detects with its bumper.

A completed TuteBot is shown in Figure 2.1. The front bumper acts as a sensor and detects collisions with obstacles in its path. A trailing caster wheel maintains stability. Above the chassis is the battery case and mounted on top of the batteries is the breadboard containing TuteBot's electronic circuitry.

All the mechanical components used here are Fischer-Technik parts: motors, gears, axles, wheels, switches and connectors. Fischer-Technik is an excellent source of parts for building robots as the designer can prototype mechanisms quickly without recourse to a machine shop. The Fischer-Technik parts and pieces are available in multiple quantities. Catalogs are available and should be requested.

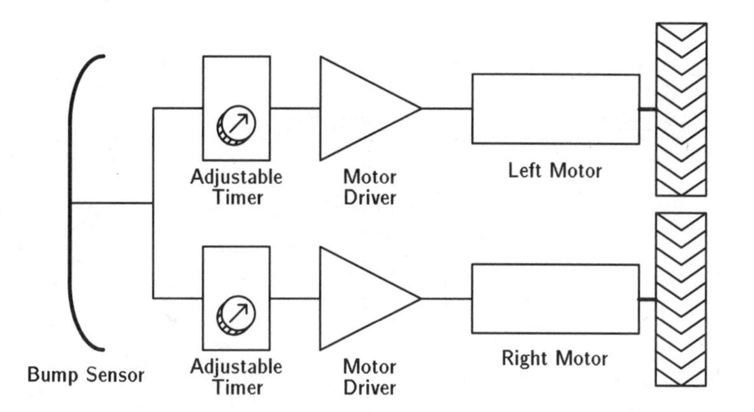

Figure 2.2. The essence of TuteBot. Two motors, two wheels, a bump sensor, two potentiometers for programming, and two motor drivers are enough to create a concrete example of a simple robot—an intelligent connection of perception to action.

An address and phone number for Fischer-Technik is listed in Appendix C. Other types of mechanical building block kits are also quite usable and widely available, including LEGO, LEGO Technik and Meccano.

TuteBot's brain is entirely analog circuitry. No integrated circuits are required and almost all of the components, including the breadboard, can be found at a Radio Shack store. The only tools required to put TuteBot together are wire cutters, wire strippers, and possibly a soldering iron for making connectors. An oscilloscope is not necessary although having one always makes debugging easier. A multimeter should suffice for debugging TuteBot.

A block diagram of TuteBot, shown in Figure 2.2, illustrates how the bump sensor is connected to the actuators. The signal created when the bump sensor detects contact is sent to the motor-driver circuitry for each wheel, signaling the robot to back up. Adjustable timers associated with each motor driver determine for how long each wheel should reverse.

2.2 TuteBot Behaviors

With a minimal amount of hardware, obstacle avoidance can be implemented on TuteBot. Figure 2.3 depicts the sequence of actions

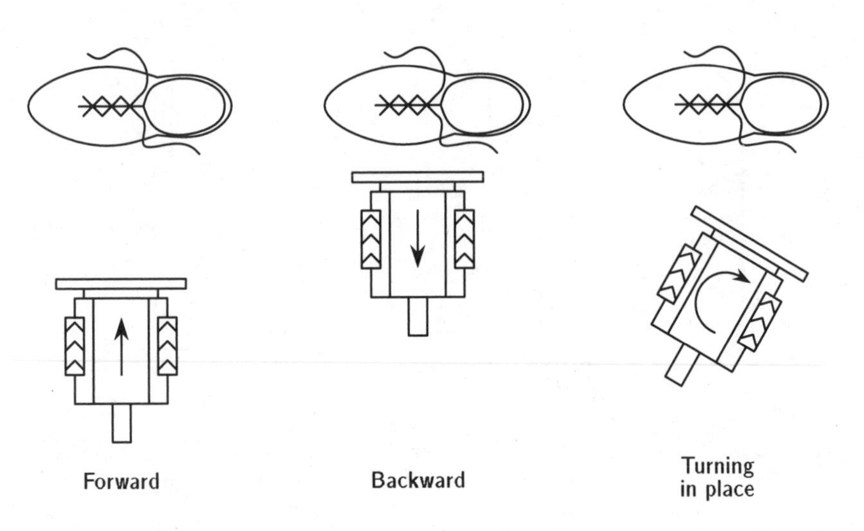

Figure 2.3. TuteBot's basic operation. When TuteBot is powered up, the robot moves forward until it encounters an obstacle. TuteBot then backs up, turns in place, and resumes its forward motion. The time spent backing up and turning in place is programmed by the user.

that occur when TuteBot strikes an obstacle. The robot is initially moving directly forward toward the shoe. As it strikes the shoe, both motors are switched to reverse and the robot backs straight up. However, one motor stays in reverse longer than the other and the robot begins to turn; in this case, the right motor reverses for a longer time period, causing TuteBot to turn to the right. At some point, the right motor stops reversing and both motors go forward, leading TuteBot off in a new direction, hopefully with a wide enough berth to avoid the shoe. If not, the TuteBot bumps into the shoe again and the process repeats until TuteBot turns far enough to the right to finally avoid the shoe.

A timing diagram which graphs this sequence of events is shown in Figure 2.4. The top graph depicts the signal generated by the front bumper's bump sensor. The bottom two graphs illustrate the signals sent to the right and left drive motors.

Initially, both motors receive signals which direct them to go forward. If a collision occurs, the bumper sends a binary signal to the adjustable timers—low for no-contact, high when an obstacle is struck. The timers, in turn, provide a binary signal to the motor

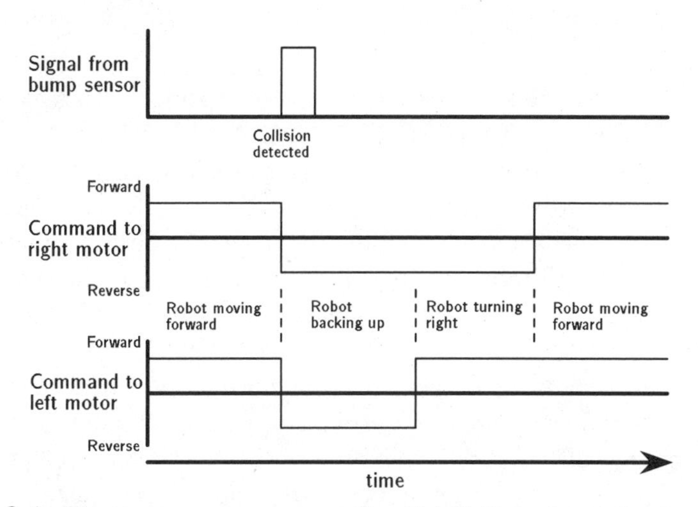

Figure 2.4. The timing sequence generating TuteBot's backup behavior. Both motors normally move in the forward direction as shown in the bottom two graphs. When the bump sensor is activated, both motors reverse. The right motor continues in reverse longer than the left, causing TuteBot to turn to the right. When both motors resume forward motion, TuteBot moves on in a new direction.

drivers—high for forward rotation, low for reverse rotation. Once activated, each timer continues to supply the low signal for a characteristic time. The motor drivers interpret this high or low signal by providing forward or reverse current to the motors respectively.

Assume that the timers are set for delays of t_r seconds and t_l seconds for the right and left motors and that $t_r > t_l$. After encountering an obstacle, the robot will backup for a time t_l. It will then turn to the right (the left motor turns forward, the right motor stays in reverse) for a time $t_r - t_l$. It will then resume moving forward in a different direction, thus avoiding the obstacle or repeating the sequence until it does avoid the obstacle.

An additional behavior can be made to emerge from the robot. If we bias the motors so that, when going forward, one motor turns faster than the other, the robot will move in an arc. This slowdown in speed can be implemented by adding a resistor in series with one motor. If, for instance, the left motor is forced to turn significantly more slowly than the right, the robot will arc to the left. By combining this forward arcing behavior with the earlier back-and-turn behavior, TuteBot can be coerced to follow a wall as was illustrated in Figure 1.2.

To demonstrate this, one would place the robot with a wall to its left, and adjust the timers so that, after encountering a bump, the robot backs and turns a bit to the right. Now when going forward, the robot arcs to the left until it hits the wall; then it backs up, turns right, and then heads forward in an arc until it bumps the wall again. For suitable settings of the parameters, it should be able to turn through a doorway and negotiate either inside or outside corners.

It is an important point here, that nowhere in TuteBot's simple brain does it have knowledge of what a wall is or what is required to follow a wall. Rather, the superposition of a simple set of reflex actions allows a more complex behavior to emerge. This idea of seemingly complex behaviors emerging from a collection of simple rules is the underlying notion of behavior control, which we introduced earlier. We will see more complex examples when we get to the microprocessor-controlled Rug Warrior.

2.3 Building TuteBot

TuteBot senses the world through a front bumper. It steers by individually changing the direction of its drive wheels, while a trailing caster wheel supports the robot. A simple relay, transistor, and capacitor circuit provide all the computational power TuteBot needs.

Figure 2.5 lists the parts needed to construct TuteBot. Most of the parts are available from Fischer-Technik. The remaining parts are easily obtained at Radio Shack or another electronics store. We will begin describing the construction of TuteBot by stepping through the mechanical layout of how to mount the motors and attach the wheels.

Motors for TuteBot

The Fischer-Technik motors have an attached worm gear, transfer box and large axle-mounted gear. Direct current (DC) motors usually spin too fast and have too little torque to drive the loads of the wheels. "Gearing down" a motor causes a motor to spin more slowly

DESCRIPTION	F/T PART NO.	QUANTITY	AREA USED
BOTTOM HALF OF HINGE	31426	2	BUMPER
TOP HALF OF HINGE	31436	2	BUMPER
4W 4P/F	38464	2	BUMPER
2W 2P/F	38242	1	BUMPER
CASTER WHEEL PIVOTHOLDER	32321	1	CASTER WHEEL
CASTER WHEEL PIVOT	31124	1	CASTER WHEEL
CASTER WHEEL WHEELBLOCK	32085	1	CASTER WHEEL
CASTER WHEEL AXLE	31690	1	CASTER WHEEL
CASTER WHEEL TIRE	36573	1	CASTER WHEEL
1W P/C	37237	2	FRONT END ASSEMBLY
1W2L P/C 4C	32879	2	FRONT END ASSEMBLY
1W1L P/5C	32881	4	FRONT END ASSEMBLY(2) AND WHEELS(2)
LEFT AND RIGHT MOTOR	32293	2	MOTOR
LEFT AND RIGHT GEARBOX	31078	2	MOTOR
1W1L P/P +4C	32882	1	MOTOR
2W C/P	35049	4	POWERBLOCK
1W P/P	37238	2	POWERBLOCK
POWERBLOCK TOP	35986	1	POWERBLOCK
POWERBLOCK BOTTOM	36165	1	POWERBLOCK
WIREHOLDER	35969	1	POWERBLOCK
ANGLE C/P	38423	2	SWITCH
SWITCH	37783	1	SWITCH
GEARHUB	35031	2	WHEELS
GEARHOLDER	35033	2	WHEELS
GEAR	31021	2	WHEELS
BIGWHEEL HUB HOLDER	31058	2	WHEELS
BIGWHEEL HUB	32883	2	WHEELS
1W2L 5C	32880	2	WHEELS
60 mm SHAFT (METAL)	31032	2	WHEELS
TIRES	32913	2	WHEELS
ELECTRIC PLUG (GREEN)	31336	4	WIRING
ELECTRIC PLUG (RED)	31337	4	WIRING
WIRE	31360	2	WIRING

W = WIDE; L = LONG; P = PIP; C= CHANNEL; F = FLAT

Figure 2.5. TuteBot can be constructed from these or similar parts.

Figure 2.6. Assembly of the motor, transfer box and wheel gear.

but with more torque at the output of the gear stage. Thus, the wheel can push against the floor with more force.

This worm gear, transfer box and axle-mounted gear which are unique to Fischer Technik give the TuteBot a total speed reduction of about 30:1 and wheel revolutions of about one per second. A speed reduction of between 20:1 and 30:1 is an appropriate goal when using Fischer-Technik or other motors and gears. Gears and motors are explained in more detail in the later chapter on motors.

The first step is to build the left-side motor, transfer box, and wheel and gear assembly as shown in Figure 2.6. (Follow the same steps for building the right-side motor assembly.) The left and right sides are the same except that one of the transfer boxes is upside down so that both 10-tooth gears are facing inward.

The two motor-sides are joined by a connector block as shown in Figure 2.7(a). Notice that this connector block is actually made from three Fischer-Technik pieces (2.7(b)).

The caster wheel which is the rear wheel for the TuteBot is assembled from five Fischer-Technik pieces. They are shown in Figure 2.8 both apart and assembled. This caster wheel assembly fits in the back of the chassis between the two motors. It should slide in between them and allow the chassis to stand on three wheels.

Figure 2.7. Joining the motors with a connector block.

Figure 2.8. Assembling the caster wheel.

The TuteBot chassis can be constructed by following the sequence of steps outlined in Figure 2.9.

Front-end assembly

The front end of the TuteBot is made from four black pieces and two red pieces. Slide one of the black double-long blocks into the other and slide two black single blocks onto either end, half the distance of each block. Then, holding the joined blocks with the two single blocks facing up, slide the two red pieces onto the same side of the joined double blocks, as shown in Figure 2.9(a). This side is now the back of the front-end assembly.

Switch assembly

The switch slides into two red L-pieces; this switch-assembly slides onto the front side of the front-end assembly just constructed. (See Figure 2.9 (b).)

Bumper assembly

The actual bumper is composed of seven pieces, shown in Figure 2.9 (c).

The completed bumper assembly slides onto the front side of the two single blocks on the top of the front-end assembly.

To complete the chassis slide this entire unit onto the front of the wheel gear-motor assembly.

Connecting power to the chassis

Now we will connect the power block (battery pack) to the TuteBot chassis. The power block requires six AA alkaline batteries. Attach the wire-holder connector to the front right-hand side of the power block. (The power holes are on the back side). Using one short and one long red piece each (see Figure 2.10(a)) construct two "legs" and attach them to the rear of the power block. Slide the power-block assembly onto the motors at the rear of the chassis. The front of the power block should simply rest on the front-end assembly (see Figure 2.10(b)).

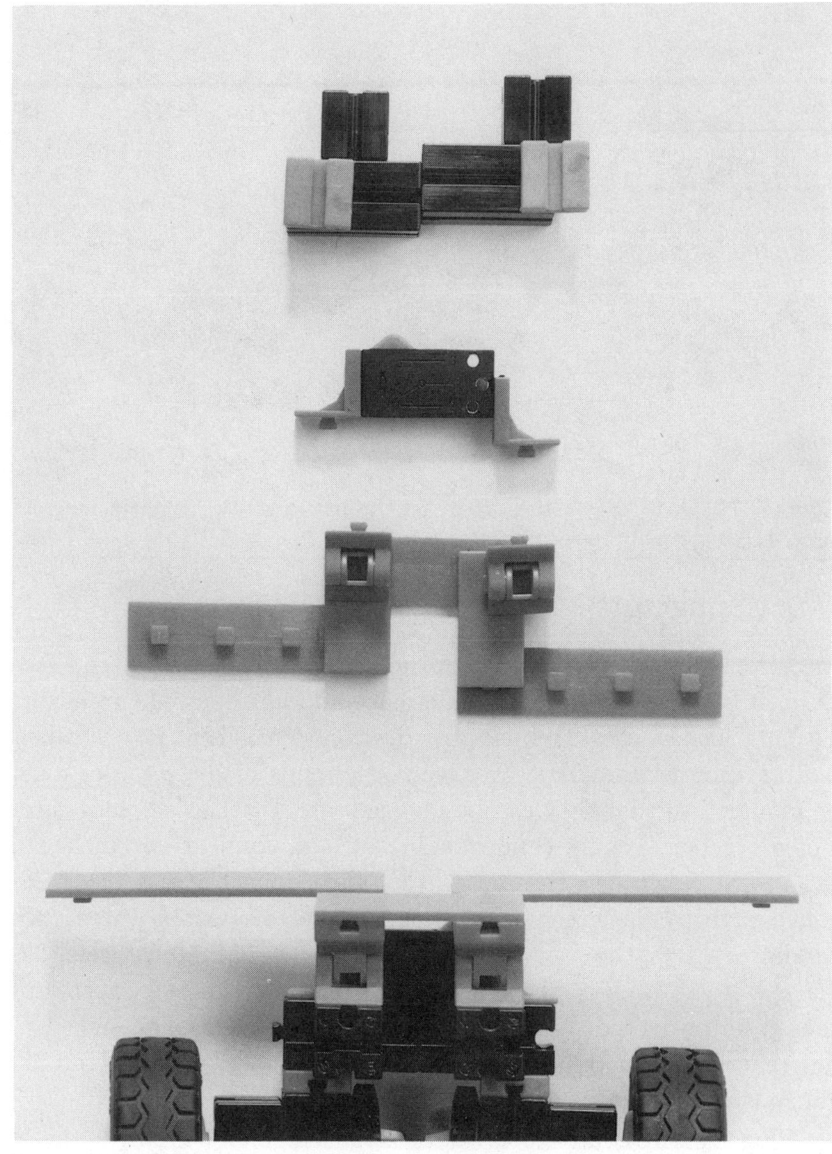

Figure 2.9. Step-by-step instructions for building the TuteBot chassis. (a) Front-end assembly. (b) switch assembly (c) bumper assembly (d) connection of bumper assembly to front-end assembly.

Figure 2.10. (a) Constructing "legs" for the power block. (b) Attaching the power-block assembly to the motors.

Wire Connectors

Next, we will make connectors for the motors and bump switch. Two Fischer-Technik connectors should make all four required connectors for TuteBot. Fischer-Technik provides connectors that fit with their motors and switches. However, the other ends of these cables must be modified so that they can be plugged into TuteBot's breadboard. Cut a Fischer-Technik connector in half and connect two inches of red 22-gauge wire to the red wire of the connector and black 22-gauge wire to the green wire of the connector. Make three such connectors. For the fourth connector connect a two-inch green 22-gauge wire to each of the red and green wires. These connections should be protected with electrical tape. Although color makes no difference electrically, you can avoid confusion by using a green 22-gauge wire to connect to the switch.

Plug the wire connector with the attached green wires in the holes labeled 1 and 3 on the switch. This sets the switch up as normally open (NO). It is this switch that will cause the TuteBot to back up when it collides with an obstacle.

Plug two of the three remaining connectors into the left and right motors at the rear of the chassis.

Figure 2.11. The TuteBot is now assembled. All that is left is to mount the breadboard to the top of the battery pack.

Plug the remaining connector (red to +, green to -) into the corresponding holes at the back of the power block.

Now your TuteBot should look similar to that in Figure 2.11. The final step is to mount the breadboard to the top of the battery pack.

In the next section, we will discuss building the electronic circuitry for TuteBot's brain. Once this has been assembled, mounting it on top of the constructed TuteBot should produce a robot resembling that in Figure 2.1, shown at the beginning of this chapter.

2.4 Electronic components

Before we get into the specifics of the control system for TuteBot, we take a moment here to describe the basics of a few common electronic components such as relays, transistors, capacitors, diodes, etc. Figure 2.12 illustrates the relationship between the physical components we will use on TuteBot and their schematic symbols.

First, the *relay* shown in the upper left-hand corner of Figure 2.12 is a type of electrically controllable switch. TuteBot uses relays to switch the polarity of the voltage applied to its motors and thus reverse their direction. The idea behind a relay is that a small current flowing in the relay's coil can allow much larger currents to flow through its contacts. The way a relay works is that, when different voltages are applied to the two lines marked coil, the resulting current creates a magnetic field inside the device. This field attracts a metal lever to which the internal switch contacts are attached. Activation of the lever in turn disconnects one circuit and connects the other. With no voltage applied, the line marked com or *common*, is connected to nc , the *normally closed* pin. When a voltage is applied across the coil, com is disconnected from nc and connected to no, the *normally open* pin.

Next come bipolar *transistors.* Bipolar transistors have three terminals: a *base, b*, a *collector, c*, and an *emitter, e.* For a particular transistor case design, the correspondence between these symbols and the physical leads can be found in the manufacturer's data book. Transistors can be used as amplifiers or switches. TuteBot employs transistors to supply a current sufficient to activate the relay. There are a great variety of transistors. Two of the important parameters that differentiate among them are amplification factor and maximum power-handling ability.

A *diode* is a device which allows current to flow in one direction but not the other. If the "+" end of a diode, the *anode*, is connected to the "+" terminal of a battery and the "-" end of the diode, the *cathode*, is connected to the "-" terminal of the battery, a large current will flow through the diode, enough to damage the diode or battery. Usually a resistor is placed in series with a diode to limit current to a safe level. If the connection is reversed, no current flows. Diodes are rated according to the amount of current they can handle without damage and the maximum reverse voltage they can sustain. A band on the diode usually marks the "-" end. The triangle on the diode's schematic points in the direction current is allowed to flow. TuteBot uses diodes to isolate parts of the circuit and short out induced voltages of the wrong polarity.

A single-pole, single-throw (SPST) switch is shown at the left of the second row in Figure 2.12. Switches are characterized both by the number of connections that can be made or broken by moving the switch lever and by the number of different lever positions that make contact. A single pole, single throw (SPST) switch is the simplest type of switch. With the switch lever in one position, connection between its two leads is broken. In the other position, connection is made. An SPST switch might serve as the power switch for TuteBot, if desired.

To detect collisions, TuteBot uses a *momentary contact switch.* These types of switches have an internal spring that endeavors to keep the switch in one state. As long as the switch lever or push button is pressed, the switch circuit is closed. When the lever is released, the circuit opens. Momentary contact switches with the opposite sense (open when pressed, closed when not pressed) are also available.

Resistors impede the flow of current. Their ability to do this is measured in ohms, Ω; kilohms, $K\Omega$; or megohms, $M\Omega$. The current, I, that will flow through a resistor with resistance, R, given an applied voltage, V, is $I = V/R$. This is known as Ohm's Law. When current flows through a resistor, it must dissipate power. A resistor's capacity for dissipating power is measured in watts. In general, a resistor with a higher wattage rating will be physically larger than one with a smaller wattage rating.

To block direct current but allow the passage of alternating current, one uses a *capacitor.* Once connected to a voltage source, such as a battery, current flows into the capacitor until it has accepted as much charge as it can. This ability to accept charge is usually measured in units of micro- or picofarads (μF or pF). If the voltage supply is removed from the capacitor, the stored charge keeps the voltage across the capacitor constant. Shorting the leads together causes a current to flow until the charge is depleted and the voltage across the capacitor goes to zero. TuteBot uses capacitors as memory cells. The presence or absence of stored charge represents the robot's recent history, or state.

There are many different capacitor technologies. Most capacitors can be connected into a circuit without regard for polarity. One type for which polarity is important is the electrolytic capacitor. The leads on these capacitors are marked "+" and "-" so that it is clear which way they should be inserted into the circuit. Electrolytic capacitors can generally store more charge in a smaller volume than other types of capacitors. The maximum voltage that can be applied to a correctly connected capacitor before damage occurs is listed as the WVDC (Working Voltage, Direct Current).

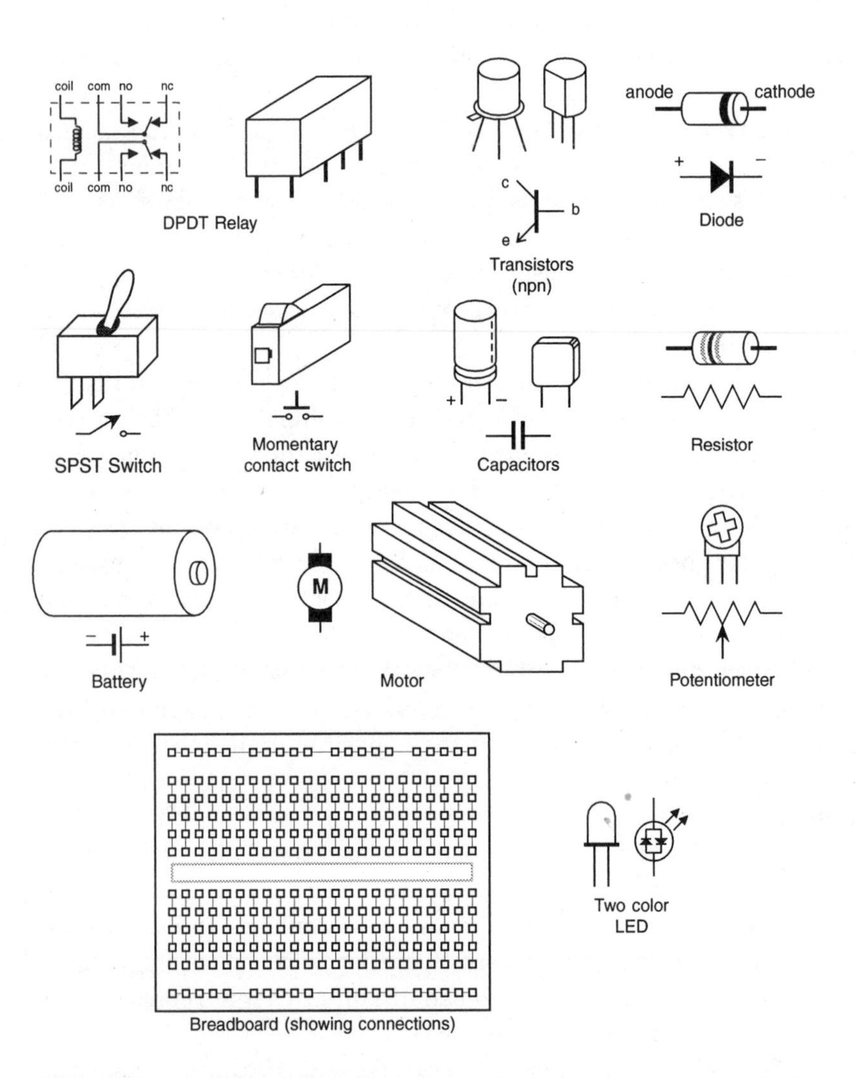

Figure 2.12. The relationships between schematic symbols and the physical components they represent. All of these components are used in TuteBot's brain. No other components are necessary, and the entire circuit will fit in a 6-inch-long breadboard mounted on top of TuteBot's chassis.

A *potentiometer* is simply a resistor whose resistance is adjustable. As with fixed resistors, there are a large number of resistances and maximum power ratings from which to choose. A potentiometer allows the user to manually alter some parameter of a circuit. We will use potentiometers in TuteBot to control its response to collisions—how long it backs up and how long it turns in place before proceeding forward again.

The first item found in the third row of Figure 2.12 is a *battery*. Batteries supply currents required at some characteristic voltage. The nominal voltage rating of a battery is normally stamped on its case. TuteBot, for instance, uses six 1.5 volt (V) alkaline batteries. Many toys and portable appliances use nickel cadmium (NiCd) batteries which come in 1.2 volt cells.

Motors convert electrical energy to mechanical energy. FischerTechnik motors were chosen for TuteBot because they are easy to integrate into the chassis and they happen to provide sufficient power for this application.

The last component in Figure 2.12 is an electronic *breadboard*. Internal connections among its sockets are shown. A breadboard allows us to quickly connect components into a circuit and to make changes easily. Vertical columns are connected together, as are the top and bottom horizontal rows. Typically, one would connect these rows to power, the positive side of the battery pack in this case, and ground, the negative side of the battery pack. The space between the columns in the center is the correct width to accommodate standard integrated circuit chips. The relays are the same width as standard chips.

Later on in this book when we discuss Rug Warrior, we will introduce a number of other components such as power MOSFET transistors, crystals, operational amplifiers, photoresistors, light emitting diodes, logic gates, microprocessors, memories, etc.

2.5 Electronic Construction

With those device descriptions as background, now let us look at the circuit for TuteBot's brain. Figure 2.13 gives the schematic. A schematic illustrates the topology of how all the electronic components are connected into a circuit.

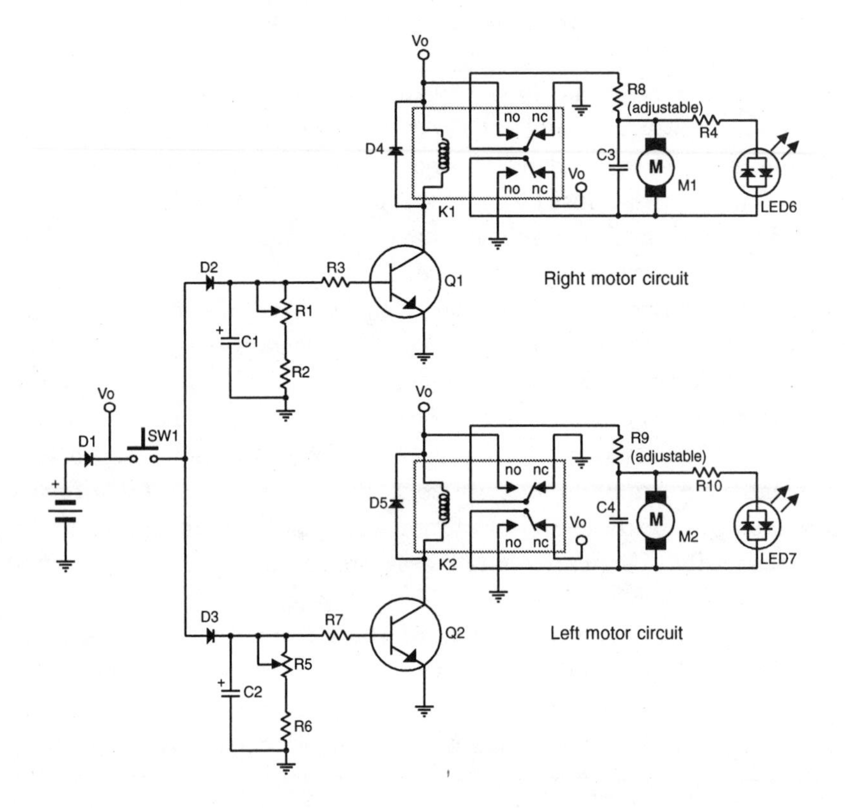

Figure 2.13. Schematic for TuteBot's brain.

In the circuit for a robot's brain, there are typically transducers connected on either side. For instance, on the input side, batteries and sensors act as input transducers. A battery converts chemical energy into electrical energy, and a sensor converts a physical phenomena from a mechanical form (say, the force acting on a bump switch) to an electrical form. On the output side, motors, speakers, lights, etc., act as output transducers. The motors on TuteBot convert electrical energy into mechanical energy. In between the input and the output transducers is the electrical circuit, which does the information processing. The time variation in the signals, the voltages and currents in the circuit, provides information transfer.

In describing a circuit's behavior, one usually speaks of voltage across a device and current through it. One bit of confusion can arise due to a verbal shorthand of speaking of such things as "the voltage at point A". What is meant and what would be more precise would be to speak of "the voltage across the network between points A and ground." The verbal shorthand comes about because ground is usually taken to be the reference, 0 volts.

The basic idea of TuteBot's circuit is that the front bumper switch (SW1 in Figure 2.13) generates a signal that tells the robot to back up. This bump signal is sent to each half of the circuit. The diodes D2 and D3 act to separate the circuit driving the left motor from the circuit driving the right motor so that they can independently have specifiable time constants for how long each wheel should back up. The time constants are implemented with resistor-capacitor (RC) circuits that hold a voltage for a given amount of time, depending on the values of the resistor and capacitor. The timing signals from these RC networks then direct the motors to reverse direction for the specified amount of time. Some driver circuitry to condition the signal to provide enough current to drive the motor has to be added at this point. This motor-driver circuitry is implemented with transistors and relays. A bank of resistors may be added in series with one motor to regulate its speed in comparison to the other motor.

There are two ways to proceed at this point. One is to go ahead and just build the circuit and not worry about understanding how it works. Simply build it, mount it on TuteBot's chassis, plug in the connectors and start playing with various behaviors by tweaking

Electronics:	
1	Breadboard - Experimentor 350, Radio Shack 276-175
2 K1, K2	DPDT 5 volt relays - Omron G5V-2-H-DC5 (DigiKey Z770-ND)
2 Q1,Q2	Transistors - 2N2222A
2 C1, C2	2200 uF Capacitors, electrolytic, 10 WVDC, radial lead - DigiKey P6219-ND [Physical size of cap is important, substitute must not be bigger than above but may be smaller e.g. 1000 uF]
2 C3, C4	01 uF Capacitors, 50 WVDC, radial leads - Radio Shack 272-109 Substitutions OK, but try to keep as small as possible]
2 R1, R5	1 K potentiometers, Bourns 3352T, DigiKey 3352T-102ND [Must be finger adjustable]
5 D1, D2, D3, D4, D5	Diodes, 1N914
2 R8, R9	33 Ohm Resistors, 1/2 watt
2 R2, R6	82 Ohm Resistors, 1/4 watt
2 R3, R7	270 Ohm Resistors, 1/4 watt
2 R4, R10	220 Ohm Resistors, 1/4 watt
2 LED6, LED7	Dual-color LEDs, Radio Shack 276-012 [Substitution may require adjusting the 220 Ohm resisitor]
3	12" lengths 22 AWG solid conductor wire: Red, Black, Yellow

Figure 2.14. Use these parts to construct TuteBot's brain. A good understanding of how the circuit functions will allow the builder to make substitutions. Radio Shack part numbers are given in parentheses. Where no part number is given, any component with the listed parameters can be used.

potentiometers and adding resistors in series with the motors. The other way is to convince yourself you understand every last detail of the circuit configuration before you start stripping wire.

We recommend a quick skimming of the circuit description and then directly putting the circuit together. The parts list for the circuit is given in Figure 2.14, and because the purpose of this chapter is designed to overcome the inertia of getting started, an exact layout on a Radio Shack breadboard is shown in Figure 2.15. Build the circuit just like this one and TuteBot should work. One can then go back through the circuit observing voltage signals across various portions of the network with an oscilloscope to compare traces to graphs for a better understanding.

A photograph of a finished breadboard is shown in Figure 2.16. One detail to note in assembling this circuit is that all the parts and pieces must make firm connection in the breadboard. Relays may be too short to make good contact when inserted into the breadboard and may need to be plugged into a dual inline pin (DIP) socket before being plugged into the breadboard. A 16 pin DIP socket ought to be sufficient. Use care when installing the diodes and the electrolytic

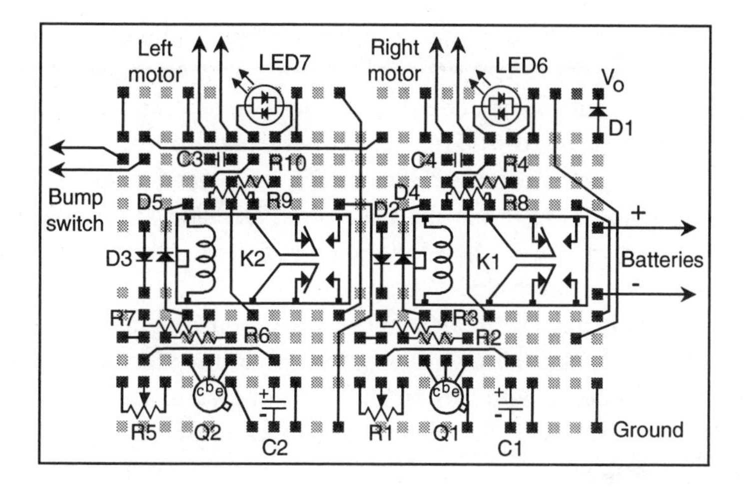

Figure 2.15. One possible layout of the TuteBot circuit.

Figure 2.16. Details of the breadboard.

capacitors. These devices are polarized. If they are installed the wrong way they may be damaged.

It is a good idea to test the circuit as you go. Build only half of it first and check to see that it drives the motors as desired. With power applied and the motor not connected, check to see that pressing the bumper switch activates the relays. If operating properly, a click will be heard. The bias resistors, R3 and R7, may need to be adjusted if relays or transistors other than the ones specified are used. If the relay does not operate, choose smaller resistors until it does (but don't go below about 100 Ω.)

In general, it pays to be neat when breadboarding a circuit. Any time saved in quickly throwing together a sloppy circuit is usually more than wasted in debugging time. Cut and strip wires to appropriate lengths so they lie flat on the breadboard. Buy lots of different colors of hookup wire and stick to conventions for power and ground. If you use red for +9 volts and black for ground, then it becomes easy to visually check your breadboard, because all wires connected to the top horizontal row should be red and all the wires connected to the bottom horizontal row should be black.

Another important tip before connecting power is to always "ohm out" power and ground–that is, check with an ohmmeter that power and ground have not been inadvertently connected together on your breadboard. This prevents smoke from streaming out of your circuit. Never remove components with the power on. Power down first. If the circuit does not work, first check with a voltmeter that all points in the circuit that should be connected to power are actually at +9 volts, and that all points which should be at ground actually read 0 V. While this all sounds rather obvious, you would be surprised at how many problems are caught by these few simply steps.

2.6 Operation

For a more detailed exposition of the TuteBot circuit of Figure 2.13, we break the system into modules and explain each piece. The circuit is divided into two nearly identical halves. For simplicity, we describe only one half, the half which controls the right motor.

As soon as power is applied by plugging in the black (green) wire to the negative battery socket after having plugged in the red wire to the positive battery socket, both motors begin to turn forward and TuteBot is able to move straight ahead. If we look at the portion of the schematic showing the right motor's connection to its relays, we see that there is a lever arm that can switch between normally open and normally closed connections. This type of relay is a double pole, double throw (DPDT) relay. It is the only electronic component not available from Radio Shack in a 9-volt variety. We can see for the right motor that the normally closed connection applies 9 volts across the motor. If the motor moves in the reverse direction, switch the leads going to the motor and it will then move forward. This is true for the left motor also. Notice the light emitting diode (LED) is green when it is going forward and red when going in reverse.

Again, looking at the right motor-portion of the circuit, if TuteBot strikes an obstacle and the bumper switch is closed, a current flows through diode D2 charging capacitor C1. Simultaneously, current flows through resistor R3 into the base of transistor Q1. The base current causes Q1 to conduct—pulling current through the coil of the DPDT relay. When the current is provided to the relay, it switches from the normally closed state to the normally open state. The motor terminal previously connected to +9 volts is now connected to ground, and the other terminal which was previously connected to ground is now connected to +9 volts. This causes current to pass in the opposite direction through the motor, making it spin in reverse. The LED should be red while motors are in reverse.

As the reversing motors cause TuteBot to back up, its bumper is no longer pressed against the obstacle and the bumper switch, SW1, is no longer closed. With the switches open, the RC circuit is no longer connected to +9 volts. However capacitor C1 continues to supply current for a while to the base of the transistor and the motor continues its reverse rotation. The capacitor discharges at a rate controlled by the resistors. At some point, Q1 ceases conducting, the relay opens and the motor resumes its forward rotation. Diodes D2 and D3 isolate the circuits so that the capacitor can discharge at the desired rate (so that current cannot drain off C1 and begin charging the left motor's RC circuit).

Figure 2.17 illustrates how the voltage across the right motor's RC network changes with time. With the switch closed, the battery charges the RC circuit (this voltage is taken as between point A and ground) up to V_o. When TuteBot backs away from the obstacle and the switch is opened, the voltage across the capacitor falls at a rate determined by the values of the resistor and the capacitor. To be more precise, this relationship is $V = V_o e^{-t/RC}$, where V_o is the power supply voltage. Figure 2.17(b) illustrates the RC network connected to the left motor. The smaller resistance in (a) causes the current to drain away more quickly, keeping the robot's right wheel in reverse for a shorter time period than the left wheel. This causes the robot to turn to the left.

The right motor turns in reverse for a period of time, which is determined by the following factors:

- The size of capacitor, C1.
- the value of bias resistor, R3.
- The amplification factor of transistor, Q1.
- The resistance of the potentiometer, R1.
- The current level needed to activate relay K1.

A very brief motor reversal may be selected by setting the potentiometer to its smallest value. A reversal longer than the one available in the circuit is most easily achieved by increasing the value of C1, as it is actually the product of R and C which sets the time constant. You can see this time lag as the duration during which the two LEDs (6,7) are different colors (one red, the other green).

We can see how the changing currents set up by the RC network are able to activate and deactivate the transistor Q1 by referring to Figure 2.18. Depending on the characteristics of the particular transistors and associated circuit components, a transistor can be used as either an amplifying device or as a switch. The TuteBot circuit requires the transistor to act as a switch as shown in Figure 2.18(a). When base current is supplied, the switch closes and the load draws current because it is connected between power and ground (see Figure 2.18(b)). Our very simple model of how a transistor switch works shows that as long as the current flowing into the base of transistor Q1 is greater than or equal to $i_{b,sat}$, the switch is on and current

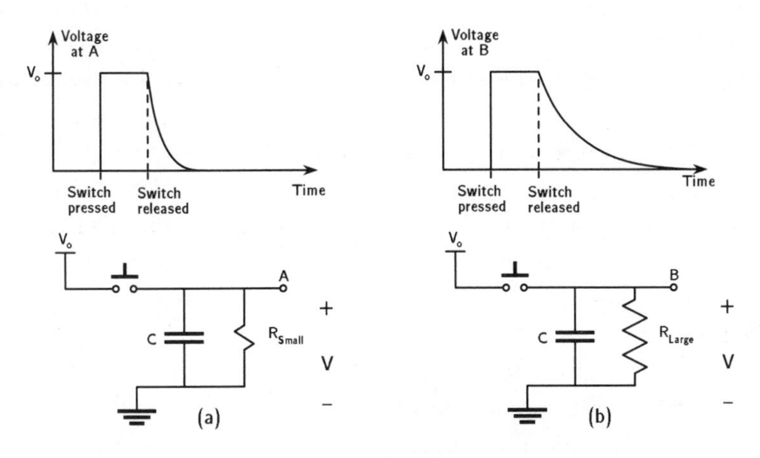

Figure 2.17. As long as the momentary contact switch is pressed, the voltage between point A and ground or point B and ground will be equal to V_o. When the switch is released, charge begins to drain from the capacitor through the resistor. The small resistance in (a) drains the capacitor more quickly than the large resistance in (b).

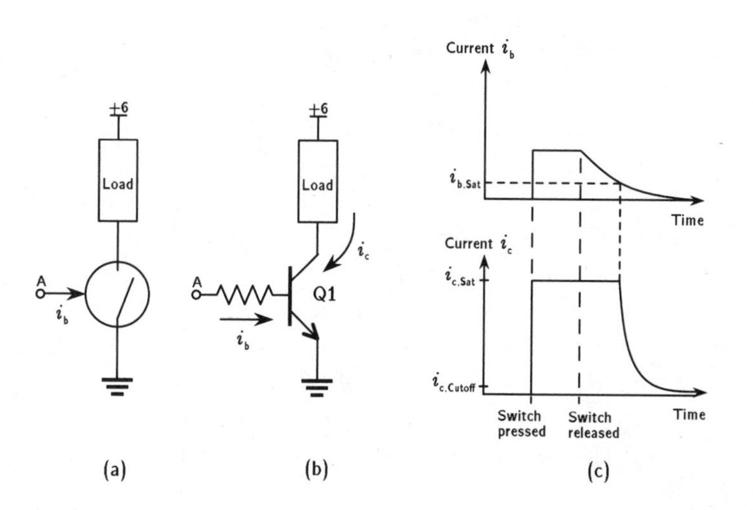

Figure 2.18. (a) A transistor is modeled as an ideal switch. (b) In reality the base current is set by the base resistor's value for a given voltage applied to terminal A. (c) The base current must be large enough to put the transistor into saturation (turning it fully on).

flows through the load. When the transistor's base current falls below $i_{b,sat}$, the transistor switches off and no current flows through the load. A small base current is able to control whether or not a large load current is allowed to flow. In Figure 2.18(b), we see that a base resistor is needed to set the base current for the transistor switch. The timing signals of the current flowing through the base resistor are shown in Figure 2.18(c). For the duration of time that TuteBot is in contact with the obstacle and the RC circuit is charged up to V_o, the base current is large enough that the transistor is completely on and saturated—that is, the collector current has reached its maximum possible level, $i_{c,sat}$.

As TuteBot backs up from the obstacle, the bumper switch opens and the voltage drains off the RC network, the current through the base becomes smaller. Eventually, it falls to $i_{b,sat}$, where the transistor begins to come out of saturation. The collector current falls to 0 and the load becomes open circuited. Actually, a small amount of current does continue to flow for a while even when the transistor is "off". The transition from "on" to "off" is not quite as sharp as in a real switch.

When the transistor switches on, it draws current, i, through the coil of the relay as shown in Figure 2.19(a). Current through the coil creates a magnetic field which forces the relay lever to move. The relay lever then switches the common connection (attached to one terminal of the motor) from normally closed to the normally open pin. This happens in the relay associated with each motor, reversing the polarity of the voltages applied across them. For all the time that Q1 is on, current is pulled through the relay causing the motor to switch from forward motion to reverse motion. The LED subsequently switches from green to red.

The essential difference between the left and right motors is the relative times at which they turn off their reversing behaviors. In Figure 2.19(b), we can see the timing diagrams of the current through the relay and the resulting voltage applied between one motor terminal and ground.

First, as the transistor Q1 turns off, it causes load current to stop flowing. This takes some amount of time after the bump switch is released due to the time delay set up by the RC circuit. When the current through the relay falls to a level which can no longer

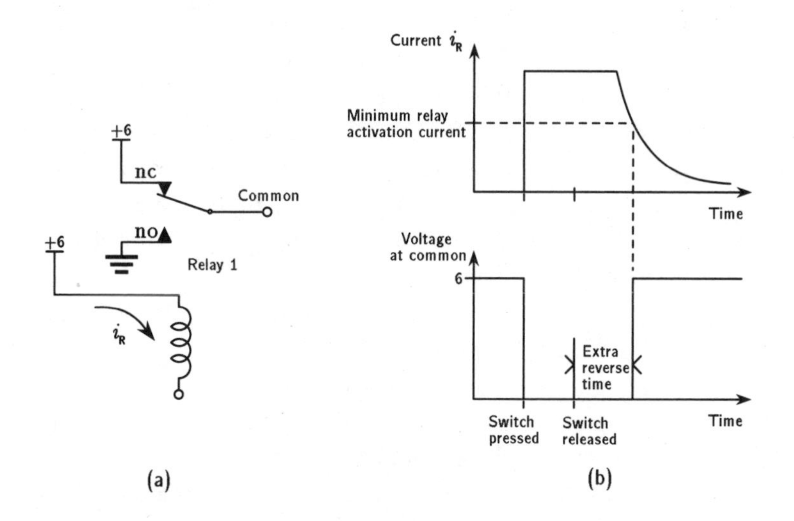

Figure 2.19. (a) The amount of current, i_R, flowing through the coil of the relay determines whether its common terminal is connected to its nc, normally closed, or its no, normally open terminal. When i_R falls below minimum activation current the state of the relay changes. (b) The "Extra reverse time" is the extra amount of time the motors run in reverse after the bumper switch has been released.

sustain the necessary magnetic field to keep the lever attracted to the normally open pin, the relay switches back to its normally closed configuration. This occurs to the relay attached to the right motor.

The lower graph in Figure 2.19 (b) shows the resulting voltage change over time for one of the right motor's terminals. The other motor terminal, normally at 0 volts, switches to +9 volts when the bump switch hits an obstacle and reverts to 0 volts again (after the time lag set up by the RC network) after the bumper is released.

A similar mechanism is implemented on the left motor except that its potentiometer, R5, is tuned to give a different time delay than for the right motor. The robot can thus be programmed to turn more or less sharply by adjusting the potentiometer settings for each wheel.

Four other points are worth mentioning concerning the right motor circuit of Figure 2.13. The first is the appearance of diode D4 across the DPDT relay. The reason for adding this device to the relay is that the diode protects the circuit from the large voltages that are induced by collapsing magnetic fields in the relay coils when the transistor turns off. If diode D4 were not there, the inductance of the coil would try to force the current flowing through it to keep flowing down through transistor Q1. Because Q1 has been opened, current through the coil results in an increase in voltage at the collector of Q1. If this voltage exceeds the maximum rating that the cutoff transistor can withstand, it becomes damaged or blows up. The diode alleviates this problem by providing a return path for the coil current when the transistor turns off.

The second point to note in the final circuit is that the capacitor C3 has been placed across the terminals of the motor. This capacitor attenuates the voltage spikes produced by the motor. Typically, these capacitors are soldered directly to the motor terminals rather than placed back at the circuit board.

Third, note that the directional LED turns green when the TuteBot is going forward and turns red when the TuteBot is going backward. Likewise note that when TuteBot is turning, one LED is red and the other green. The LED is parallel to the motor in the TuteBot circuit.

Finally, note that a "resistor bank" could be connected in series between the relay and the right motor. This bank is for matching the speeds between the two motors. Which motor should be connected to the resistor bank is something which must be determined by experiment. Although the motors and geartrains are supposed to be identical, in reality they are not.

These differences manifest themselves as mismatches in the speed at which the wheels turn. To make the adjustment, power-up the TuteBot and allow it to roll across the floor. It will make a long arc in one direction or the other. If it turns to the left, then the right motor is turning faster; attach the right motor to the resistor bank in series. Otherwise, attach the left motor. With n resistors wired in parallel, the total resistance R_T, of the resistor bank increases as each resistor, R, is removed: $R_T = (1/n)R$. The more resistance we place in series with the motor, the less current will flow and the slower the motor will turn. Add or remove resistors until both motors rotate at the same speed.

TuteBot is now complete and ready to go. Try running it in a few different environments. Try adding the wall-following behavior discussed earlier to bias the motor speeds so the TuteBot travels forward in an arc by inserting a resistor bank. If TuteBot goes too fast and falls apart when it crashes into things, electric tape, double sticky tape, Velcro^{TM}, and glue work wonders with breadboards and with Fischer-Technik components.

Have fun!

2.7 Exercises for the Reader

With the wall-following behavior implemented as described above, the robot will simply turn in circles if it is set in motion far from a wall. As an exercise, try to devise an additional behavior (possibly requiring another component or two) which will cause the robot to go straight until it encounters a wall and then begins to follow the wall.

Think about all the different ways you might add one or more photoresistors (response to light) to the TuteBot circuit. How about a thermistor (response to temperature)? What behaviors are produced in each case? Can you make a TuteBot that follows a light such as a flashlight?

2.8 References

While the TuteBot exercises in this chapter were designed to be simple examples to get started, it might be the case that many people feel more at home with a computer-controlled robot than with the analog electronics of TuteBot. If so, proceed to the next chapter describing Rug Warrior's microcontroller brain. However for background in electronics, the bibles for robot builders are Horowitz and Hill (1989) and the associated student manual (Hayes and Horowitz 1989), which give extensive practical information on analog electronics in very readable presentations. *The ARRL Handbook for the Radio Amateur* (Kleinschmidt 1990) is another very good source for the beginner new to electronics. For articles and reports on simple robots and how to build things, a few pieces have trickled out of the MIT Mobile Robot Lab over the years. Jonathan Connell (1988) describes Photovore, shown in Figure 2.20, a light-eating, dark-avoiding, relay-driven robot using three photoresistors and a Radio Shack toy car

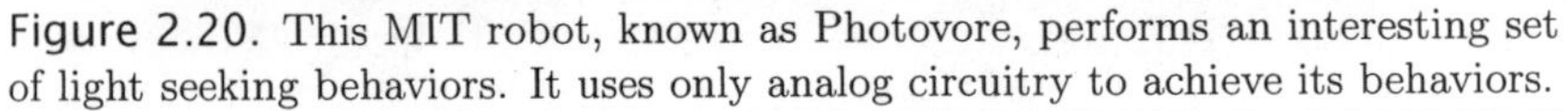

Figure 2.20. This MIT robot, known as Photovore, performs an interesting set of light seeking behaviors. It uses only analog circuitry to achieve its behaviors.

base. Photovore is also described in *The Olympic Robot Building Manual*, (Flynn et al 1988), from which this book grew. A picture book of the resulting talent show robots is contained in Flynn (1989). Another minimalist mobile robot is described in the August 1991 issue of *Popular Electronics*, (Connell 1991). Kits and printed circuit boards for building your own version of Photovore can be purchased from Johuco, Ltd. See Appendix C for addresses and phone numbers in the list of manufacturers. Also, please note that kits are available from A K Peters, Ltd. for TuteBot and Rug Warrior.

3

Computational Hardware

The elementary circuit that controls TuteBot serves its purpose well. Using only relays, potentiometers, bump switches, and some discrete components, TuteBot is able to avoid obstacles and follow walls. Adding a few more sensors and continuing in the same vein of using hard-wired logic for the intelligence system, many other interesting behaviors can also be designed. Rather than pursue this route, however, we now introduce a more sophisticated control element, the *microprocessor*. It has a number of advantages over hard-wired logic in terms of versatility, power consumption, size, and ease of use.

Most importantly, however, the microprocessor introduces a significant new tool in solving the robot control problem: software. To change the behavior of robots of TuteBot's nature, we must adjust potentiometers, rewire circuits, and add or alter components. The behavior of a software-based robot, in contrast, can be changed by typing at a keyboard.

Hardware determines a robot's ultimate potential, but realizing that potential is the job of software. There is an intimate relationship between these two elements which we will try to make clear as we proceed. Organizing the software in the proper way is also important for simulating intelligent behaviors. The low-level interface between hardware and software will be the subject of this chapter; this discussion will continue in Chapters 5 and 7 pertaining to sensors and motors. The organization of higher-level software and intelligence will be addressed in Chapter 9 on robot programming.

3.1 Rug Warrior's Design Strategy

We designed Rug Warrior as a teaching aid for this book in order to support generic discussions of subsystems with real examples of computer hardware, software, sensors, and actuators that fit together. Rug Warrior has many more subsystems than TuteBot, and complexity could easily have gotten out of hand. To avoid this, our approach has been to create a robot that was as simple as possible while still portraying the breadth of technologies we deemed important to understand.

Our design strategy toward this end has been to choose one of the cheapest yet most versatile microcontrollers available (the MC68HC11A0 from Motorola) and to essentially "max it out." By this, we mean using every pin of the chip to attach as many sensors and actuators as possible. Furthermore, we have endeavored to use all of the built-in hardware features of the MC68HC11A0, such as the timer-counter system and the analog-to-digital converters, to minimize any external interface circuitry to sensors and motors. In effect, our goal has been to strive for a *single-board robot.*

Figure 3.1 illustrates the microprocessor board we put together for Rug Warrior II, sitting atop Rug Warrior's tank-tread base. This board contains all the computer hardware, peripheral circuitry, and sensors that we used for Rug Warrior. Rug Warrior has not quite reached the goal of being a single-board robot, but we have managed to incorporate most of the computer electronics, interface circuitry, and sensors on this board, which keeps the number of connectors and cables manageably small.

The point of this book, though, is not just to describe how to build Rug Warrior but to convey general knowledge about what it takes to build a robot so that our readers can go on to build bigger (or maybe smaller) and better machines. Consequently, in this chapter, we discuss microcontrollers: what's inside them, how they work, what features they have for handling peripherals, and how to

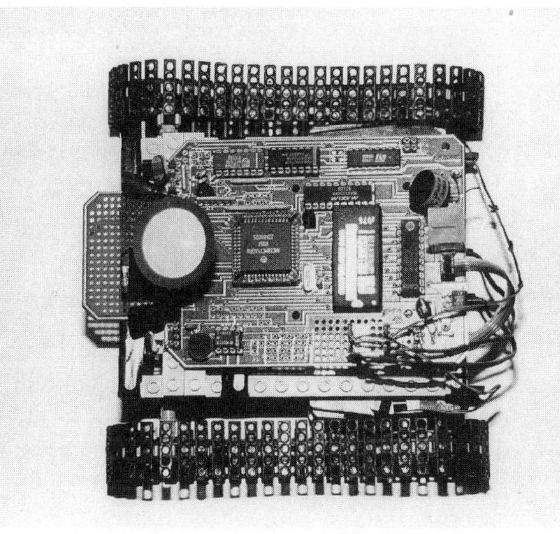

Figure 3.1. A top view of Rug Warrior, displaying its computer and sensors, which we will discuss in the next few chapters.

program them. While we use the specific example of the Motorola MC68HC11A0 throughout, the text is generally applicable to other microprocessors because while the instruction sets and particular hardware attributes for other microprocessors may be different, the underlying principles are the same as those described here.

The specific example that we will explain in this chapter is illustrated in Figure 3.2. This circuit is the computational heart of Rug Warrior. When building Rug Warrior from scratch, the reader should acquire copies of the Motorola reference manuals for the MC68HC11A0, as these are the final source for documentation and are obviously more detailed than our discussion here. Motorola data books can be ordered from the Motorola sales office. The phone number is given in Appendix E. The complete schematic for Rug Warrior, which includes the sensors and actuators in addition to the microprocessor circuitry shown in Figure 3.2, is given in Appendix A.

3.1.1 Interactive C

In addition to choosing a specific piece of hardware for the microprocessor, we also had to pick some specific pieces of software in order to produce our examples, which are threaded throughout the book.

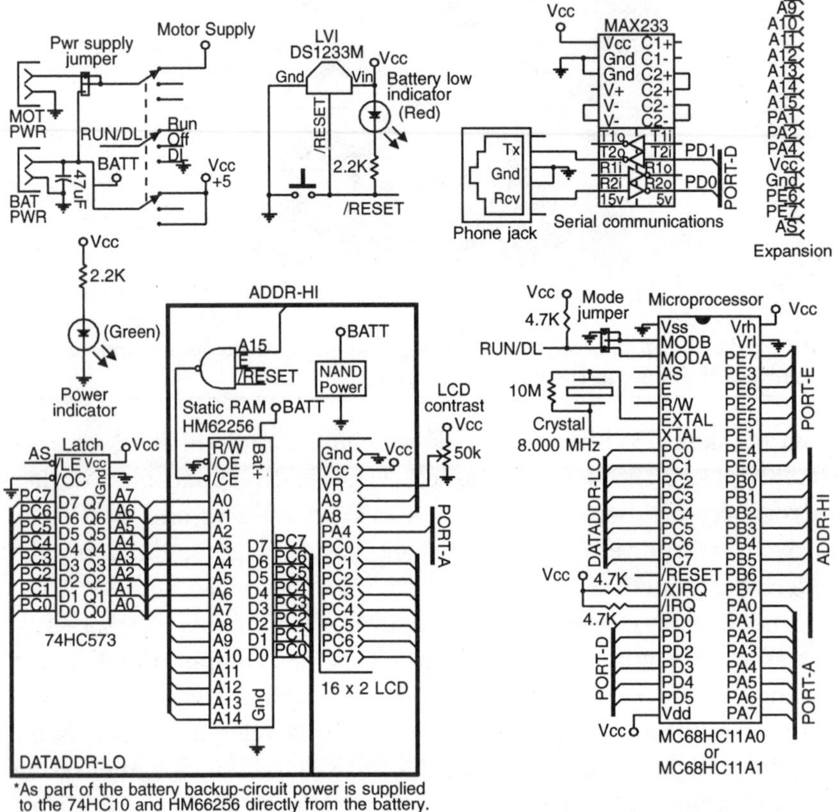

Figure 3.2. The schematic for Rug Warrior's computational hardware. The MC68HC11A0 is attached to 32K bytes of memory through a 74HC573 latch. The MAX233 chip does the level conversion for the serial port, and the DS1233M low-voltage inhibit circuit prevents problems by resetting the MC68HC11A0 when battery voltage gets too low.

Building a robot usually involves both some assembly language programming and some higher-level language programming. Assembly language programming consists of writing code in the machine-specific instruction set designed for the microprocessor you choose. Typically, the programmer writes code using a set of mnemonics for the machine instructions and then runs a program, called an *assembler*, which creates the bit-level sequences that can be downloaded to the microprocessor. Although higher-level language programming is convenient for many tasks, assembly language is often necessary when building a robot in order to direct the microprocessor to read the robot's sensors or drive its motors. Because we have chosen the MC68HC11A0 microprocessor for Rug Warrior, we use the MC68HC11 family's assembly language in our examples.

For creating robot behaviors, a higher-level language such as **C** or Lisp is often used. The user programs a higher-level language in its syntax, which is usually more concise than that of assembly language, and then translates that code to the assembly language for a specific machine using a program called a compiler.

In the research laboratory, we often use Lisp, but for this book and for Rug Warrior, we have chosen to use **C**, as more readers will likely be familiar with its syntax. Specifically, the version of **C** we have chosen to use for creating code that will run on the MC68HC11A0 is *Interactive C* (or simply **IC**). **IC** was developed by Randy Sargent and Fred Martin of the MIT Media Laboratory for an MIT undergraduate design course. **IC** runs on several versions of the MC68HC11 microprocessors and includes such useful features as the ability to initiate and terminate processes and to execute **C** statements immediately—without the need to first compile, link, and load. The interactive nature of **IC** is extremely useful when debugging a robot program.

IC runs on PCs, Macintoshes, and Unix machines. This development has helped make *Mobile Robots* tractable for a wide audience. We write examples throughout the book in both assembly language and **C**, and readers are free to acquire their own copies of **IC**. If you have access to the Internet, this involves logging in anonymously to the MIT Media Laboratory server (Internet Address 18.85.0.47 or cherupakha.media.mit.edu) and using the FTP file transfer protocol to download the **IC** compiler. An updated, expanded, and supported version of **IC** is also available for a modest fee from Newton Research Labs. Visit the website www.newtonlabs.com for more information.

Other **C** compilers are also available for the MC68HC11. Motorola (among others) maintains a library of freeware for the MC68HC11 including **C** compilers. Follow the links from Motorola's site: www.mcu.motsps.com/freeweb/areas.amcu.html A number of commercial products also exist. Dunfield Development Systems (www.dunfield.com), sells a **C** compiler for the MC68HC11 that is likely to be more stable and better supported than the freeware software available over the web.

Now let us turn our attention to a general discussion of microprocessors and everything you ever wanted to know about computers that might be helpful in designing your own robot.

3.2 Microprocessors

Programming an inexpensive, bare-bones microprocessor, such as the one we use in Rug Warrior, differs in some important ways from programming more familiar personal computers, workstations, and mainframe computers. The differences generally relate to the microprocessor's limited computational resources. Typically, such a microprocessor can utilize only a small amount of memory, has no mass storage, and runs at a slower cycle time than its more capable counterparts.

On a large computer, several layers of abstraction (such as the operating system, a high-level programming language and an application program) stand between the user and the underlying machine. These layers are useful because they obviate the need for the programmer to understand the details of the particular processor implementation and its low-level interaction with the peripheral hardware. Unfortunately, the computational overhead required to maintain such abstraction barriers is usually unacceptable for the simplest microprocessors. In most cases, it is necessary for the programmer to fully understand the bit-level interaction between the processor and the devices it controls. The only abstractions available will be those constructed by the programmer.

Recently, an important subclass of microprocessor has become available, the highly integrated *microcontroller*. A microcontroller

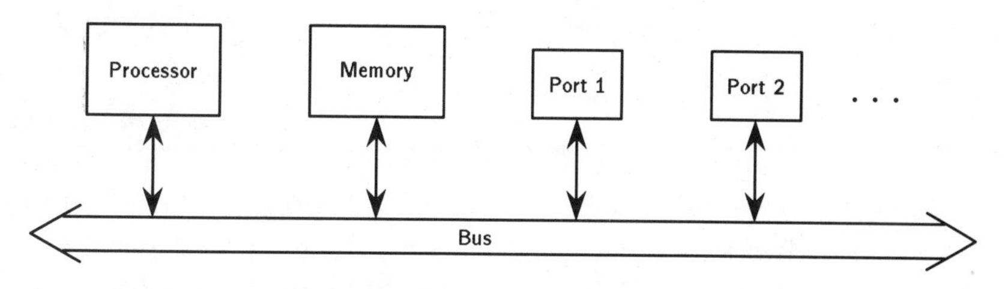

Figure 3.3. The essential elements of a computer are its processor, memory, input/output ports, and bus. The bus provides a communication pathway by which the processor can access and control the peripherals.

combines the small size, low power consumption, and computational abilities of an inexpensive microprocessor with the signal-processing proficiency of discrete circuits. In particular, microcontrollers commonly include such built-in amenities as a serial line (for communicating directly with a terminal or host computer), analog-to-digital converters, timers (for capturing events or activating hardware), and pulse counters. These features greatly simplify system design. Before the advent of the microcontroller, to achieve the sensing and actuation requirements of a robot, it was necessary to construct a system consisting of numerous printed circuit boards connected together. One or more cards were devoted to the processor and the memory; separate cards were required for each sensing and actuation function. Today, the size, complexity, power consumption, and cost of such a system can be reduced by using a microcontroller to perform all the processing tasks in one chip.

In spite of a myriad of variations, computers are basically similar. Figure 3.3 shows the block diagram of a generic computer, reduced to its essential components. A computer consists of a *processor* which executes instructions; *memory*, which stores instructions and data; *ports* which interface the computer to its peripherals ("the outside world"); and a *bus* which provides the communication pathway among processor, memory, and ports.

3.3 The Canonical Computer

It will be instructive as we go along to compare this abstract view of a computer (Figure 3.3) with two other illustrations. The first is the block diagram of the MC68HC11, shown in Figure 3.4, and the second is the schematic of Rug Warrior's logic board, shown earlier in Figure 3.2.

Most microprocessors come in families, and family members are designated with a similar numbering pattern. The MC68HC11 family of microprocessors all come with the same basic features that make them convenient processors for controlling things. Individual members of a family may differ slightly in how much memory or what types of memory they have on the chip. All members would have the same instruction set and use the same assembly language. For instance, the MC68HC11A0 and the MC68HC11E2[1] are two members of the MC68HC11 family. The MC68HC11A0, which we have chosen for Rug Warrior, is at the low end of the line. Individual members of a family also have suffix designations that differentiate the package types available. The suffix FN on the MC68HC11A0FN designates a 52-pin square version. This is the square chip situated in the center of Rug Warrior's board, illustrated in Figure 3.1.

3.3.1 The Processor

The *processor*, or *central processing unit* (CPU), is the controlling element of the computer. Its function is to execute instructions, one after another. The execution of an instruction effects some change in the state of the microprocessor. This may be reflected as an alteration of the value of a memory cell, the contents of an internal register, or the voltage on a line connected to a port.

Instruction execution occurs at a rate fixed by and synchronized with the system clock. This internal clock is driven by an external circuit that includes a high-precision crystal oscillator. In the case of the MC68HC11A0, the output of an 8.000 megahertz (MHz) crystal, connected to lines XTAL and EXTAL (as shown in Figure 3.2), is divided by 4 to produce a clock frequency of 2 MHz. The chip outputs this synchronizing signal on its E line to be used by external circuitry. The number of clock cycles required for an instruction to be completed is a characteristic of the particular instruction, but all MC68HC11 instructions require at least two cycles. Thus, each instruction takes a minimum of 1 microsecond. The longest instruc-

[1] In Motorola's labeling scheme the 8 following the HC in MC68HC811E2 indicates that the chip posses on-chip EEPROM memory.

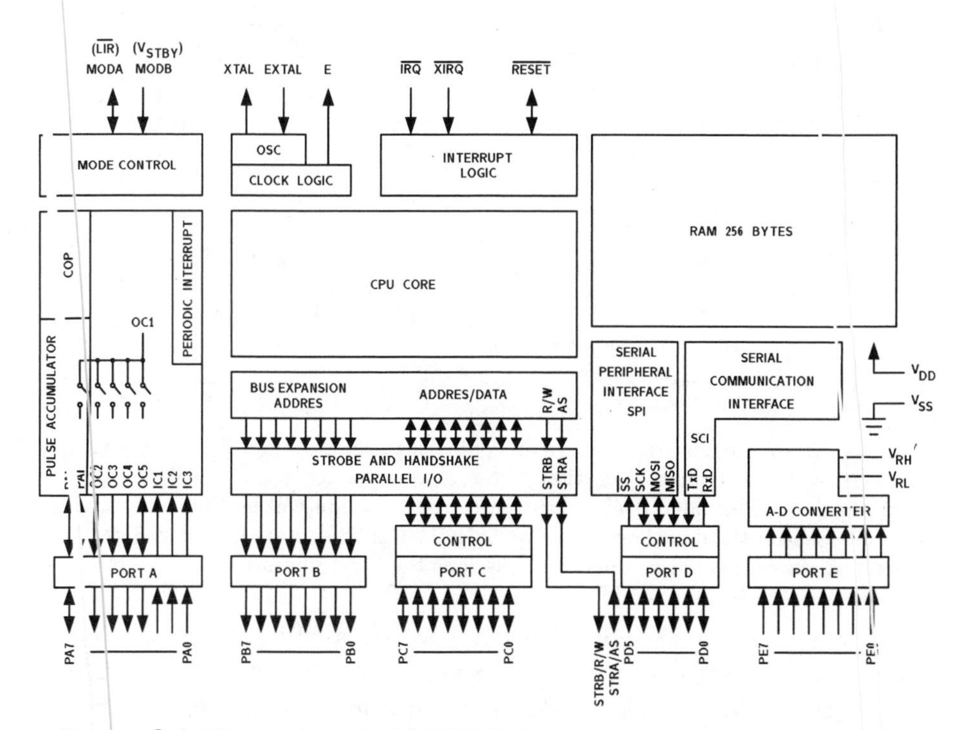

Figure 3.4. We saw how the MC68HC11A0's external pins were connected to the rest of Rug Warrior's circuits. This is the block diagram for the internals of the MC68HC11A0 chip itself. It only comes with 256 bytes of memory, but has eight analog-to-digital converters attached to port E and a timer-counter system associated with port A. Copyright of Motorola, used by permission. (1989)

tions (which do division) take 20.5 microseconds. In order to execute an instruction, the microprocessor must first fetch the instruction and any required data over the bus from its memory.

3.3.2 The Bus

A binary value stored at a particular location in memory is accessed when the CPU places the address of the location on the bus. The range of addresses available, known as the *address space*, is fixed by the width of the bus. In this case, width refers to the number of bits (usually carried by parallel wires) in the address.

The MC68HC11 has a 16-bit-wide address bus and is thus able to select any one of 2^{16}, or 65,536, different locations (also known as 64 K). At each of these locations an 8-bit (=1 byte) data value

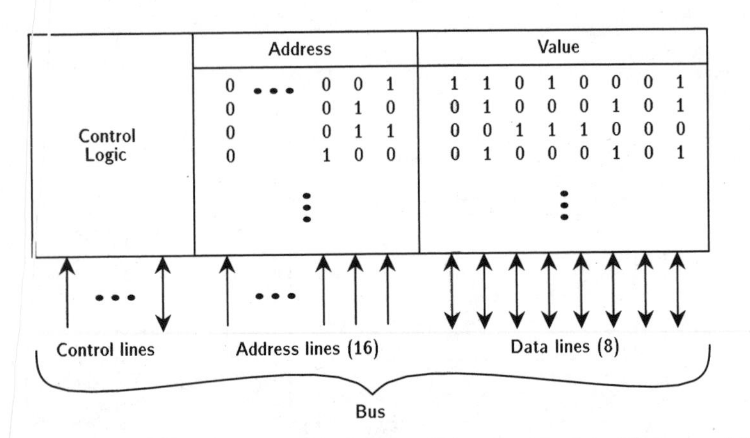

Figure 3.5. of memory. Depending on control signals, a value will be read from or written to the memory location whose address matches the signals on the address lines. The value read or written to that location will be the value that is presented on the data lines. The MC68HC11 has 16 address lines and 8 data lines. (Eight of the address lines are reused as the 8 data lines.)

is stored. The MC68HC11A0 multiplexes data and address signals. When it wishes to read or write a value to memory, it must first assert the address on all 16 address lines; it must then write data to or read data from the 8 lines that previously corresponded to the low 8 bits of the address. Whether address or data is present on these lines is specified by the state of control signals on other lines. Figure 3.5 illustrates how the address lines, data lines and control lines are organized on the bus in order to enable the reading or writing of values to memory.

Elements other than memory locations can be accessed via the bus. A port, which allows interaction between the microprocessor and external devices, may be present. Depending on its nature, the port appears to the microprocessor as a memory location that can be read from and/or written to. To the outside, the port consists of a set of lines to which a voltage can be applied and/or from which a voltage can be generated.

3.3.3 Memory

Computer memory is divided into classes based on whether or not the contents of the memory can be altered, and if so, and how that alteration occurs. The major classes of memory are:

- *random access memory* (RAM),
- *read-only memory* (ROM), and
- *programmable read-only memory* (PROM).

The desirable characteristic of RAM is that it may be read or written at will; such operations are very fast. The contents of RAM, unfortunately, are usually volatile. That is, whatever data is stored vanishes when the power goes off. It is also possible to buy nonvolatile RAM which is simply normal RAM encased in a package that contains a battery. ROM, on the other hand, is nonvolatile but once encoded at the factory cannot be changed.

Finally, PROM memory is nonvolatile and possesses a mechanism that allows the user to program it at least once and possibly to erase it. An important subclass of PROM is EEPROM (electrically erasable programmable read-only memory). EEPROM allows both read and write operations but with some restrictions. The memory may fail if altered more than a specified (large) number of times, and writing may take much longer than with RAM (milliseconds as opposed to nanoseconds).

A more common type of erasable PROM, called EPROM, (erasable programmable read-only memory) can be cleared using ultraviolet light. Such chips have small windows built in so that the physical memory cells can be exposed to an ultraviolet light source.

An important feature of the MC68HC11 family of microprocessors is that versions are available with all three types of memory on the chip. This makes it possible to design applications that need almost no components other than the microprocessor chip itself.

In particular, the MC68HC11A0 chip employed by Rug Warrior has 256 bytes of on-chip RAM but no general purpose EEPROM or ROM. The MC68HC811E2 version has 256 bytes of RAM, 2K of EEPROM, and no ROM. And the MC68HC11E9 has 12K of ROM, 512 bytes of EEPROM, and 512 bytes of RAM.

3.3.4 Ports

A port is the microcontroller's connection to the outside world. A computer for which a port is just a memory location is said to have *memory-mapped input/output*, (I/O). Other architectures are possible. One commonly encountered architecture uses special lines and instructions for accessing peripherals. The venerable Z80 microprocessor uses such a scheme.

Figure 3.6 illustrates how the memory-mapped I/O is arranged for the MC68HC11A0 used on Rug Warrior. The MC68HC11A0 has five ports, labeled A through E. Typically, a microprocessor, as opposed to a microcontroller, has either no ports or ports that support only digital inputs or outputs. Ports on the MC68HC11, however, perform a rich variety of functions.

Port A has eight lines, three of which are dedicated to input, four to output, and one to either function. (Please refer to Figure 3.4 throughout this discussion of ports) A timer-counter system is associated with port A. The input lines, PA0 through PA2, can be used to capture events. When the line changes state, the time of that occurrence is automatically latched into an internal counter. The output lines, PA4 through PA7, can initiate external events.

When the current time matches a preset time, the state of the line can automatically change. One port A line, PA7, can be configured as a pulse accumulator. Each time an externally applied voltage changes state (from high to low or low to high), an internal counter is incremented. These operations, handled by the hardware of the microcontroller, are truly automatic. Once the hardware has been set up in the proper way, no instructions need be executed to perform these functions.

The MC68HC11 has four modes of operation. The actions of some ports depend on which mode has been selected. In the expanded multiplexed mode, the microcontroller uses ports B and C as a part of the bus. In single-chip mode on the other hand, the microcontroller assumes that no external memory is available, so the operation of an external bus is not supported. In this case, port B operates as a digital output port, where each line is a dedicated binary output, and port C operates as a digital I/O port, where each line may be individually configured as input or output. The

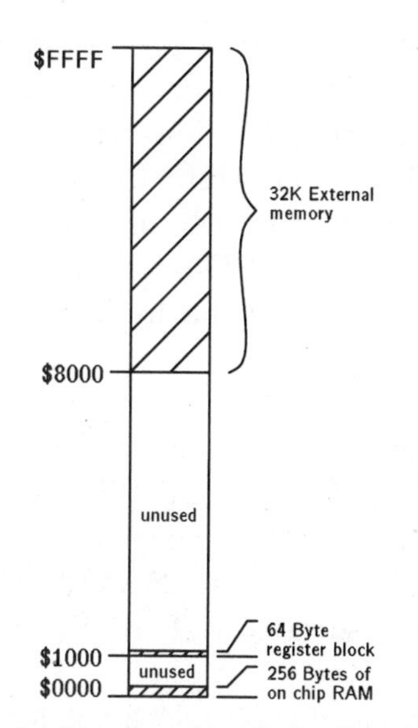

Figure 3.6. A memory map describes the relationship among addresses and the functions associated with each address. Shown here is the map used by Rug Warrior. The $ indicates that the address is given in hexidecimal (base 16) format.

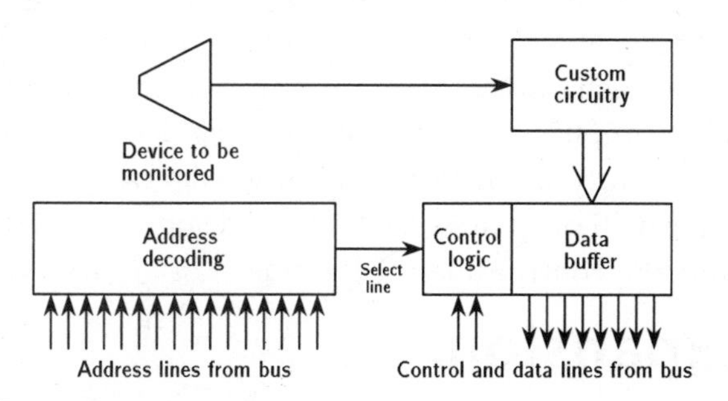

Figure 3.7. It is possible to build ports as desired to enhance the capabilities of a microprocessor. This is done by adding some discrete logic (integrated circuit chips) for address decoding.

remaining two operating modes are special bootstrap mode, used for loading programs into the microprocessor and special test mode, used mostly for factory testing.

Port D has six lines. Each may be configured as either a binary input or output. The lines of this port serve two other important functions, as well. The low-order lines, PD0 and PD1, are part of the communication system. Using these two lines, it is possible to connect the chip to a terminal or host computer. The high-order lines, PD2 through PD5, form a high-speed synchronous data-exchange facility that can be used to network a number of MC68HC11s.

Finally, port E can be used either as a general purpose 8-bit digital input port or as an 8-channel analog-to-digital (A/D) converter. Each channel has 8 bits of resolution. When the A/D converter feature is activated, voltages in the range of 0.0 to 5.0[2] are converted to binary numbers in the range of 0 to 255. Applying, say, 2.5 V to pin PE0 and reading the associated A/D result register would return a value of 128.

If the microprocessor of choice does not have enough ports or if the existing ports have the wrong functions for a particular application, it is possible to build a port of any desired type. We will describe this in detail later, see Section 3.4, in the meantime Figure 3.7 shows how we would go about adding, in this case, an input port. First, design custom circuitry to perform the required interface to the sensor or actuator. Next, build a circuit to decode an address. Chips such as the MC74HC688, which can compare two sets of 8 lines, simplify construction address decoders. Choose an address not currently mapped to any other device. Finally, build a data buffer that will output its contents in response to signals on the select line and the control lines of the bus.

3.4 Expansion

In Section 3.3.4 we noted that it is possible to construct additional memory-mapped ports. In this section we will elaborate on one simple method for accomplishing this expansion. Port expansion circuitry is divided into two components: address decoding and data buffering. The method we describe simplifies the problem by performing only partial address decoding.

[2]In fact, the range depends on the reference voltages at pins VRL and VRH. Most commonly these are set to 0.0 V and 5.0 V respectively.

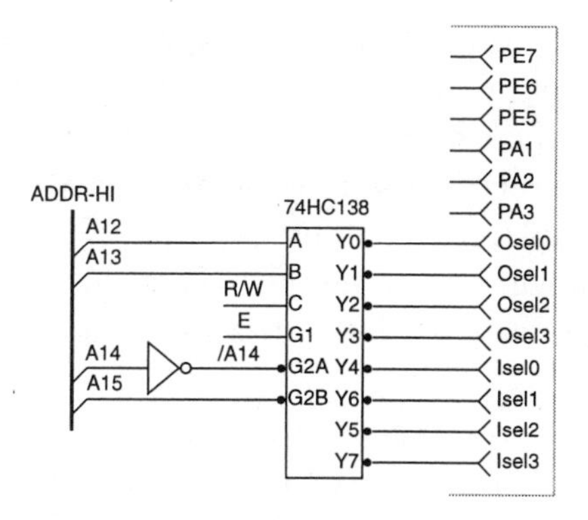

Figure 3.8. Address decoding circuitry is implemented on Rug Warrior Pro™ and on Rug Warrior's RugUp™ upgrade board. Four address lines plus the R/W line are decoded to produce four input select lines and four output select lines.

3.4.1 Address Decoding

Port expansion circuitry connects to the microprocessor's address and data bus. It is the responsibility of the expansion circuit to read data from the bus or write data to the bus. These operations must happen only when a preselected address is referenced and then during just the proper portion of the clock cycle when the microprocessor is ready to send or receive the data. We begin by considering the address decoding portion of the circuit (see Figure 3.8).

To map a byte of data from an I/O port into a particular spot in the microprocessor's memory space of 65536 total bytes, 16 address bits must be specified. This mapping feat requires a circuit to compare two 16 bit quantities—the address currently on the bus and the address of the data port. When the addresses match, the address decoder activates a select line. The second part of the expansion circuit, the data buffer, reads data from the bus or writes data to the bus during the short time when the select line is in the proper state.

Typically, address decoding is accomplished using at least two 18 or 20 pin chips. There is, however, a way to get by with a much simpler one chip circuit if we can accept a partially specified address. That is, our circuit will examine (decode) only some of the address bits and ignore others.

In our example, (see Figure 3.8), expansion circuit address decoding is done using a 74HC138 chip. This chip is called a 1-of-8 decoder because a three digit binary code applied to the chip's A, B, and C inputs will select (cause to go low) exactly one of the chip's 8 outputs. For example, if the inputs A, B, and C all have the value 0, then output Y0 will be low and all others will be high. For inputs 0, 0, and 1, respectively, output Y1 is low and all others high, and so on.

Three other control signals are also required to activate the 74HC138's outputs. Inputs G1 must be high and G2A and G2B must be low. If we connect the MC68HC11's A15 address line to G2B then any time A15 is high, outputs from the 74HC138 chip will be disabled. This is the behavior that we require, because when A15 is high the microprocessor is addressing high memory (addresses of $8000 and higher). (See Figure 3.6.) This is where external memory is located and peripheral devices must not respond to addresses in that range.

Also, we do not want peripheral devices to collide with bytes in the lowest part of memory, $0000 to $00FF or $1000 to $103F, because those are the addresses of the microprocessor's on-chip RAM and the control and status registers, respectively. By requiring A14 to be high, we will avoid that part of memory. We do this by inverting A14 and connecting it to G2A. According to the manufacturer's literature, only during the high part of the clock cycle should peripheral devices read or write to the bus—this can be ensured by connecting the clock signal, E to input G1.

Now we can use A12, A13, and R/W to further select a peripheral device. Using the R/W signal makes certain that an input and an output device will never try to access the bus at the same time. Four outputs of the 74HC138 chip select output devices and four others select input devices.

Putting this all together and representing the binary value of A13 by A and the value of A12 by B we see that one of the outputs of the 74HC138 chip will be active any time the microprocessor selects an address of the form %01ABxxxxxxxxxxxx. Where x represents "don't care," and % indicates a binary number.

For example, if the microprocessor writes a value to the binary address: %0100000000000000 (that is $4000 hex) then the signal

```
Address Range      Inputs  Outputs
===============    ======  =======
0x4000 - 0x4FFF    Isel0    Osel0
0x5000 - 0x5FFF    Isel1    Osel1
0x6000 - 0x6FFF    Isel2    Osel2
0x7000 - 0x7FFF    Isel3    Osel3
```

Figure 3.9. Relation between addresses and active signals for the address decoder portion of the expansion circuit.

Osel0 will become active for a portion of the bus cycle. The data (one byte) that the microprocessor writes can be read by a peripheral device if that device latches (loads and saves) the data present on the data bus at that time. Or, if the microprocessor reads the address $4000, a peripheral device should write its data to the data bus while the signal Isel0 is low. In fact, since only the highest four bits of the address are decoded, Isel0 and Osel0 will become active for any address in the range $4000 to $4FFF.

Any peripheral input device must place its data on the bus when one of Isel0 to Isel3 is low (the device must choose one). Any output device must latch data from the bus only when one of Osel0 to Osel3 is low (again the device must pick one of the signals).

3.5 Data Buffer

The data buffer circuitry must respond in a particular way when the address decoder decides that the microprocessor wishes to access the expansion port. Please refer to Figure 3.10. The circuit shown is that of the RugIO™ Stackable Expansion Module, SEM.[3]

Consider first an input operation. We wish to interpret voltages appearing on the IN0 - IN7 lines of connector J4 as eight bits of data to be input into the robot. IN0 - IN7 are connected to the inputs of

[3]RugIO™ adds eight additional inputs and eight additional outputs to the Rug Warrior Pro™ robot. RugIO™ attaches directly to Rug Warrior Pro™ but to connect RugIO™ to Rug Warrior™, a RugUp™ board is required. Both RugIO™ and RugUp™ are supplied by A K Peters.

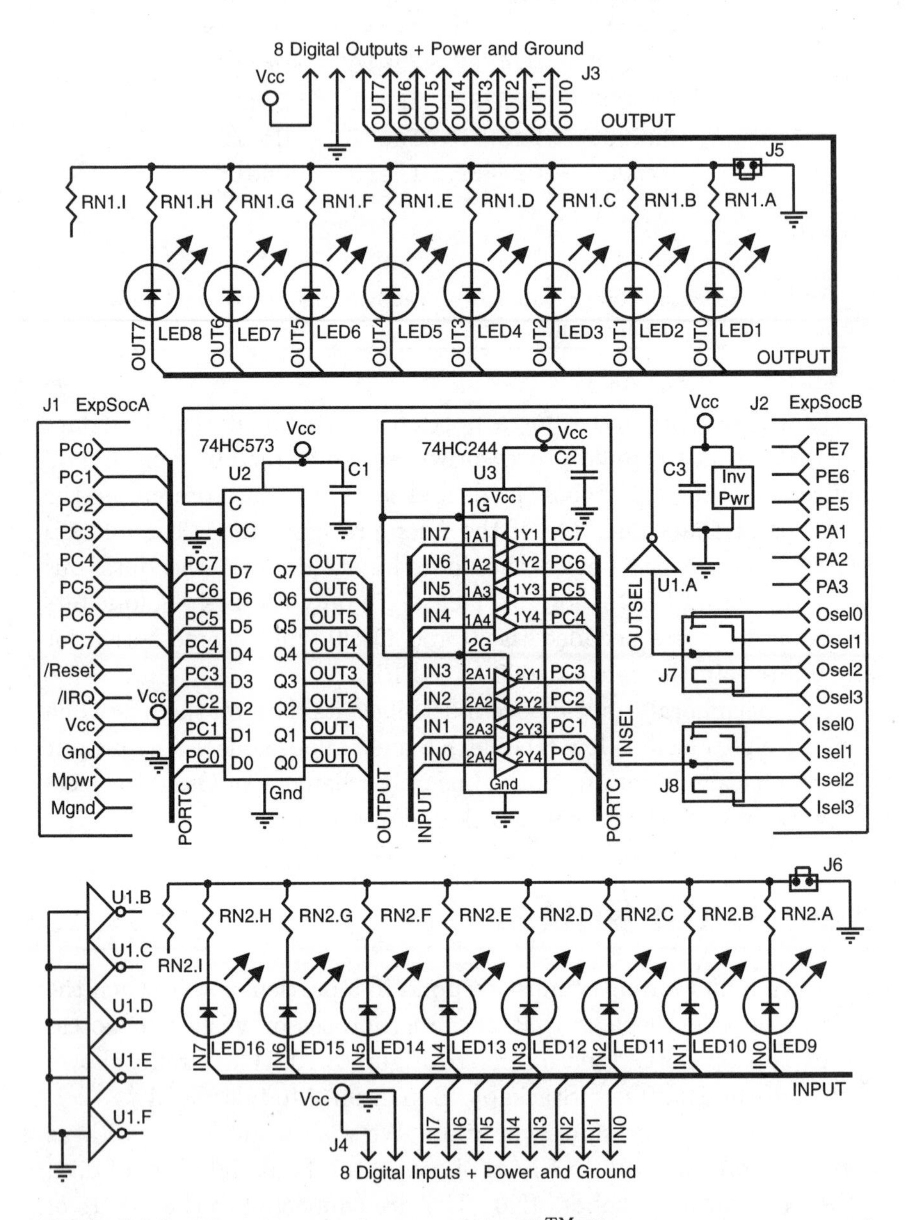

Figure 3.10. The schematic diagram of RugIO™ illustrates memory mapped input and output circuitry. Each of the 16 I/O bits is monitored by an LED. Knowing the state of each line aids in debugging attached circuits.

the 74HC244 buffer chip. Normally, the outputs of the 74HC244 are driven neither high nor low. As long as pins 1G and 2G are high, the 74HC244's outputs are in a high impedance state. This is a vital feature of any chip connected to the data bus as is the 74HC244. If the 74HC244 tried to impose a voltage on the bus at times when it did not have control, signals from the 74HC244 would conflict with what ever other signals were on the bus at that time. This condition is known as *bus contention.*

At the moment when the microprocessor is ready to read data from a particular memory location, the select line on jumper block J8 corresponding to that location, goes low. One of these lines, chosen by the user, is jumpered to INSEL (the dashed line in Figure 3.10 indicates that Isel0 has been chosen). When INSEL goes low the enable lines (1G and 2G) of U3 go low forcing the chip's 8-outputs to leave their high impedance states. In the low impedance state, the logical values present at U3's inputs appears at U3's outputs. Thus, the microprocessor reads whatever values are imposed on IN0 through IN7.

Memory-mapped output is handled in a way similar to memory-mapped input. When OUTSEL goes low, the inverter U1.A causes U2 line C (the latch enable line) to go high. This latches into U2's outputs whatever signals are present at U2's inputs at that moment. Thus, a value written by the microprocessor appears on the OUT0 through OUT7 lines. Builders who wish to construct this circuit should note that standard LEDs drawing 20 mA of current each cannot be used. These levels exceed the maximum current output specifications of the buffer and latch chips. High efficiency LEDs drawing 2 mA or less are required. (Use an HLMP-4700, for example.)

3.6 Rug Warrior Logic

Now that some of the inner mysteries of the MC68HC11 have been divulged, we can present more details of the logic components that run Rug Warrior (see Figure 3.2).

3.6.1 Power

The power switch on Rug Warrior turns on or off power to the microcontroller and sensor circuits while separately controlling the power going to the motor driver chip. It also selects run mode versus download mode. Power is supplied to the MC68HC11A0 through its VDD pin. Ground is connected to VSS.

A three-pole, three-position switch controls power to the circuit. The center position is off. In the down position, programs may be downloaded to the microprocessor from a host. In the up position, all circuit components receive power and a previously stored program will run.

3.6.2 The Clock

An 8.000 MHz crystal provides an accurate and stable time base for Rug Warrior. Such a circuit is critical to the proper functioning of any microprocessor because every operation is synchronized by the clock.

3.6.3 Reset

Pressing the reset button pulls the $\overline{\text{RESET}}$ line low. (A signal name written with an overbar means that the signal is asserted when low.) When this happens, the microprocessor halts—it stops executing instructions. After the button is released and the $\overline{\text{RESET}}$ line goes high again, the microprocessor restarts its program from the beginning.

3.6.4 Mode Selection

As stated above, the MC68HC11 has four operating modes: *single-chip mode*, *expanded multiplexed mode*, *special bootstrap mode*, and *special test mode*. Of these, only the special bootstrap and expanded multiplexed modes are of interest to us.

A particular mode is selected according to the voltages placed on the MODA and MODB lines. When the power switch is in the Download position, MODA and MODB are both low. This places the chip in the special bootstrap mode, where it is possible to load a program via the serial line into the microcontroller's memory. With the power switch set to Run, the program just loaded will begin to execute (after a reset).

3.6.5 Low-Voltage Inhibit

The MC68HC11 is designed to operate at voltages no lower than 4.5 V. However, when the power is switched off, the voltage falls below this level through an illegal range before reaching 0.0 V. In this nether region between 4.5 and 0.0 V, the MC68HC11 exhibits some unmannerly behavior; namely, it may write random values into memory locations. The chip can be inhibited from doing this if the $\overline{\text{RESET}}$ line is held low as power is switched off. This is the purpose of the DS1233M low-voltage inhibit chip.

3.6.6 The Serial Line

In order to program a microcontroller, we must communicate with it in some way. The MC68HC11 facilitates this with a built-in serial line. On a host computer, programs can be typed, edited, and assembled to a form understandable to the microprocessor. Then the machine language form of the program is downloaded to the MC68HC11 through its serial line. Unfortunately, there is an incompatibility between the most common communication standard, RS232, and the microcontroller's format. RS232 specifies that 0's and 1's are represented by voltage swings of -15 to $+15$ V while the MC68HC11, a CMOS (complementary metal oxide semiconductor) device, represents binary digits using 0.0 and 5.0 V. Fortunately, this common problem has a ready solution: Several clever circuits will perform the interface function. We have chosen to use a MAX233 chip for this purpose because it allows full-duplex operation (it can transmit and receive at the same time) and no components besides the chip itself are needed.

3.6.7 External Memory

The HM62256LP-12 RAM chip holds Rug Warrior's 32K-byte external memory. This is exactly half the total memory that a MC68HC11 can directly address. The 32K block fills the upper half of memory, the address space from the addresses $8000 to $FFFF, as illustrated in Figure 3.6.

In single-chip mode, the MC68HC11 assumes that no external memory is available and so it is free to configure ports B and C as general purpose I/O ports. In the expanded multiplexed mode that we use for Rug Warrior, however, the MC68HC11 must use ports B and C to implement the address and data lines needed to access external memory. In this case, these ports cannot be used for I/O. This is the design choice made for Rug Warrior. There is, however, a special chip called a *port replacement unit*, the MC68HC24. When added to the circuit, this chip makes ports B and C available even while operating in the expanded multiplexed mode.

Each byte of the 32K memory space can be addressed by using only 15 address lines. Together, ports B and C provide 16 lines, so one line, PB7, is left over. This line is used to select the memory chip itself.

The high part of the 15-bit address is formed using port B lines PB0 through PB6. Port C lines PC0 through PC7 form the low part. Line PB7 selects the HM62256LP-12 memory chip. Any address of $8000 or above has the highest-order line asserted; that is, PB7 outputs a 1. Thus, the memory chip is selected and will respond only when the microcontroller asserts an address of $8000 or more. Addresses below this number are ignored. The signals from PB7, the low-voltage inhibit chip, and the E pin of the MC68HC11 are combined in a triple-input NAND gate whose output goes to the memory chip's $\overline{\text{CE}}$ (chip enable) line. (The output of a NAND gate is low, if, and only if, all its inputs are high.) The RAM chip is selected only when there is sufficient voltage to operate, when the system clock is in the proper part of its cycle, and when an address of $8000 or higher is specified. If we wished to expand Rug Warrior's memory by filling in the addresses below $8000, we could wire in a second 32K RAM chip. This chip would be selected by inverting the sense of PB7 and connecting it to the new RAM chip's $\overline{\text{CE}}$ line. The new chip would be selected only when PB7 output a 0. This would deselect the first memory chip.

At the beginning of a memory read or write cycle, port C outputs the low part of the address (bits 0 through 7) and port B, the high part (bits 8 through 15). Control signals then cause the low part of the address to be latched by the 74HC573 chip. After latching has been enabled, the 74HC573 chip will continue to output to the memory chip the signal first sent to it by port C, even when data on port C later changes. Thus, during the second part of the read/write cycle, the lines of port C are free to be used as data lines either to write data to or read data from the memory chip. (The AS and RW lines from the microprocessor determine this.) This dual use of the port C lines is known as *multiplexing*.

3.6.8 Battery Backup

As the contents of the external memory chip are volatile, some extra mechanism is required if we wish for the robot to remember its program after the power is turned off. We have chosen a scheme of battery backup for the RAM chip. A very helpful property of chips using CMOS technology is that they require only tiny amounts of current to maintain their state.

Thus, we have routed power from the battery directly to the supply pin, VDD, of the memory chip (bypassing the power switch). This chip continues to be powered, even when the switch is off. This choice has essentially no impact on how long the batteries will last, however, as the current required to maintain the contents of the RAM is only about one microamp. The 74HC10 triple NAND gate is part of the enabling circuitry for the memory chip. By always providing power to the NAND chip, we can make sure that RAM is disabled whenever the power in the main circuit is switched off.

Another alternative is just to buy a nonvolatile RAM chip which is a normal RAM that has a small lithium battery in the case. A nonvolatile RAM chip is only a few dollars more expensive than a normal one. Dallas Semiconductor and Greenwich Instruments sell nonvolatile RAMs.

3.7 Hardware-Software Interface

Software controls hardware and hardware supports software. The nature of this relationship is the topic of this section. In what follows, we will assume the reader has some familiarity with programming in a higher-level language. After an aside concerning number formatting, we will begin with an example of what actually goes on when a program runs.

3.7.1 Representing Numbers

When programming a microprocessor at the lowest level, it is useful to be able to easily refer to numbers in bases 2, 10, and 16 (known, respectively, as binary, decimal, and hexadecimal, or hex). Unfortunately, every programming language establishes its own standard for specifying the base. Unless the base is clear from the context, we will use the convention shown in the following table for representing numbers in assembly language programs and in **C** language programs. The decimal number 123 is used as an example.

	Assembly language		**IC language**	
Base	**Prefix**	**Example**	**Prefix**	**Example**
2	%	%01111011	0b	0b01111011
10		123		123
16	$	$7B	0x	0x7B

The base 2 representation for the decimal number 123 is 01111011. The syntax of our assembler requires us to specify this as %01111011 so that it understands that we mean the binary number 01111011 and not the decimal number 1,111,011. The syntax for **IC** would have us write 123 in binary form as 0b01111011. Similarly, hex numbers are specified using the $ prefix for our assembler and a 0x prefix for **IC**. The prefix 0b for representing binary numbers is part of **IC** but is not included in standard **C**.

3.7.2 An Example

The details of writing a workable program and loading it into the microcontroller will be explained later. For the moment, we will assume that a simple three-instruction program has already been loaded. We will watch what happens as the program runs.

Figure 3.11 illustrates the changes that take place in two of the microprocessor's internal registers and an output port when the following fragment of a program runs:

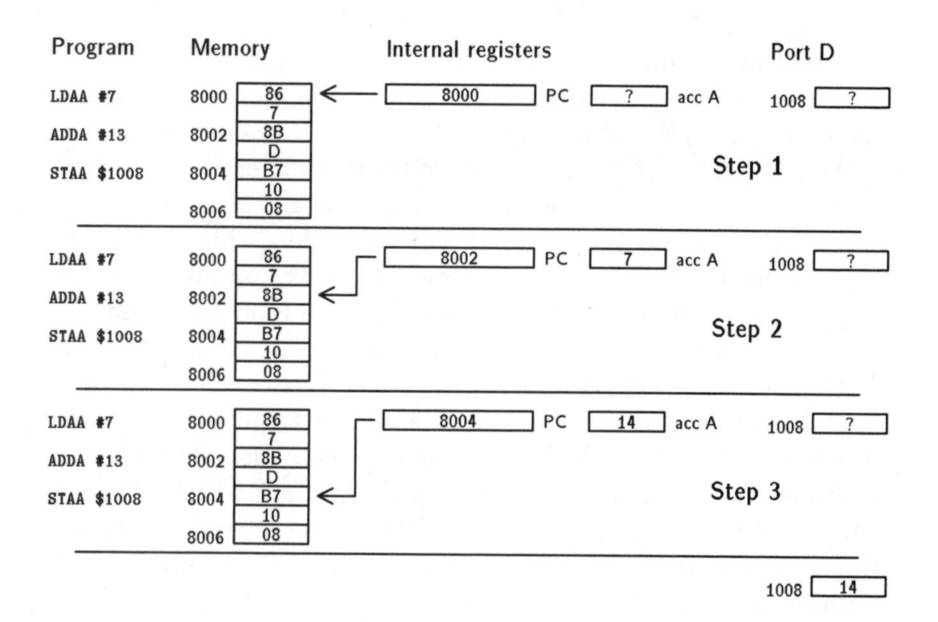

Figure 3.11. The program counter (PC) keeps track of which instruction the CPU will execute next. As each instruction is processed, the address of the next instruction is placed in the PC. The contents of internal registers and memory cells are altered as a result of instruction execution. Here, three steps in a program are shown. The final contents of address $1008 is $14 (hex) or 20 (decimal).

```
LDAA #7        ;Load 7 into accumulator A, # means immediate
ADDA #13       ;Add 13 to accumulator A
STAA $1008     ;Store contents of A to port D
```

The left column of Figure 3.11 (labeled Program) shows the code written by a programmer. In this case, the program consists only of the names of instructions and arguments for those instructions.

The second column, Memory, displays the contents of memory (in hexadecimal) after the program has been loaded. To translate the code supplied by a programmer into the internal representation (the machine code) used by the microprocessor, another program called an *assembler* is required. The `LDAA` instruction has been converted into its machine language code, which happens to be the number $86. This `LDAA` instruction is stored at memory location $8000. The numbers into which the instruction mnemonics are converted are also known as *opcodes*. Following $86 in memory is 7, the argument that will be used by this instruction.

The third column reports the state of two special registers internal to the microprocessor. The program counter, or PC, is the microprocessor's way of keeping track of where it is; the value stored in the PC is the address in memory of the instruction the microprocessor is about to execute or the argument it is about to fetch. Note that the box representing the PC is twice as wide as those representing memory locations and other registers. This indicates that the PC holds a 16-bit address while the others hold 8-bit data values.

The MC68HC11, like many other microprocessors, requires nearly all computations to be performed in a special register called the *accumulator*. For example, it is not possible to directly add the contents of one memory cell to that of another. Rather, one value must be loaded into the accumulator and then the next must be added to the contents of the accumulator. The MC68HC11's accumulator A, one of its two 8-bit accumulators, is shown beside the program counter Figure 3.11.

Finally, port D, which resides at location \$1008 in the memory map is shown in the last column. The purpose of the program is to change the value stored at memory location \$1008 and thus the voltages on the lines connected to port D.[4]

Step 1 of Figure 3.11 shows the state of the microprocessor before any computation has taken place: The program has been loaded, the program counter is pointing to the first instruction, and the contents of accumulator A are arbitrary and unknown. When the program begins execution, the microprocessor uses the address stored in the PC to get the first instruction opcode, \$86.

It then increments the PC. Interpreting this instruction tells the microprocessor two things: how to find the instruction's operand and what to do with the operand. In this case, the value fetched from the memory location pointed to by the PC, location \$8001, is the *operand*, 7. (An operand is a data value that is processed by an instruction in some way.) `LDAA` further instructs the microprocessor to place this value into accumulator A.

By the beginning of Step 2, accumulator A holds 7 and the PC points at the next instruction, `ADDA #13`. Again, we use the PC to

[4]To simplify the example we assume that register DDRD, the data direction register for port D, has already been set correctly, enabling the lines of port D as outputs.

fetch the operand, 13, but the **ADDA** instruction causes its operand to be added to the contents of accumulator A. Step 3 shows the result: $7 + 13 = 20$ decimal or $14 hex.

The last statement, **STAA $1008**, finally effects a change in the world outside the microprocessor. This command causes the contents of accumulator A to be transferred to port D. The argument of **STAA** is the address where the data is to be stored. The binary representation of $14 is %010100. This is interpreted by the hardware of port D as a set of voltages to be output. In particular, pins PD0, PD1, PD3, and PD5 are set to 0 V, while pins PD2 and PD4 are set to 5 V. From the schematic of Rug Warrior's sensors and actuators (see Appendix A), we observe that this will make LEDs (light emitting diodes) 1 and 3 glow.[5]

3.7.3 CPU Registers

The MC68HC11 has several registers internal to its CPU, besides the two introduced in the preceding example. Figure 3.12 offers a graphical representation of the register set we will describe more fully later.

Accumulator A and its twin, accumulator B, are both 8-bit registers used for performing arithmetic computations. Some instructions treat these registers as if they were a single 16-bit accumulator. In this case, accumulators A and B are referred to collectively as the double accumulator, accumulator D.

The register known as the stack pointer (SP) is used to hold a 16-bit address. The operation of this register will be explained later in the context of the stack (see Section 3.7.9).

Registers that hold 16-bit values IX and IY are known as *index registers*. They are used by the indexed-addressing mode to access instruction operands. Additionally, register IX is used by the division instructions.

The condition code register (CC) is an 8-bit register that holds information about recent CPU operations. Each bit of this register

[5]The LEDs described in the example are present on Rug Warrior but were eliminated from Rug Warrior Pro™. Rug Warrior Pro™'s LCD made separate debugging LEDs superfluous. Running the example code on Rug Warrior Pro™ will activate the left IR emitter and select positive rotation for the left motor.

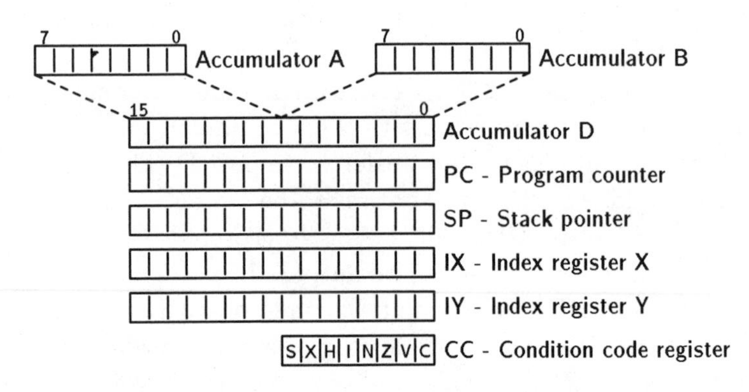

Figure 3.12. The internal registers of the MC68HC11 describe the processor's state.

has a special purpose (described in the MC68HC11 documentation). For example, when the **STAA** instruction stores a 0 value to memory, the Z bit of the condition code register is set to 1. If any number other than 0 is stored, the Z bit is 0. If two numbers are summed to zero by, say, the **ADDD** instruction, the effect on the condition code register is the same.

Other bits signify other conditions. If the most recently processed instruction produced a negative number, then the N bit is set. The occurrence of an arithmetic carry causes the C bit to be set.

An arithmetic overflow affects the V bit. Branch instructions, which are discussed later, (see Section 3.7.7), examine the state of the bits in the condition code register to determine whether or not to transfer control to another part of the program.

We can think of these bits as flags. When some condition is met, the corresponding flag is raised—the bit is set to 1. When the condition is not met, the flag is lowered—the bit is cleared to 0.

None of the MC68HC11's registers appear in the memory map. This means that the only way to access these registers is through the use of special instructions. Certain microprocessors do, however, map their internal registers to memory.

3.7.4 Instructions and Operands

The instruction set of a microprocessor is the set of primitive operations that it can carry out. Figure 3.13 lists a majority of the instructions in the MC68HC11 family's instructions set. Most instructions require one or more operands. In the above example, the operand of the LDAA instruction was 7. The instruction stored that number in accumulator A. The operand of the ADDA instruction was 13. Executing ADDA added this number to accumulator A.

An instruction can locate its operand in several ways. In the example; LDAA and ADDA both used a form called *immediate addressing*. With this method, the operand itself is stored in memory following the instruction code. Figure 3.14 illustrates an example of immediate addressing.

In an assembly program, the programmer specifies how the operand is to be found by the way the instruction's argument is written. The # sign in front of the numbers 7 and 13 in the program in Figure 3.11 indicates to the assembler program that these numbers should be referenced using immediate addressing. The following list summarizes the operand-addressing schemes used by the MC68HC11.

Immediate: The operand itself follows the instruction code in the program stream. The argument is prefixed by #. *Example*: `ADDA #$2F` means that the hex value $2F should be added immediately to the value of accumulator A.

Extended: The argument is the address of the operand. Two bytes are required to form the address (given the 64K address space of the MC68HC11). The argument has no prefix. *Example*: `JSR subr_foo`.

Direct: Direct addressing is similar to extended addressing except that it takes one less byte to specify the operand. The first 256 bytes of the address space are sometimes called the *zero page*. Because the high-order byte is always 0, this portion of the memory (which corresponds to the MC68HC11's on-chip RAM) can be addressed with only 1 byte. The argument again requires no prefix. *Example*: `LDAA variable_1`.

Mnemonic	Operation performed
ADDA	Add argument to acc A
ADDD	Add double; add argument to acc D
BCLR	Bit clear; clear specified bits of memory location
BEQ	Branch if result = 0
BGT	Branch if result is > 0 (signed)
BHI	Branch if higher (unsigned)
BLO	Branch if lower (unsigned)
BLT	Branch if result is < 0 (signed)
BNE	Branch if result ≠ 0
BRA	Always branch
BRCLR	Branch if specified bits are clear
BRSET	Branch if specified bits are set
BSET	Bit set; set specified bits of memory location
CLI	Clear I flag of CC register, enable interrupts
COMA	Complement; bitwise negation of acc A, $FF − argument
IDIV	Divide one 16-bit integer by another
JMP	Jump to an absolute address
JSR	Jump to subroutine
LDAA	Load a value into acc A
LDAB	Load a value into acc B
LDD	Load double; load argument into acc D
MUL	Multiply two 8-bit numbers, return 16-bit number
NEG	Two's complement argument, 0 − argument
NOP	No operation; this instruction makes no changes
PSHA	Push contents of acc A onto stack
PULA	Load acc A with value at top of stack
RTI	Return from interrupt
RTS	Return from subroutine
SEI	Set I flag of CC register; disable interrupts
STAA	Store acc A to memory
STAB	Store acc B to memory
STD	Store double; store acc D to memory
SUBA	Subtract argument from acc A
SUBD	Subtract double; subtract argument from acc D
TSTA	Test acc A; set condition codes accordingly
TSX	Transfer Stack pointer to IX register

Figure 3.13. Selected instructions from the MC68HC11 instruction set. Code written using this instruction set, typed in and edited on a host computer, would be assembled into machine code and then downloaded to the robot via a serial cable.

Indexed: The argument is added to the contents of the IX or IY register to compute the address of the operand. This addressing scheme is useful for accessing items within blocks of data. If, for example, the address of the 0th element of an array of values is loaded into the IX register, then any other element can be found by giving an instruction only the index of the desired element. *Example*: Suppose that the number $9000 has been loaded into the IX register. The instruction `LDAA 3,X` will then place the value stored at location $9003 into accumulator A.

Relative: The argument is added to the program counter to compute the operand. This is the scheme used by branch instructions to pass control forward or backward in the program. The argument requires no prefix or other indicator. *Example*: `BRA label_1`.

Inherent: No explicit operand is required for instructions using inherent addressing. *Example*: The `TSX` instruction takes no argument. It transfers the contents of the stack pointer to the IX register.

3.7.5 Arithmetic

In the discussion so far, you may have noticed the absence of numbers other than integers. This is no coincidence. Unless the microprocessor comes equipped with special hardware for dealing with floating-point numbers or numbers containing exponents, integer arithmetic is all the microprocessor is able to do. If floating point computations are required, the programmer must write routines that implement such computations entirely from the integer-based instructions native to the microprocessor. Floating-point operations typically require much more time and storage space than integer operations.

An integer represented by a binary value can be interpreted in one of two common ways. An 8-bit byte, for example, can be seen as an unsigned integer in the range of [0, 255] or as a signed integer in the range [−128, 127]. Sixteen-bit quantities can also be regarded as either signed or unsigned.

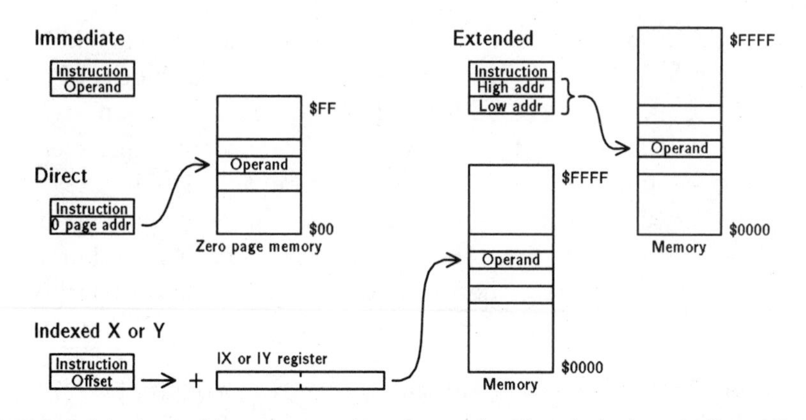

Figure 3.14. A graphic representation of several addressing schemes. Immediate addressing finds the operand itself stored following its instruction in memory. The extended and direct addressing schemes have the address of their operands stored in the memory location following the instruction. The operands are found by fetching from these locations. Indexed addressing computes the address of the operand by adding the number following the instruction to the contents of a special CPU register. The operand can be fetched once this calculation has been performed.

The unsigned representation, in which each bit of the byte corresponds to a power of 2, is straightforward. If the bits of byte B are designated b_n, where n ranges from 0 to 7, then the integer I represented by byte B is:

$$I = \sum_{n=0}^{7} b_n 2^n$$

For example, $\%00001011 = 2^3 + 2^1 + 2^0 = 11$ decimal.

Negative numbers are represented in 2's complement form. Suppose we wish to construct a byte that holds the 2's complement of, say, -5. First, take the 8-bit representation for 5, %00000101, and complement it, obtaining %11111010. The 2's complement operation replaces every 0 bit with a 1 and every 1 with a 0 and then adds 1 to get %11111011. This representation has the correct arithmetic property: If we add -5 and $+5$, we know we should get 0, and, properly, $\%11111011 + \%00000101 = \%00000000$. (This operation sets the carry bit, C, in the condition code register.)

The value stored in a memory location is just a string of 1's and 0's. Whether %11111011 is to be interpreted as −5 or +251 is left to the programmer. Different instructions of the MC68HC11 are used to select one interpretation or the other. It is also necessary to use different instructions depending on whether we want to manipulate 8-bit or 16-bit quantities.

3.7.6 Control and Status Registers

As stated earlier, several of the MC68HC11's ports have multiple functions. How does the microprocessor select one function as opposed to another? The answer is that special memory-mapped registers (not to be confused with the CPU's internal registers) control these functions.

In the example program in Figure 3.11, we used port D to control some external devices, four LEDs. The instructions listed in that program will not have the desired effect. That is, they will not turn on the LEDs unless we first configure the pins of port D as outputs. To effect this change, we must use the DDRD register.

DDRD	Bit 7					Bit 0
$1009	–	–	D5	D4	D3	D2	D1	D0
			1	1	1	1	0	0

Register DDRD will contain the value shown if we first execute these instructions:

```
LDAA #%111100          ;Set PD2,3,4,5 for output, PD0,1 for input
STAA $1009             ;Store data to memory-mapped register DDRD
```

The MC68HC11A0 has a total of nearly 50 control registers, status registers, and ports. The purpose of each bit of each register and the default state of each bit (that is, whether the bit is a one or a zero after a system reset has occurred) is specified in the MC68HC11A0 documentation.

3.7.7 Jumps and Branches

Flow control in a microprocessor is implemented by *branch* and *jump* instructions. Consider the following program. Its purpose is to compute the absolute value of the 8-bit signed integer that has previously been loaded into accumulator A.

```
ABS           ;ABS is the label of this section of code
  TSTA        ;Test value in acc A (maybe set CC register N flag)
  BLT ABS-NEG ;If value is less than 0 branch to label ABS-NEG
  BRA ABS-END ;If the above branch was not taken then go to ABS-END
ABS-NEG       ;The dash is not a minus sign, just a part of the name
  NEGA        ;Negate the quantity in acc A
ABS-END
```

`ABS`, `ABS-NEG`, and `ABS-END` are not instructions but rather symbolic labels created for the convenience of the programmer. Such labels make it easy to refer to specific points in the instruction stream.

The first instruction, `TSTA`, examines the contents of accumulator A. It sets the condition code bits appropriately. In particular, if the number in accumulator A is negative, `TSTA` will set the N bit of the condition code register. The next instruction, `BLT`, is the "Branch if Less Than Zero" instruction. If the N bit of the condition code register is set, then this instruction will cause control to be passed to the instruction following the label `ABS-NEG`. If the N bit is clear, then control passes to the instruction following `BLT`. That is, control flow will pass to the `BRA` instruction.

The `BRA`, or "BRanch Always," instruction, is an unconditional branch. It always forces program control to jump to the address specified by its argument. The effect of these branches is that if the contents of accumulator A is positive, then flow control jumps to the end, leaving accumulator A unchanged. If accumulator A holds a negative number, the `NEGA` instruction following the `ABS-NEG` label is executed, negating the contents of accumulator A. The MC68HC11 offers many additional branch instructions for testing other arithmetic conditions.

There is one subtlety to be aware of when using branches. The operand of each branch instruction is only 1 byte long. This means the instruction cannot specify the absolute address of the location to which it will pass control. Rather, a branch causes a jump forward or back in the instruction stream. The 1-byte operand can specify a displacement of 127 locations forward or 128 locations backward from the memory location in which the branch instruction is stored. To go further than that, we must use a `JMP`, or "JuMP," instruction. This instruction takes a 2-byte operand and can pass control to any location in the MC68HC11's memory space.

3.7.8 Subroutines

The previous example illustrated how we might implement an absolute value function by writing it directly into the instruction stream of the program. If an absolute value were needed at another point in the program, the same code could be repeated. The labels **ABS**, **ABS-NEG**, and **ABS-END** would have to be changed (perhaps by calling them **ABS-1**, **ABS-2**, etc.) to eliminate ambiguity.

We can make more efficient use of the available memory if we implement, as a *subroutine* any piece of code that is used repeatedly. This is also true with higher-level language programs. By adding one instruction to the previous example, we can turn the code fragment into a subroutine.

```
ABS                 ;Subroutine named ABS
  TSTA              ;Test value in acc A (set CC register flag)
  BLT ABS-NEG       ;If value is less than 0 branch to ABS-NEG
  BRA ABS-END       ;If the above branch was not taken go to ABS-END
ABS-NEG
  NEGA              ;Negate the quantity in acc A
ABS-END
  RTS               ;Return to the place where the subroutine was called
```

The essential difference between this subroutine and the *in-line code* (code in the main body of the program) in the previous example is the inclusion of the **RTS**, "ReTurn from Subroutine," instruction. This instruction causes control to switch back to the point elsewhere in the program where the subroutine was called.

Subroutine **ABS** assumes that the argument (the quantity whose absolute value is to be computed) has been stored in accumulator A. It also returns the result in accumulator A. To call **ABS**, we could say:

```
  LDAA Value        ;Load acc A with some value
  JSR ABS           ;Jump to subroutine ABS
  STAA Value        ;Store the result
```

The assembler program replaces the label **ABS** with the address of the first instruction in the subroutine. This allows **JSR**, the "Jump to SubRoutine" instruction, to determine where to transfer program control. But how does the microprocessor find its way back to the instruction following the **JSR ABS** after the subroutine has been completed? The answer is that, before transferring control to subroutine

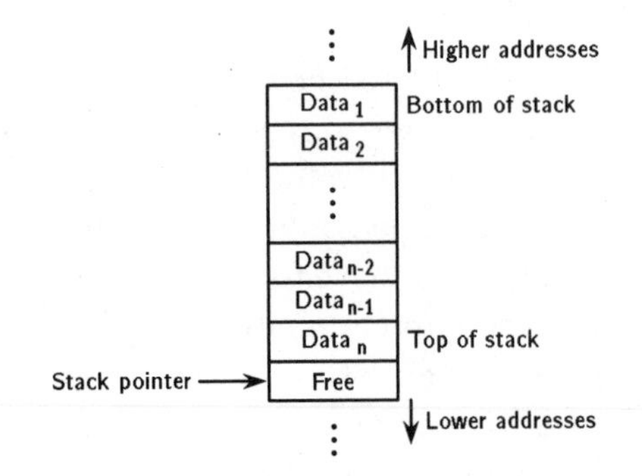

Figure 3.15. A value is added to the stack by storing it at the location pointed to by the stack pointer and then decrementing the pointer. A value is retrieved from the stack by incrementing the pointer and returning the item at that memory location. The stack illustrated here grows by inserting values at decreasing memory addresses. However, stacks that grow toward increasing addresses are also often implemented. In either case, the location indicated by the stack pointer is still considered the "top."

`ABS`, the `JSR` instruction pushes the address of the next instruction, `STAA Value`, onto the stack.

3.7.9 The Stack

Modern computers make use of a *stack* to transfer control to and from subroutines, to pass information, and to store local variables. Figure 3.15 illustrates implementation of a stack. A stack is implemented as a contiguous set of addresses in RAM memory. The stack pointer, or SP, (usually an internal register of the microprocessor) holds the address of the next free location. When a value is "pushed" onto the stack, that value is written to an address specified by the stack pointer. The stack pointer is then decremented. To "pop" a value from the stack, the stack pointer is incremented. The value of the stack pointer is then used as the address of the operand to be fetched.

The sequence shown in Figure 3.16 illustrates how the stack is used to transfer control between in-line code and a subroutine. When the `JSR` instruction is executed, the 2-byte address of the next instruction in the instruction stream, `STAA Value`, is placed on the stack. After `ABS` has finished, the `RTS` instruction loads the program

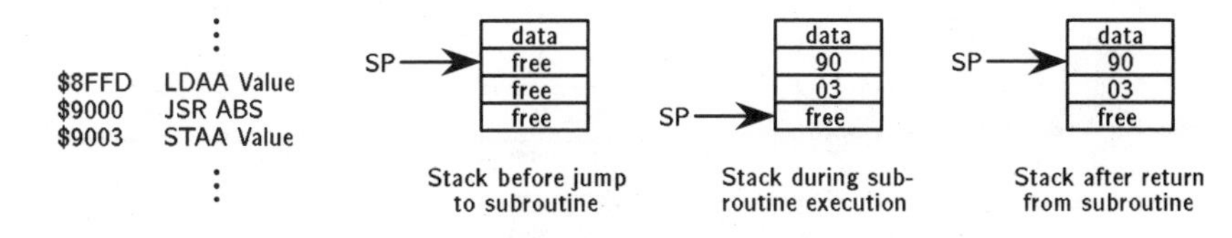

Figure 3.16. This sequence of steps shows how the stack is used to pass control to and from a subroutine. The values placed on the stack, the return address bytes, remain after subroutine ABS has been completed. However, the next number written to the stack will overwrite these values.

counter with the top 2 bytes on the stack. Control is thus returned to the instruction following the JSR instruction.

Nothing prevents a programmer from "nesting" subroutines, that is, having one subroutine call another. Although there is no advantage to doing so in the simple example given here, it is even possible for a subroutine to call itself. This very powerful concept of subroutines calling themselves is known as *recursion*.

3.7.10 Passing Arguments

A key issue in using subroutines is determining how to pass arguments and results between the calling code and the subroutine. In the preceding example, this posed no problem. Since both the single argument and the result were only 1 byte long, each fit neatly into accumulator A.

One common way to pass large chunks of data to a subroutine is to send the address rather than the data itself. If the data to be processed are stored at a block of successive memory locations starting at DATA-ADR, we can enable the subroutine PROC-DATA to access them by:

```
LDX #DATA-ADR    ;Load the address of the data into the IX register
JSR PROC-DATA    ;Jump to the data-processing subroutine
```

The same subroutine can now be used to process any number of different blocks of data. The subroutine can, for example, look at the first item in a block by saying LDAA 0,X. It could acquire the third with LDAA 2,X.

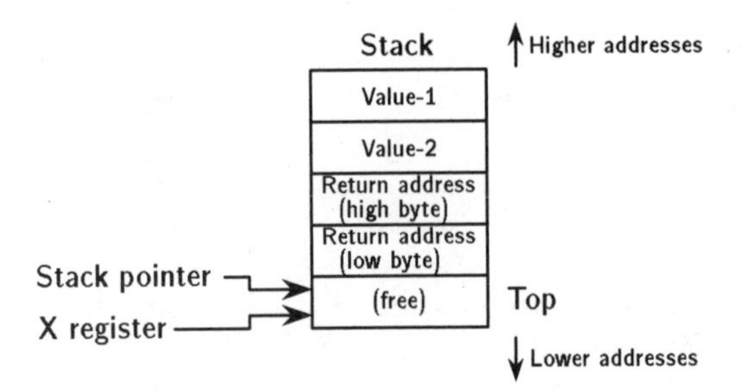

Figure 3.17. Data can be passed to a subroutine via the stack. VALUE-1 and VALUE-2 are pushed onto the stack and a Jump to SubRoutine (JSR) instruction is performed. One result of the JSR instruction is to leave the return address on the stack.

This procedure is adequate if it is possible to allocate storage for all data in advance and keep such storage space around indefinitely. There is, however, another more clever way to create a block of data "on the fly" and reclaim the memory space used when the data are no longer needed. We can create local variables by storing temporary data on the stack. Suppose, for example, we wish to pass two 8-bit values, `Value-1` and `Value-2`, to a subroutine. The calling code might say:

```
LDAA Value-1     ;Get the first value
PSHA             ;Push that value onto the stack
LDAA Value-2     ;Get the second value
PSHA             ;Push it onto the stack
JSR SUBRTN       ;Jump to the data-processing subroutine
```

Figure 3.17 shows the situation after the jump to the subroutine has occurred. Notice that jumping to the subroutine has placed additional data on the stack. The two bytes that comprise the return address appear following the data of interest. In order for the subroutine to access the stored values, all it must do is point the IX register to the same address as the stack pointer (the `TSX`, Transfer SP to IX instruction will accomplish this) and bypass the return address bytes to access the data. To get `Value-1`, we can say `LDAA 4,X`; `Value-2` is accessed with `LDAA 3,X`.

Values can be passed back to the calling code by storing them on the stack, as well. If it is important not to overwrite the calling data already on the stack, then the calling code should push extra dummy values onto the stack so that the subroutine has space to store its results.

Use great care when manipulating the stack in this way. If the return address stored there is accidentally overwritten, the microprocessor will almost certainly crash when the return from subroutine instruction is executed.

3.8 Real-Time Control

To this point, we have reviewed the fundamental components of microprocessor software and described how they are supported by the hardware. Next, we discuss how to assemble these building blocks into strategies for real time control.

There are three strategies for writing software that can respond to external events in real time. *Polling* is a method where the software loops, continuously checking an input pin. Polling ties up the processor, keeping it busy even when no external events are happening. *Interrupt*-driven software is more efficient. In this method, the external event creates a signal that directs the processor to postpone whatever it is doing and respond to that event immediately. The third strategy, *input capture*, can be used if the processor has special hardware, known as input capture registers. By taking advantage of this special purpose hardware, the processor is never interrupted. Instead, event handling is taken care of in the background. We will expand on these concepts with some examples.

3.8.1 Polling

Suppose that we wish to monitor an input closely in order to take action immediately after some event of interest has occurred. We might, for example, wish to measure the time of flight of a sonar pulse. Figure 3.18 illustrates an example of sonar ranging. It is important to measure precisely the difference between the time the pulse was sent out and the time it returns. In the following example,

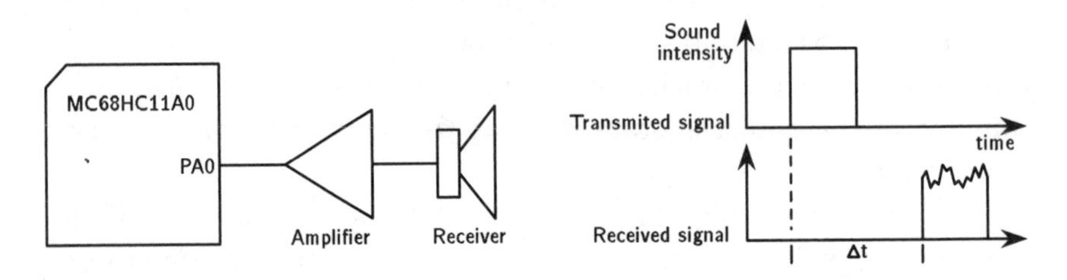

Figure 3.18. The time of flight of a sonar echo can be measured using polling, interrupts, or input capture.

we assume that the output of the ultrasonic receiver is connected to PA0. This line goes high when an echo is detected.

The following subroutine will measure the time difference between initiation of the sonar ping and detection of the returning echo, with an accuracy of a few microseconds. This subroutine must be called immediately after the ping starts. It will store the measured time in `Sonar-tof`. The timer-counter system associated with port A is used to measure the time of flight of the sonar pulse. The timer counter is a 16-bit register called TCNT, which is a free-running timer. With every clock cycle, the hardware automatically adds 1 to the contents of a 16-bit register called `TCNT`. The following `TIME-SONAR` code uses this feature to advantage:

```
TIME-SONAR              ;Measure the time of flight of a sonar echo
  LDD TCNT              ;Get the starting time from the system timer
  STD Sonar-tof         ;Save start time
WAIT-FOR-ECHO
  BRCLR PORTA, %00000001, WAIT-FOR-ECHO ;Keep checking PA0
  LDD TCNT              ;Echo detected so get current time
  SUBD Sonar-tof        ;Acc D now holds current time -- start time
  STD Sonar-tof         ;Store the 16-bit time to location Sonar-tof
  RTS                   ;Return to the calling code
```

`TIME-SONAR` begins by loading the value of `TCNT` into accumulator D. Next, it executes the `BRCLR PORTA, %00000001, WAIT-FOR-ECHO` instruction. This instruction tests the state of the lowest-order bit of port A—the bit corresponding to line PA0. If the value of this bit is 0, a branch is made to `WAIT-FOR-ECHO`; that is, the same instruction is executed again. Program control thus stays in a tight loop, repeatedly testing the state of PA0. When an echo finally returns and PA0 goes high, control passes to the next instruction. This instruction, LDD TCNT, and the one following it, SUBD Sonar-tof, compute the difference between the time the WAIT-FOR-ECHO routine started and the time the echo returned. Finally, this time-of-flight value is stored in the variable Sonar-tof. The main code that calls and responds to the TIME-SONAR subroutine could be written:

```
JSR Turn-on-sonar      ;This subroutine initiates the sonar ping
JSR TIME-SONAR         ;Jump to the example code
JSR Compute-distance   ;Use the measured time to compute distance
```

This strategy of repeatedly checking for a condition is known as polling. For the sake of simplicity, an important safeguard has been left out of this example. If it happens that the sonar ping fails to return, then control will never advance beyond the tight loop. The program will be stuck indefinitely. A robust program would include within the loop some sort of time-out feature to exit the routine should the echo take too long.

3.8.2 Interrupts

Polling offers an effective way to respond quickly to real-time events. The problem is that this method can use up all the microprocessor's resources, waiting for just one event. While the microprocessor waits, it cannot do anything else. Much more efficient use could be made of the microprocessor's time if there were some automatic way of responding to an event. The microprocessor should only have to take action (execute instructions) when the event actually occurs. Such a mechanism exists and is called an *interrupt.*

An interrupt is an event that triggers an automatic response in the microprocessor. The code that responds to that event is called the *interrupt service routine.* Interrupt service routines are quite similar to subroutines except that they are called by the occurrence of an event rather than by a JSR instruction.

Interrupts are asynchronous; the microprocessor cannot anticipate when an interrupt will occur. Thus, when an interrupt does happen, the microprocessor will be executing some piece of unrelated code. To respond to the interrupt, it will first have to stop executing the current code and save the state of the ongoing computation on the stack. Then it must locate the proper interrupt service routine and transfer control there. After servicing the interrupt, the microprocessor must be able to restore its pre-interrupt state and return control to the code that was running originally, before the interrupt.

We will now rewrite the sonar-ranging example from the previous section, demonstrating event handling using of an interrupt service routine:

```
TIME-SONAR-ISR         ;Sonar timer interrupt service routine
  LDD TCNT             ;Get the time at which the interrupt occurs
  SUBD Sonar-tof       ;Difference is echo time of flight
  STD Sonar-tof        ;Save difference
  LDAA #1              ;Clear interrupt flag by
  STAA TFLG1           ; by writing 1 to register
  RTI                  ;Return control to the interrupted code
```

This code assumes that, at the time the sonar ping was initiated, the current time was stored into `Sonar-tof`. When the returning echo triggers the interrupt, the difference between the time the ping was initiated and the time it returned will be stored in `Sonar-tof`.

Writing the interrupt service routine is only one of the things we must do to make the interrupt happen. There are two others.

The ultrasonic receiver is connected to pin PA0. This pin is associated with IC3, the MC68HC11's input capture register number 3. Several registers must be initialized to have IC3 generate an interrupt when a signal appears on PA0. To enable IC3 to generate an interrupt, we must set a *mask register*. (Mask registers enable certain microprocessor operations.) Setting the lowest order bit of TMSK1 enables the IC3 interrupt:

TMSK1	Bit 7							Bit 0
$1022	OC1I	OC2I	OC3I	OC4I	OC5I	IC1I	IC2I	IC3
	x	x	x	x	x	x	x	1

We want the interrupt action to occur when the state of pin PA0 changes from 0 to 1. Following the MC68HC11 documentation, this choice is realized by setting the lowest-order bits of register TCTL2 to %01:

TCTL2	Bit 7							Bit 0
$1021	–	–	EDG1B	EDG1A	EDG2B	EDG2A	EDG3B	EDG3A
	x	x	x	x	x	x	0	1

After an interrupt has been generated, a flag will be set in the TFLG1 register. This flag must be cleared, once the interrupt service routine has been entered, or else, when action returns to the main code, it will think another interrupt is pending and service it again. The processor will do that forever if the flag is not cleared. To clear the flag, we must write a 1 to the corresponding bit, IC3F, in the TFLG1 register:

TFLG1	Bit 7							Bit 0
$1023	OC1F	OC2F	OC3F	OC4F	OC5F	IC1F	IC2F	IC3F
	x	x	x	x	x	x	x	0

The following code implements these choices, enabling the interrupt to occur when PA0 goes high:

```
LDAA #%01               ;Setup IC3 to generate an interrupt
STAA TCTL2              ; on rising edge
LDAA #1                 ;Clear IC3 flag
STAA TFLG1              ;Clear the bits of the register by writing 1's
LDAA #1                 ;Enable the IC3 interrupt
STAA TMSK1
CLI                     ;Global intrpt enable, intrpt system now ready
JSR Turn-on-sonar       ;Initiate sonar ping
LDD TCNT                ;Get the time the sonar was turned on
STD Sonar-tof           ;Save turn on time
      ⋮
                        ;The microprocessor is free for other uses
JSR Compute-distance ;At some later time compute the distance
```

One more operation must be performed before the interrupt can be successfully initiated. The microprocessor must be told how to find the interrupt service routine code. For each interrupt facility that the MC68HC11A0 provides (there are 21) a location is specified in memory where the address of the associated interrupt service routine is stored. For the IC3 interrupt, this address is $FFEA. When the program is loaded, it must fill this location with the address of TIME-SONAR-ISR.

Setting up an interrupt is clearly much more complicated than setting up a simple polling operation. But the increased efficiency of

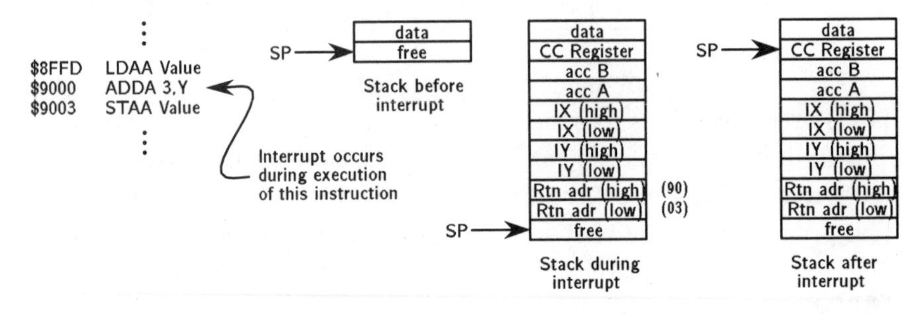

Figure 3.19. When an interrupt occurs, the instruction currently underway continues to completion. The state of the microprocessor is then saved. Preserving the state requires saving all of the CPU registers on the stack as well as the address of the next instruction to be executed following the interrupted instruction. After the interrupt service routine completes, the preinterrupt state is restored. Using the data saved on the stack, all CPU registers are reloaded with the values they had before the interrupt occurred.

an interrupt usually more than makes up for the increased complexity. Figure 3.19 shows how the microprocessor saves and restores the state of ongoing computations. The interrupt service routine is free to use whatever CPU registers it needs. The values stored in these registers are automatically restored when the routine exits.

3.8.3 Input Capture

To illustrate the point in the previous section on interrupts, we actually did more work than was necessary. The input capture facility of the MC68HC11 allows us to compute the time of flight of the sonar pulse without resorting to an interrupt routine.

When properly set up, the timer-counter hardware can capture the time when PA0 goes high. We must make use of one more built-in 16-bit register, TIC3:

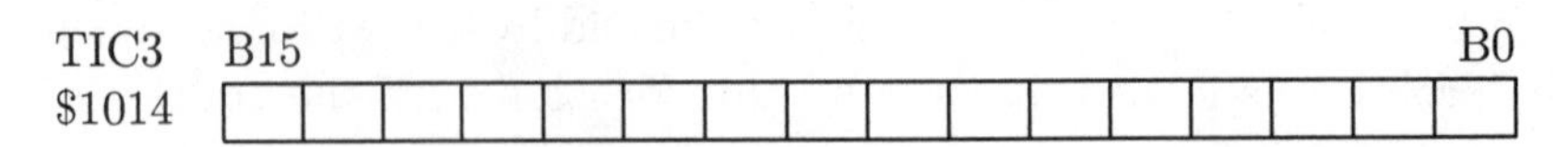

Now when input capture IC3 occurs, the time of that event (the instantaneous value of TCNT) will automatically be latched into register TIC3. To set up this feature, the following code would be required:

```
Setup-IC3                ;Code to activate input capture
  LDAA #%01              ;Trigger IC3 capture on rising edge.Each such
  STAA TCTL2             ; capture latches the time into register TIC3
```

At any later time, when a sonar ping is initiated, the time of that event will be saved in `Sonar-start`:

```
  JSR Turn-on-sonar      ;Initiate sonar ping
  LDD TCNT               ;Get the time at which the sonar was turned on
  STD Sonar-start        ;Save turn on time
```

At any point after the sonar echo has returned, the distance can be computed from the elapsed time:

```
  LDD TIC3               ;Get the time the echo returned
  SUBD Sonar-start       ;Subtract the time ping started
  STD Sonar-tof          ;Store difference for distance computation
  JSR Compute-distance   ;Compute the distance
```

Built-in features like input capture and its counterpart, *output compare*, add greatly to the power of the microcontroller. See 5.5.3 for a worked out sonar example.

3.8.4 Traps

What if something goes wrong? Perhaps an unexpected condition causes an attempt to divide by 0, or maybe a memory cell is accidently overwritten, causing the microprocessor to try to execute an opcode that doesn't exist. What will happen?

The *trap* facility gives a computer an opportunity to recover from events that would otherwise cause a crash or an arbitrary response to an unexpected condition. A trap strongly resembles an interrupt. The user writes a trap service routine and stores its address as an interrupt vector. When the microprocessor detects the error condition, it jumps to the trap code.

High-powered microprocessors and computers provide many different traps. The MC68HC11A0 has just one, the *illegal opcode trap.*

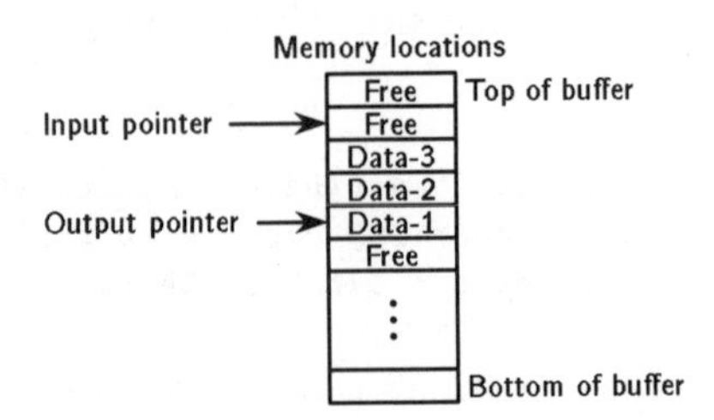

Figure 3.20. Values are added to the buffer by inserting at the place pointed to by the input pointer, then incrementing the input pointer. Values are removed from the point indicated by the output pointer. This pointer is then incremented. When input and output pointers point to the same location, output stops. When a pointer reaches the buffer's top, the next increment sends it to the bottom.

It does however, have another feature that can help it recover from a crash, the *computer operating properly* or COP, facility. When the COP feature is enabled, the user must provide a piece of code that causes a special location to be written to every so often. If this operation fails to happen (presumably, because a crash has occurred or the program is hung), then the system automatically jumps to the address specified by the COP failure interrupt vector. The user should supply (at the chosen address) code that will restart the system.

3.8.5 I/O Buffers

Frequently, it is more advantageous to move data through a buffer rather than directly from its source to destination. As an example, consider the problem of writing data to a serial line. Suppose a program must send a string of characters to a terminal. Ideally, the characters should be output as quickly as the microprocessor can move them from memory to the output port. However, terminals (or rather serial lines) typically operate much more slowly than the microprocessor itself. To accommodate a direct transfer, the microprocessor would be forced to output a character and then wait for an acknowledgment signal telling it that the terminal is ready for the next character. This would be slow.

The solution is to send the characters using an interrupt routine. To do so, the function that wishes to output characters must move the string to a buffer and activate the interrupt routine. Moving the individual characters out the serial port is then handled automatically.

Figure 3.20 illustrates a buffer structure. In an empty buffer, the input and output pointers both point to the same location. To add characters to the buffer, write a character to the location indicated by the input pointer and then increment this pointer. (If the pointer reaches the top, the next character must move it to the bottom.) Whenever the serial line is ready for another character, it initiates an interrupt. The code that handles this interrupt then takes the character designated by the output pointer and moves it to the serial line. The output pointer is then incremented. As soon as the input and output pointers both point to the same value in the buffer, all characters have been sent and output can stop.

3.9 Loading a Program

So far, we have seen only fragments of assembly language programs. What do we have to do in order to write a complete program and run it on the microprocessor?

3.9.1 The Assembly Program

The first step in writing an effective program does not involve the microprocessor at all. The first step is to work out the details of the algorithm on which the program will be based. Code is used to implement an algorithm, but the algorithm itself, the method used to solve the problem, is independent of the particular code that supports it.

In the following examples, we will loosely follow the syntax of the Motorola assembler, called AS11. This assembler allows the programmer to create symbolic labels such as the following:

```
PORTA EQU $1000
```

Here, the symbol `PORTA` has been assigned the value \$1000. In any subsequent code, we may refer symbolically to port A rather than having to remember and write out its address. Symbolic labels make code easier to understand and debug. Use them liberally.

We should point out that, in the examples in this book, we have used labels of arbitrary length. Some assemblers however, restrict the number of characters a label is allowed to have. We have also used the symbol "–" as a normal character when embedded in a label. Many assemblers treat the "–" as a special character, indicating that subtraction is to be performed.

We have learned that Rug Warrior's memory space extends from \$8000 to \$FFFF and that programs are stored in memory. How does the microprocessor decide where, within this space, to put a particular instruction opcode or other data? The Motorola assembler uses the ORG directive to determine where instructions will be placed. Suppose our program begins:

```
ORG $8000
LDAA #my-value
        ⋮
```

This construction will put the opcode for LDAA at location \$8000 in memory. Subsequent opcodes and data values will follow.

One important assembly function remains. After the code has been loaded into memory, how does the microprocessor know where to begin? When power has been turned off and then back on or the reset switch has been pressed, which address should be loaded into the program counter to begin program execution? In the MC68HC11, the last two locations in memory, \$FFFE and \$FFFF, hold the reset vector. Whenever the microprocessor is restarted (by turning the power on or hitting the reset button), the address stored in the reset vector is loaded into the program counter. If the code fragment shown above is to be the beginning of our program, then, at some point in the instruction stream, we must say:

```
ORG $FFFE          ;Next data will be stored in the reset vector
FDB $8000          ;Store location of start of program
```

FDB, like ORG, is not an instruction but rather a directive. FDB instructs the assembler to use the next 2-bytes of memory to store the given number. That is, the reset vector at address \$FFFE now has the number $8000 stored in it. Whenever the reset button is pushed, the program counter will point to address $8000 and start executing the code that begins there. Depending on the sophistication of the assembler, many other useful directives and features may be available to aid in preparing an assembly language program.

After the program code has been written, run the assembler to convert the code into machine language instructions. The next step is to get these instructions from a file on the host computer into the memory space of the microprocessor. This is done using a *downloader*, a program that takes the assembled file, the output of the assembler, and sends it to the microprocessor. In the case of the MC68HC11, assembled code is usually loaded via the serial port.

Somehow, the microprocessor must intercept the machine code instructions being sent to it over the serial line and store them in the right locations. Servicing the serial line and moving data into memory locations sounds like a job for an assembly language program. But how is the microprocessor able to accomplish this before the first program has been loaded into it? How can it load a program unless it already has a program to tell it how to do this?

3.9.2 A Bootstrap Loader

The answer to these questions is to first load a *bootstrap loader* program. Loading this initial program is assisted by a special mode of operation of the MC68HC11. The MC68HC11 has four modes of operation selected by the two lines, MODA and MODB. If both MODA and MODB are low, the microprocessor enters a state of monitoring the serial line. In this state, the first 256 bytes sent over the serial line are intercepted and stored in internal RAM (addresses $0000 to $00FF). After receiving the last byte, control jumps to the beginning of RAM, $0000, and execution of the program just received begins. All of these operations are controlled by special factory-installed code in the MC68HC11's ROM.

Thus, one way to execute the user's program would be to load it in the way just described. If the program takes less than 256 bytes, the remainder can be filled with null operations (NOPs). However, this is not a very useful method for loading a program, since the length is severely limited and each time the microcontroller is reset or switched off, the program will be lost. More typically, we use this feature to load a loader program into internal RAM. The only function of the loader program then is to load into memory the program that comes after it. When the loader program begins execution, it loads the user's program—the next set of instructions and data that come over the serial line. This code is presumably stored into on-chip EEPROM or external RAM. After switching the MC68HC11 back to single-chip or expanded mode, this new program will begin executing as soon as a reset occurs.

The simplest way to program the MC68HC11 is to use a commercial or public domain development system that solves the problems of assembly and downloading for the user. So, rather than plunge into the peripheral issues of how to write assemblers, downloaders, and loaders, we will assume that the user has acquired the appropriate software.

3.10 Getting Started

In the years since the first edition of *Mobile Robots* was published, a great number of microprocessors, single-board computers, and microprocessor-based robot kits have been introduced. These devices make life much easier for today's beginning robot builder. We can mention here only a few of the products that can help you get started.

Rug Warrior began life as a built-from-scratch robot board. In response to reader inquiries, A K Peters, the publishers of *Mobile Robots: Inspiration to Implementation* offered a Rug Warrior kit. Rug Warrior Pro,™ a second-generation robot providing improved functionality, is now available from A K Peters. You will find either commercial version of Rug Warrior highly compatible with this text.

There is now little need to build your own robot board from scratch. A number of commercially available single-board computers are suitable for use in robots. New Micros manufactures a line of single-board computers based on the MC68HC11 microcontroller. Boards come complete with a built-in programming language burned into on-chip ROM. Thus, all that is necessary to program such a board is a host computer of modest power.

Two excellent boards designed by Fred Martin of the MIT Media Lab are available from various sources. The Mini Board board uses an MC68HC811E2 and is compatible with the Dunfield C compiler. The design, which includes onboard motor-driver chips, is distributed free of charge. Plans for constructing the board are available over the Internet via anonymous file transfer protocol (FTP) from cherupakha.media.mit.edu.

The more powerful single-board computer, the Handy board, is available fully assembled and in kit form from several vendors (see the Gleason Research site, www.gleasonresearch.com). The Handy board has 32 Kb of on-board static RAM and is compatible with the **IC** programming language.

The Basic Stamp, a tiny single-board computer based on Microchip's PIC microcontroller, has become very popular. The Basic Stamp (programmable in Basic as the name implies) offers a good combination of price and functionality and the size is hard to beat (see www.parallaxinc.com).

LEGO has recently introduced a product called Mindstorms. Mindstorms is an outgrowth of work originally done at the MIT Media Lab where the concept device was called the Programmable Brick. The Mindstorms' brick contains a microcontroller and is programmed via a host computer using a visual programming system. This processor brick can be combined with other LEGO elements to construct simple autonomous robots. Mindstorms is an introductory product for younger robot builders. More advanced builders may chafe at Mindstorms' small number of inputs and outputs and the limited expressiveness of the programming system.

The workings of a microprocessor are sufficiently complex that you cannot hope to get a full understanding of the subject by reading a chapter from a book. Each device has its own set of special abilities and idiosyncrasies. As with most things, the only effective way to learn is to do. Hook up a microprocessor, and try to program it!

3.11 References

This chapter has given a very brief description of microprocessor basics along with some particulars of the Motorola MC68HC11A0 which we use in Rug Warrior. We chose the MC68HC11A0, because it was the lowest-end member of the MC68HC11 family of 8-bit microcontrollers. Even so, it allowed us to put the entire circuitry for Rug Warrior's brain (including 10 sensors, 2 motor drivers, a music maker, and a serial port) on a 3.4" × 4.5" board. All the details of the numerous capabilities of this chip cannot possibly be explained in a book of this scope, so we strongly recommend that, to follow along in the construction of Rug Warrior, the reader acquire the Motorola MC68HC11 data books (Motorola 1988, Motorola 1991) listed in the bibliography at the end of this book. The first of these data books, *Microprocessor, Microcontroller and Peripheral Data*, gives detailed hardware descriptions and specifications for all Motorola microcontrollers. The MC68HC11A0 takes up just two dozen or so pages of this set. The second data book, *Motorola M68HC11 Reference Manual*, is easier reading, goes into extended examples, and gives much more information on programming the MC68HC11.

For readers looking for a more gentle introduction to microprocessors in general, Horowitz and Hill (1989) give a clear exposition on the subject. Textbooks on computer architecture, such as Ward and Halstead (1990) cover the complete field in great depth. For lighter fare and for additional expositions on digital circuits, glue logic, and support circuitry, Lancaster (1977) and Zaks (1986) are helpful. If you wish to delve seriously into computer architecture, consult the classic Hennessy and Patterson (1996).

Another useful reference for the novice robot builder is *The 6.270 Robot Builder's Guide* (Martin 1992). This book has been used in an undergraduate MIT LEGO Robot Design course developed by Fred Martin, Randy Sargent and Pankaj Oberoi. The course provided kits of LEGO parts, a microprocessor circuit board, motor drivers, and a collection of sensors. *The 6.270 Robot Builder's Guide* describes interfacing bend sensors, infrared proximity sensors, touch sensors, and the like to the MC68HC11 board through software drivers provided with **IC**.

Finally, Fred Martin's latest book (Martin 1998) is a true tour de force for robot enthusiasts. This book contains a wealth of detailed information on robot control, sensors, actuators, LEGO construction, and much more.

Designing and Prototyping

4.1 Practical Problems

To turn a schematic into an actual circuit that can be mounted on your robot, a few basic pieces of prototyping equipment are required. There are a variety of routes to choose for constructing a circuit, but for a small mobile robot, it is important to use a technology that is light and compact, yet flexible enough to accept changes.

There are several choices for prototyping: breadboards, wire wrap, Scotchflex, Speedwire, and printed circuit boards to name a few. *Breadboards* (see Figure 4.1), are commonly used by engineers for testing new designs and have the advantages that they are relatively inexpensive and easily changed. Debugging is simplified because wires and components are on the same side of the board.

Breadboarding has several serious disadvantages, however, particularly if the breadboard will be permanently incorporated into the robot. The component density is necessarily low, and the resulting package is bulky. Stray capacitance between rows can also degrade the performance of high-frequency circuits. Probably the least obvious aspect, though, is that the wiring sockets in breadboards are easily sprung, leading to intermittent connections.

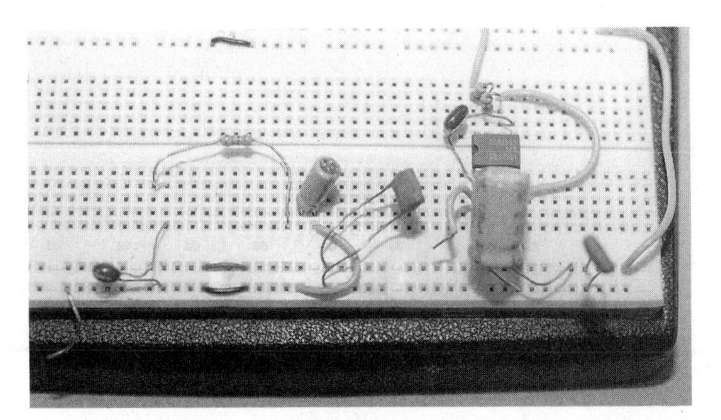

Figure 4.1. Breadboarding can be useful for initial testing. The 5 pins of each vertical row are connected together, as are each of the horizontal rows at the bottom. Discrete components and 22-gauge solid hookup wire can be pushed into the holes.

The problem is that the sockets are typically made to fit one size of solid wire (usually 22-gauge solid hook-up wire), and invariably, a prior user has jammed the next larger size wire into the hole, stretching the socket. Then when a subsequent designer attempts to prototype a circuit using correct-sized wire, the wire intermittently makes contact.

4.1.1 Attention to Detail

Intermittent connections are the most frustrating to debug. The way to avoid this problem is to build your circuit neatly and carefully the first time. When soldering, do not use globs of solder. Use heat-shrink tubing to cover exposed wires. Use connectors liberally for quick disassembly. Add strain reliefs to cable harnesses. Wire things carefully the first time. Keep in mind that a little quality goes a long way.

4.1.2 Wire-Wrap

Another widely used technology for prototyping circuits is *wire-wrap*. This method involves stripping 30 gauge solid wire-wrap wire, inserting one end into a hand tool called a wire-wrap gun, and placing the tip of the gun over a long pin of a wire-wrap socket. Triggering

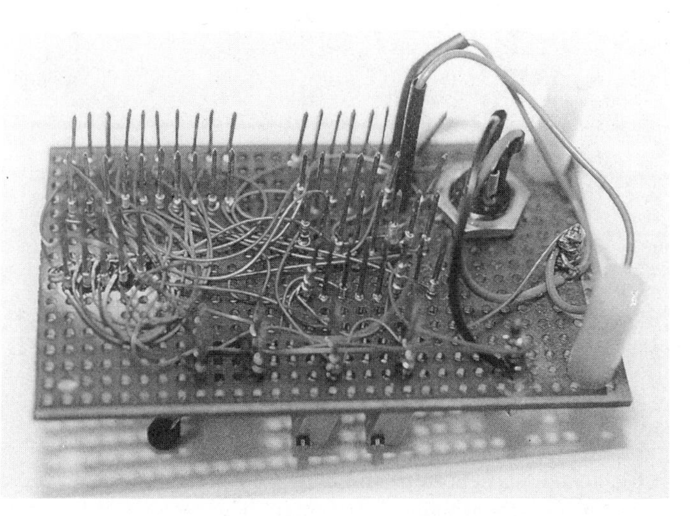

Figure 4.2. Wire-wrap pins stick up a fair distance from the back of the board. The stripped end of 30-gauge wire-wrap wire is curled around a pin with a wire-wrap gun.

the gun wraps the wire around the pin. A small wire-wrap board is shown in Figure 4.2.

The final board is thick due to the length of the pins. Also, connecting one signal (say, ground) from pin to pin to pin (this is called daisychaining) is impossible. Wire-wrap is strictly a point-to-point technology, since each portion of wire must be cut and stripped to fit into the wire-wrap gun. Also, it is rather inconvenient to make changes, as you have to uncurl the wires. This is especially inconvenient when the wire you want to change is below another wire, which is frequently the case since the wiring is point to point.

4.1.3 Scotchflex

The 3M company sells a connector line called *Scotchflex* which is convenient for quick prototyping. There are three components—sockets, plug strips, and the wiring tool. You will also need perfboard to mount the sockets. These components are shown in Figure 4.3. Gerber sells a glass-epoxy board with appropriately sized holes. See Appendix C for all suppliers mentioned in this section.

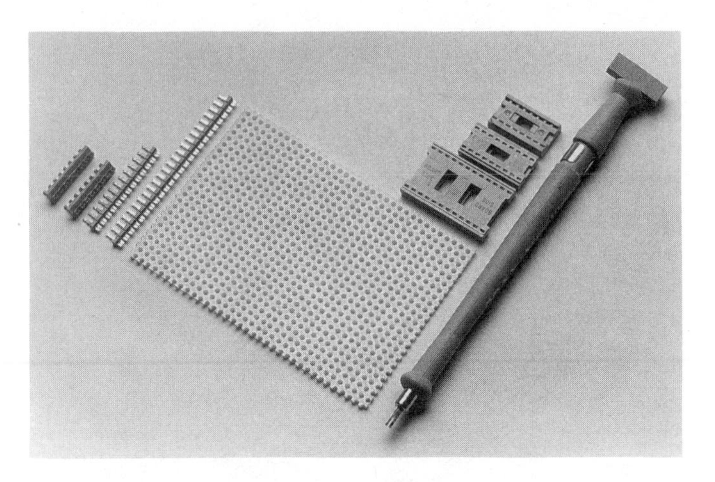

Figure 4.3. Scotchflex sockets (top) and plug strips (left) are mated through the holes in glass-epoxy perfboard shown at center. The top end of the wiring tool is used for mating the plug strips into the sockets and the other end is used for pushing wires into plug strips on the backside of the board.

Scotchflex sockets come in a variety of shapes corresponding to most integrated circuit dual-in-line packages (DIPs), such as 8-pin, 14-pin, 16-pin, and so forth. The plugs come in long strips and are broken off according to the number of pins that correspond to one line of the socket's receptacles. The socket is placed on the top side of the perfboard, and the plug strips are pushed into the sockets from the bottom side through the holes in the perfboard. The wiring tool is double ended, with one end shaped for pushing the plug strips into the sockets and the other end shaped for pushing wires into the plug strips. Wiring is very simple, as it only requires laying 30-gauge insulated solid wire-wrap wire over a plug and then using the tool to push it between the two prongs of the plug's pin. The prongs slice through just the insulation, making contact with the wire. Daisychaining is then very convenient, as you just continue laying wires across prongs and pushing the wires onto them with the wiring tool. At most, two wires can fit into the prongs of any plug, as the pins are fairly short. (You never need more than 2 wires per pin anyway, because of daisychaining.) Consequently, the final boards are thin and can be stacked close together, if necessary. Also, making a change merely involves pulling the wire out and laying in a new one.

If you need to mount discrete components such as resistors and capacitors, make headers by using 8-pin, 14-pin, or 16-pin component carriers and solder the discretes into them. Then just insert the component carriers into the matching sockets and wire in the same manner as for DIPs.

Scotchflex is a very useful technology for quick prototyping. It is portable and compact, as you can cut the perfboard to any shape you want or punch holes in it for other connectors. Do not use a band saw or a drill press on glass-epoxy materials. These materials will damage the cutting edges of the tools. Also, the dust produced by sawing or drilling may be harmful to breathe. Use a punch or a shear instead, and remember to leave room and extra holes for prototyping space on your board for future circuit additions.

The disadvantages of Scotchflex are twofold. Scotchflex does not make sockets for all shapes of chips. In particular, there is no 52-pin-grid array socket of the type needed for an MC68HC11. Another problem involves intermittent connections stemming from the way Scotchflex sockets are mated to their plugs through the perfboard. If chips must be frequently removed from their sockets (for instance, in debugging or reburning EPROMs), the sockets have a tendency to pull away from the plug strips. Eventually, they become loose and do not maintain good contact. For quick prototyping, Scotchflex is useful; for permanent circuits, other methods may be more appropriate.

4.1.4 Speedwire

The Vero Electronics company markets wiring tools and equipment known as Speedwire. Speedwire has but two components: Speedwire pins and the Speedwire wiring tool, pictured in Figure 4.4. Again, perfboard provides the substrate, but with Speedwire, individual pins are pressed through the holes so that the top portion of the pin sticks through to the top of the perfboard and the bottom portion of the pin sticks through the back. To make a 14-pin DIP socket, for example, seven pins are inserted along one row of perfboard holes and seven pins are inserted along a parallel row three columns over.

Speedwire involves more work than Scotchflex, but the advantages offset the disadvantages of Scotchflex: It is possible to make

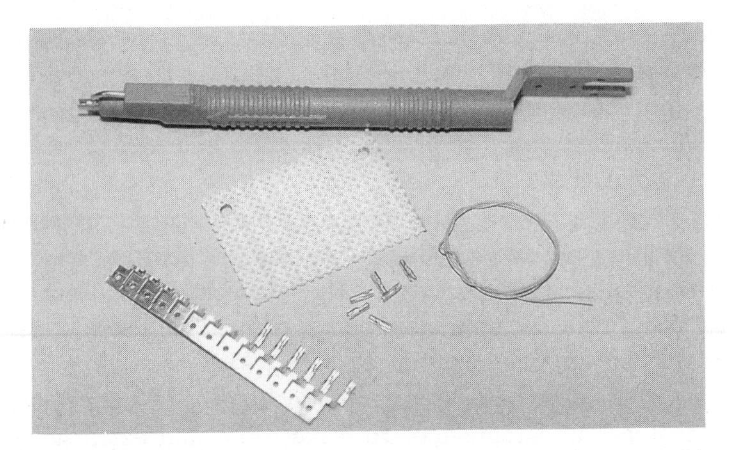

Figure 4.4. Speedwire pins come in reels of one thousand (bottom left) and are broken off and inserted in perfboard (center). The wiring tool is used to push 30 gauge wire-wrap wire through the backside prongs of the pins.

a pattern for any arbitrary pin-grid array, and there are no mating connectors vulnerable to loosening when removing DIPs. Additionally, if discrete components are required, they can be pushed directly into Speedwire pins, without the need for component carriers. Thus, the final boards can be made relatively thin.

Wiring is accomplished in the same manner as with Scotchflex. With Speedwire, you should take care to orient the pins uniformly at 45° to facilitate laying wires diagonally to the rows and columns of perfboard holes. This alleviates the problem of having the end of a wire sticking directly into a pin of a neighboring hole (leading to intermittent shorting problems). The technique is outlined in detail in the instructions that come with the Speedwire wiring pins.

For pushing pins, we have found that first reaming out the holes slightly with an X-ACTO knife makes things easier. Using a large Allen wrench that fits well in the palm of your hand is sufficient to push the pins. Just prop something underneath the perfboard (such as a slab of aluminum), and work at the edge of it to simplify inserting the pins. Another effective technique is to use long-nose pliers to hold the pin by its breakaway tab while pressing the pin into the perfboard. Figure 4.5 diagrams both Scotchflex and Speedwire technologies.

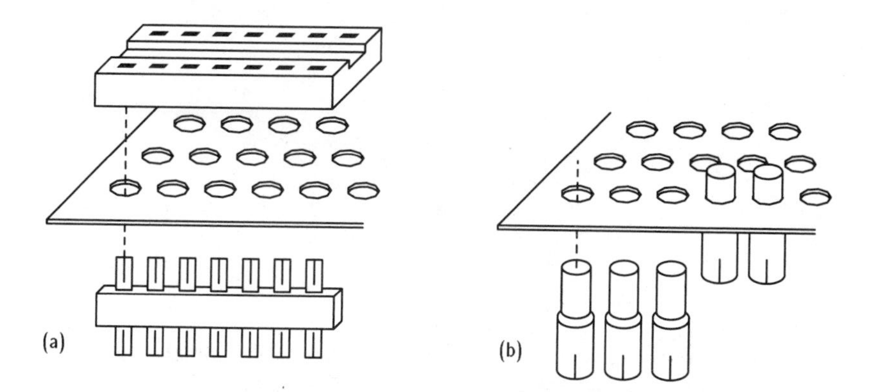

Figure 4.5. (a) Scotchflex technology uses sockets and plug strips that press in through the board and into the sockets. (b) Speedwire technology does not use sockets but rather individual pins that are pressed into the perfboard. Chips and discrete components fit into holes on the top side, and wiring is done on the back.

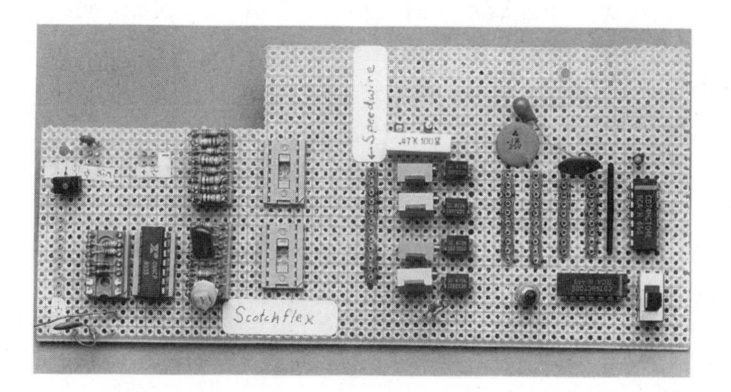

Figure 4.6. This board shows Scotchflex technology on the left and Speedwire technology on the right.

Figure 4.6 shows a board that was prototyped using both Scotchflex and speedwire technologies. On the left are Scotchflex sockets holding a 14-pin DIP and three 14-pin component carriers in which resistors have been soldered. Two empty sockets are shown above the Scotchflex label. Wiring is done on the backside. To the right are discrete or odd-sized components mounted in Speedwire pins (potentiometers, capacitors, and 4-pin DIPs). An empty row of Speedwire pins is shown below the Speedwire label.

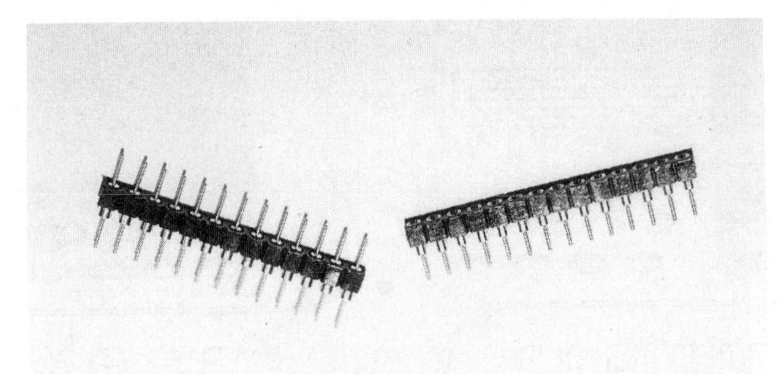

Figure 4.7. Terminal plug strips are shown on the left and terminal socket strips, on the right. The plug strips can fit into these socket strips or into Speedwire pins. Socket strips would be soldered into printed circuit boards, and Speedwire pin sockets would be used for perfboard prototype boards.

Remember that the more components you incorporate into your design, the more time you have to spend prototyping, wiring, and debugging connectors. This is why we focus on using a microprocessor controller, keeping parts count down, and getting to software as soon as possible.

4.2 Connectors

Connecting sensors, motors, and power supplies to your electronics board usually requires making cabling harnesses. Connectors are a problem. It is not uncommon to design a sophisticated, compact, and elegant processor board yet have the connectors take up most of the space on the board. To avoid this result, we put most of the sensors for Rug Warrior directly on the board.

One connector technology that we have found useful for prototyping uses terminal plug strips and terminal socket strips, as shown in Figure 4.7. They come in long lengths and can be broken off for the number of pins necessary for the corresponding number of wires needed. The pins on the terminal plug strips fit into Speedwire pins, also.

A convenient way to use these terminal strips is to assemble them in a fashion that we call mobot connectors, for want of a better name. A mobot connector is shown in the lower part of Figure 4.8. The

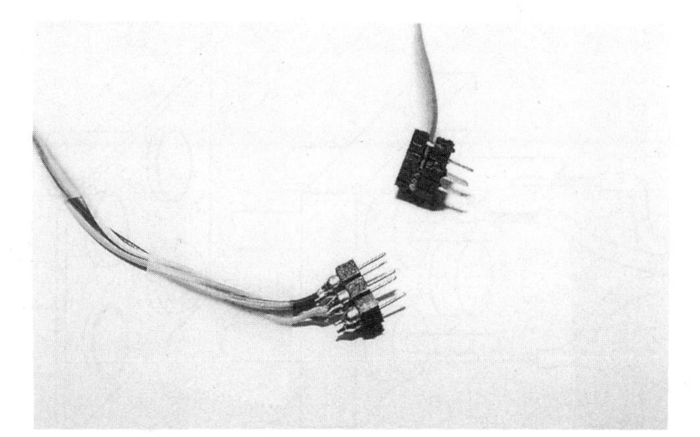

Figure 4.8. A mobot connector made from two 3-pin-long terminal strips glued together is shown in the lower portion of this photograph. The top-side pins are trimmed slightly and wires are soldered on. Ribbon cable plug connectors from Samtec are also convenient; one is shown on the right.

idea is to glue two terminal plug strips together and slightly trim the top-side pins. Then tin each pin with solder. Strip each piece of wire and tin it. Always use stranded wire for cables, as stranded wire is less likely to break. Do not strip the insulation very far back. Solder each wire onto the pin so that the insulation reaches nearly to the top. Figure 4.9 diagrams a mobot connector.

These types of connectors work well with Speedwire technology, as the pins on the terminal plug strips fit into Speedwire pins. Since you push the pins and you make the mobot connectors, the strategy grants flexibility. That is, you can make connectors for cables with any number wires without having to stockpile a vast assortment of different-sized connectors in your laboratory. Another very nice feature of this technique is that the connectors are fairly low profile, which help in keeping things small and elegant.

It is good practice to always key all connectors that you make. *Keying* is a way of making sure that you cannot put the connector in backwards. Having one extra terminal on the mobot connector facilitates keying, as is shown in Figure 4.9. The strategy is to snip off the pin on the unused terminal and drop solder in the mating Speedwire pin. This prevents the connector from fitting into the Speedwire pins in any other way except the correct one.

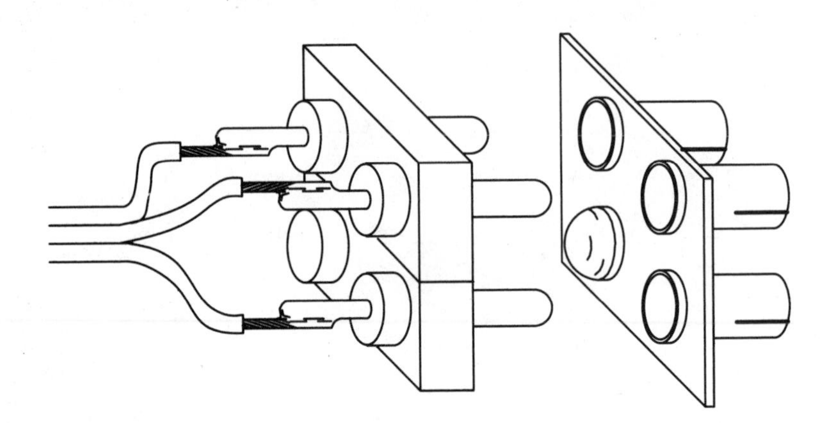

Figure 4.9. On the right is an epoxy-glass perfboard prototyping board with four Speedwire pins inserted. On the left is a mobot connector. Two 2-pin lengths of terminal plug strips have been glued together to create a 3-wire connector. The connector is keyed by snipping off one of the four pins and filling the mating Speedwire pin with solder.

Another possibility for making low-profile, compact connectors is to use cable plug assemblies, such as the one shown above the mobot connector in Figure 4.8. Cable plug assemblies come ready-made with multicolored ribbon cable and also fit into Speedwire pins. Samtec sells both terminal plugs and sockets and also cable plug assemblies.

In Chapter 3 we learned how to design a microprocessor circuit. It is easy to prototype your own microprocessor circuit using the prototyping techniques we have described here. Figure 4.10 shows an early prototype of Rug Warrior's processor board using Speedwire technology. Speedwire pins were pushed into the perfboard (after the holes were reamed out slightly with an X-ACTO knife), the backside was wired up with an assortment of colored 30-gauge wire-wrap wire, and integrated circuits and discretes were inserted into the topside Speedwire pins. The board needs both a power connector and a serial cable for downloading code to the processor. Both of these can be made by pushing Speedwire pins into the perfboard and then making matching mobot connectors. Of course, then you also have to make the connectors on the other ends of the cables. For the

Figure 4.10. An early prototype of Rug Warrior's board. Fifty-two Speedwire pins were inserted in perfboard to match the footprint of the MC68HC11A0 pin-grid array socket. The board was cut on a shear, and a large hole for the reset switch was made with a punch. Integrated circuits and discrete components fit directly into the Speedwire pins.

downloader cable, it will probably be necessary to buy a connector that fits into the back of your host computer.

4.3 Printed Circuit Boards

For stable, reliable hardware that will allow repeated programming of your robot over the long term, there is no better choice for circuit construction than *printed circuit board* technology. The trade-off is cost for reliability.

4.3.1 In-House Fabrication

A typical printed circuit board factory consists of large process lines of etching and plating baths. For designers and people who prototype constantly, it would be helpful to have a machine in house for prototyping printed circuit boards. No companies have solved this problem quite yet, but it would certainly be useful if you could send your layout to a special printer, from which would emerge an actual flexible printed circuit board.

A few companies provide partial solutions; these are rather serious investments. T-Tech makes a circuit board routing machine

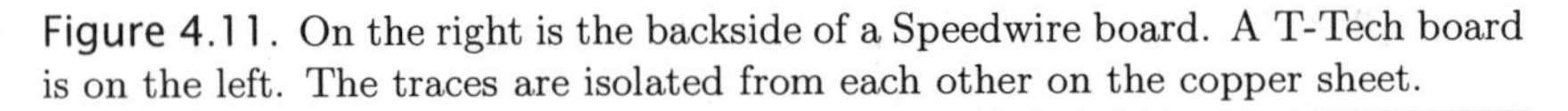

Figure 4.11. On the right is the backside of a Speedwire board. A T-Tech board is on the left. The traces are isolated from each other on the copper sheet.

that utilizes a desktop numerically controlled X-Y milling machine along with isolation software to mechanically carve your circuit from stock copper-clad fiberboard (see Figure 4.11). The advantage of such a machine is that it enables prototyping with chips that come in surface-mount packages. This is nice when you want a small compact board or if you want to use a chip in your design that is available only in a surface-mount package. The disadvantage is that this machine cannot make plated-through holes, narrow lines, or solder masks. Sockets have to be soldered on both the front and back of the board, and connecting traces from one side of the board to the other requires inserting and soldering pins.

Other companies are starting to market even more sophisticated machines. Direct Imaging offers an in-house machine that patterns conductive ink on a flexible substrate for multilayer and flexible printed circuit boards. This is movement in the right direction, but at the moment, these machines are expensive.

Finally, we should mention that it is possible to buy solid copper clad circuit boards and etching chemicals for the truly do-it-yourself approach. The technique involves transferring a printed representation of your layout to a chemically coated board. The chemicals

Figure 4.12. Commercial products, such as this circuit board from inside a Canon camera lens, use printed circuit board technology with surface-mount components. The outer diameter is 6 cm.

remove the copper that is not part of the layout leaving only the connections you desire. You must then drill all through holes by hand. This approach is messy and time consuming but remains in common practice.

4.3.2 Mail-Order Solutions

Prototyping with printed circuit boards has become easier and less expensive in recent years. Having a board fabricated by a commercial manufacturer is now a viable alternative for even budget-minded hobbyists.

To achieve rapid turn around and low cost, the least expensive vendors require that your design meet certain constraints. Multilayer boards, that is boards with one or more layers of traces buried inside, are not allowed. Rather, boards can have traces on only the bottom or both top and bottom. Boards must be rectangular (but you can cut the finished product to any desired shape yourself). Holes in your board cannot be of arbitrary diameter, you must select from a limited number of standard drill sizes. There can be no solder mask or silk screen.

The payoff for living within these restrictions is high. For example, as of this writing, it is possible to have two boards similar in size and complexity to the Rug Warrior board manufactured to your specifications and delivered to your door within three business days for well under $100.

Two companies, both located in Canada, who supply such a prototyping service are AP Circuits, (www.apcircuits.com) and EP Circuits (www.uniserve.com/epcircuits).

There are several steps to perform if you follow this route. Assuming you have access to the proper CAD (computer-aided design) software, you must first create a schematic of your circuit. This is called *schematic capture.* Second, in the *layout* phase, you specify the positioning of components and mounting holes. Third, you must specify the interconnection of all components. This step, called *routing* is tedious if done manually but requires an expensive program if done by computer.

Schematic capture/layout/routing programs are available at a very wide range of prices. Simple programs can be obtained for less than $100; the most capable programs can exceed $100,000. The effectiveness of the routing program is often the most significant determinant of the total price.

The fourth step in the process of having a circuit board made is to convert your design from the internal format of your program to a standard format readable by the board fabrication house. *Gerber* format is a commonly accepted standard. A conversion facility is often present in the CAD package or available as an option.

Finally, send your design via FTP to the board fabrication house of your choice.

If you end up sending your circuit out to have your design implemented by a commercial fabrication house, it is not necessary to use through-board DIP packages for chips. It is fine to use the smaller surface-mount packages. Figure 4.12 shows how dense consumer products can be, using surface-mount technology. The photograph is of a circuit board mounted on the inside of the lens of a Canon camera. The electronics include a DC-DC converter and motor-drive electronics to drive a piezoelectric ultrasonic motor used for autofocusing the lens.

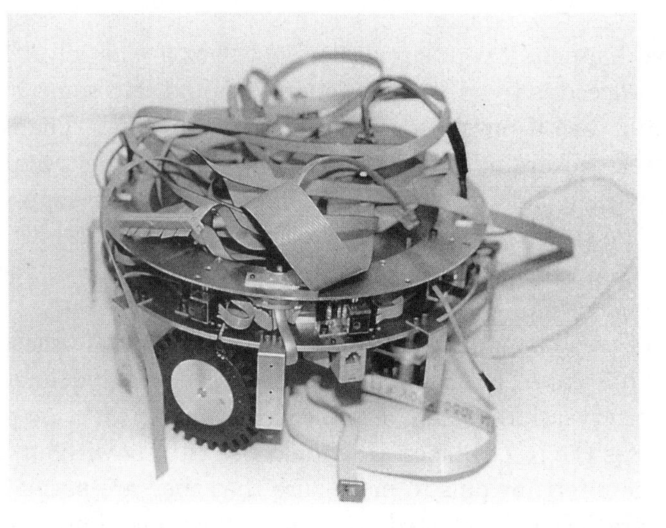

Figure 4.13. The wiring on this MIT robot has not been finished. Connector soup!

4.4 Debugging

Nothing ever works right the first time. Thus, debugging is one skill worth mastering.

You can perhaps avoid errors if you enforce certain disciplines when wiring up your board. For instance, attach stick-um labels to the back of the perfboard that name the chip and mark the position of pin 1. This alleviates having to juxtapose front to back in your head. Another trick is to make a copy of your circuit diagram and highlight each signal with a marker after adding each wire. Finally, buy lots of different colors of wire, use them liberally, and stick with a convention for power and ground. Pacer Electronics sells multitudes of different colors of wire.

Once you have finished constructing your circuit (but before inserting chips in the sockets or applying power!), check with an ohmmeter if +5 V is shorted to ground. If it is okay, test that all points that should receive power are connected together and that all points that should be grounded are similarly connected.

Next, insert chips and discrete components. Check +5 V and ground once more before pressing the "on" button.

What happens if your circuit does not behave properly? The best way to proceed is to go back to square one and find something that does work. See if power is getting to all your chips. The batteries could need recharging or possibly an IC's pin was bent and missed mating with a hole in a socket. Check the power supply on an oscilloscope to make sure it is not corrupted with noise. It is good practice to add capacitors across the supply and across the power and ground pins of digital ICs.

If you are debugging a MC68HC11 circuit, start by checking the clock. Pin 5, called E, should be a square wave at a frequency of one-fourth the crystal frequency. Next, check the reset pulse on pin 17 as you depress the reset button. It should rise cleanly without glitches. Check the interrupt pins to make sure that they are normally high. If they are left unconnected, they may float and initiate random interrupts. Check that the processor is set up in the correct mode by observing the signals on pins 2 and 3, MODA and MODB, while you press the reset button. These signals are valid and read by the processor for just a few cycles after reset.

If other parts of your circuit are misbehaving, try the technique of "divide and conquer." Remove any load from the pin and check again. Be systematic and thorough; always start by finding a point in your circuit that is behaving as designed and gradually debug subsequent portions of the circuit.

Finally, think about connectors. The more subsystems you add to your robot, the more interconnecting is required. Complexity can increase quickly if many sensors and actuators are geographically scattered around the perimeter of the robot, as can be seen in Figure 4.13. (This is a photograph of work in progress on a vacuum cleaner robot built by Masaki Yamamoto at the MIT Mobile Robot Lab.) Make your first robot simple, and key your connectors so they cannot accidentally be inserted backward. One of the most common sources of problems are loose or flakey connections. Be neat, and build reliable connectors!

5

Sensors

5.1 Achieving Perception

As humans, we often take for granted our amazing perceptual systems. We see a cup sitting on a table, automatically reach out to pick it up and think nothing of it. At least, we are not aware of thinking much of it. In fact, accomplishing the simple task of drinking from a cup requires a complex interplay of sensing, interpretation, cognition, and coordination, which we understand only minimally.

Thus, instilling human-level performance in a robot has turned out to be tremendously difficult. A computer program has now beat the reigning world champion at chess but a program that reliably recognizes, say, a chair in an arbitrary scene still does not exist. The parallel computer inside each of our heads devotes large chunks of grey matter to the problems of perception and manipulation.

5.1.1 Transducing versus Understanding

While we would like our robot to understand and be aware of its environment, in actuality, a robot is limited by the sensors we give it and the software we write for it. Sensing is not perceiving. Sensors are merely transducers that convert some physical phenomena into electrical signals that the microprocessor can read. This might be done by using an analog-to-digital (A/D) converter onboard the microprocessor, by loading a value from an input/output (I/O) port, or by using an external interrupt. Typically, there needs to be some interface electronics between the sensor and the microprocessor to condition or amplify the signal.

5.1.2 Levels of Abstraction

With software, we can create different levels of abstraction, or abstraction barriers, to help us as programmers think about sensor data in different ways. At the highest level, the intelligence system, in order to seem clever, needs to have some variables to juggle: Is it dark in this room? Did a person just walk in? Is there a wall to the left?

However, the only questions the robot is able to ask are ones such as: Did the resistance fall in the photosensor? Did the voltage from the pyroelectric sensor connected to the fourth A/D channel go above threshold? Did the output of the near-infrared proximity detector change from low to high?

Nevertheless, it is possible to instill many capabilities in a mobile robot. Figure 5.1 shows a five-foot-tall mobile sentry robot called Robart II, built at the Naval Ocean Systems Center. Robart II serves as a mobile sentry robot (patrolling a building, avoiding obstacles, watching for intruders) and is able to find its recharging station and plug itself in. This robot contains a very large number of sensors, such as near-infrared proximity detectors for obstacle avoidance, sonar rangefinders for localization, microwave sensors for motion sensing, pyroelectric sensors for detecting intruders and temperature, and earthquake and flood sensors for disaster identification.

Another mobile robot covered with sensors is Attila, shown in Figure 5.2. Attila is a shoebox-sized, six-legged robot designed as a rough-terrain explorer. Sensors on the legs are used for detecting obstacles and stepping over them. There are strain gauge force sensors along the shins for detecting collisions, potentiometers on the joint motors for position calibration, and contact sensors on the feet for ascertaining stable footholds. A number of sensors are also mounted

Figure 5.1. Robart II, from the Naval Ocean Systems Center, is a five-foot-tall mobile sentry robot laden with sensors—sonar sensors, infrared sensors, bump sensors, microwave motion sensors, burglar alarms, a surveillance camera, even earthquake and flood sensors!

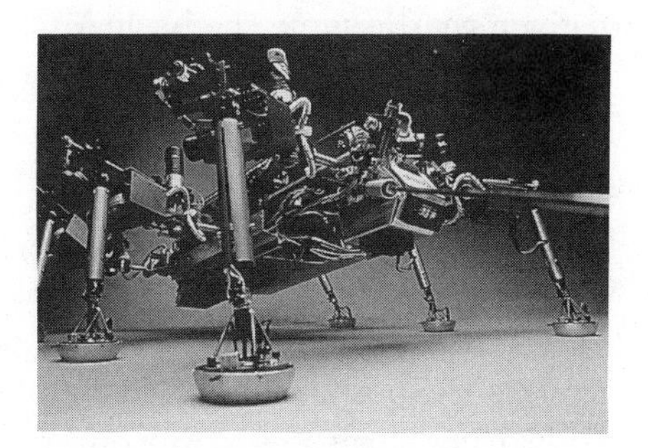

Figure 5.2. Attila, an MIT robot, is a six-legged rough-terrain explorer robot with over 60 sensors, 23 motors and 11 computers. Sensors up and down the legs include force sensors, touch sensors, color sensors, and potentiometers for measuring motor position. Other sensors are mounted on the chassis, such as a force-sensing whisker, a gyroscope, a pitch-and-roll sensor, a near-infrared rangefinder, and a small camera.

on the chassis. Whiskers protrude from the front for collision detection, a long-range, near-infrared sensor measures clear space, and a small camera gathers images.

5.2 Interfacing Sensors

In this chapter, we will focus on many types of simple sensors and how to interface them to a microprocessor. Threaded throughout the chapter are various examples of sensor-interface electronics and sensor-driver routines. A variety of sensors (such as photosensors, bump switches, microphones, pyroelectric people sensors, near-infrared proximity sensors, sonar rangefinders, bend sensors, gyroscopes, accelerometers, force sensors, compasses, and cameras) can be inexpensively acquired and interfaced to a small mobile robot.

By the end of this chapter, you will be able to understand most of the second half of Rug Warrior's "brain," which is illustrated in Figure 5.3. This brain constitutes the sensors and their interface electronics that fit (along with the computer described in Chapter 3) onto Rug Warrior's 3.4" × 4.5" board. Part of Figure 5.3, the motor-driver circuitry, will be discussed later in the chapter on motors.

Throughout this chapter, as each type of sensor is explained, partial schematics are given that assume the basic MC68HC11A0 circuit is already built. The interface electronics are shown connected to a specific MC68HC11A0 I/O pin, analog-to-digital port, or counter-timer pin, and software fragments illustrate how to convert sensor readings into internal variables. If you would like to see the entire Rug Warrior schematic all in one place, refer to Appendix A.

Most of Rug Warrior's sensors are mounted directly onto the circuit board, which is left exposed. This is to circumvent the need to make connectors and wiring harnesses to any outer cover of the robot. Many of the sensors can be seen in Figure 5.4. The pyroelectric sensor, with a cone-shaped holder for its plastic fresnel lens, points upward in the center of the board. The square aluminum package just in front of it is a Sharp near-infrared detector. Two near-infrared LED emitters are mounted on either side of the Sharp detector. Just to the outside of both LEDs are cadmium sulfide photoresistors for light detection.

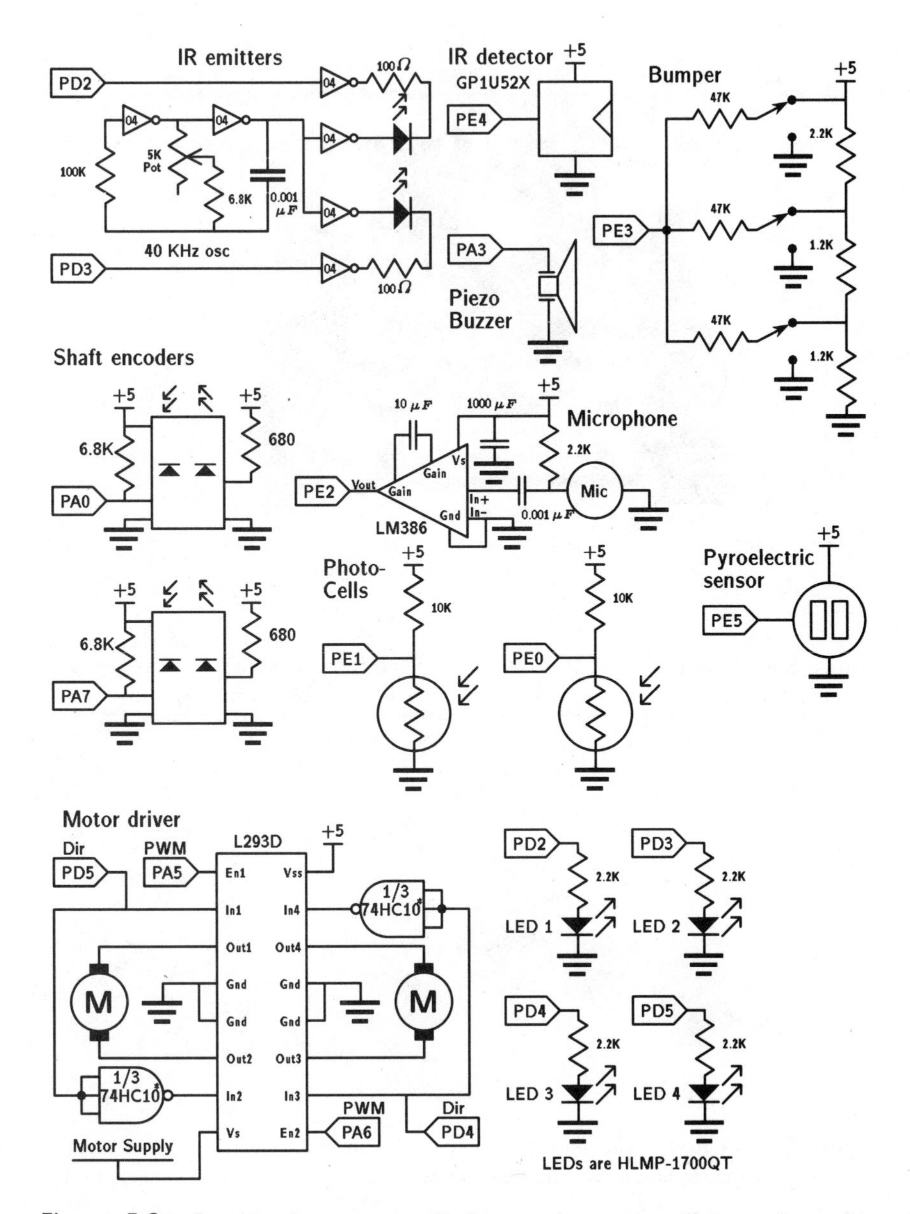

Figure 5.3. In this chapter we will discuss the sensors illustrated on this schematic of Rug Warrior's sensors and actuators: the near-infrared proximity sensors at top left, the three bumper sensors at top right and the shaft encoders, microphone, photoresistors and pyroelectric sensor shown in the center.

Figure 5.4. A front view of Rug Warrior, which shows a number of the sensors. The extra board space in the front with holes in it is spare prototyping room.

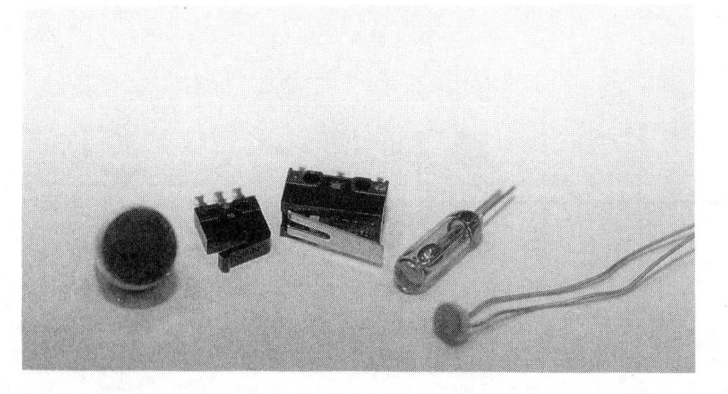

Figure 5.5. A few of the sensors incorporated in Rug Warrior. Left to right are shown a microphone, two microswitches, a mercury tilt switch sensor, and a photocell.

A few of Rug Warrior's sensors can be seen more clearly in Figure 5.5. The microphone on the left is available at Radio Shack. The microswitches in the center are of the type used on a bump skirt to detect collisions. Just to the right of the microswitches is a mercury tilt switch, which is not actually used on Rug Warrior. If the bulb is tilted, the mercury flows to cover two contacts, thus acting as a switch. Such a sensor is useful for detecting if your robot is climbing a ramp. At the far right is a Radio Shack cadmium sulfide photoresistor.

5.2.1 Software Drivers

Once a set of sensors has been selected and the proper interface circuitry has been designed to connect your sensors to a microprocessor, the microprocessor has to be programmed to read the sensors. These pieces of code are often written in assembly language and are known as software drivers.

Software drivers are pieces of code that provide a well-defined interface between a hardware device and a program that needs to use the device. We will describe here several examples of driver code that make the hardware simple to use. Where it is instructive to do so, we will implement our examples of software drivers in both assembly language and the **C** language. The syntax we use for our sample assembly language programs closely follows Motorola's AS11 assembly language. One notable exception is that, in our syntax, unless set off by spaces, we use "–" as a normal character rather than the subtraction operator.

Software drivers deal with the hardware-software interface. These routines might constantly poll an A/D pin, waiting for the trigger from a pyroelectric sensor, or they might be implemented as interrupt handlers that are only called when the return signal from, say, a near-infrared proximity sensor goes high. Sensor-driver code might take this data and store it in a memory location. Used in this way, the output from the sensor can be thought of as the value of a variable or as a flag. This data then becomes fodder for a higher-level abstraction. For instance, another part of the intelligence system might use such a flag or variable to trigger a behavior or perhaps combine it first with other information into a type of virtual sensor.

Keep in mind the different levels of abstraction, as sensors seldom reach the degree of perfection we would like.

5.2.2 Sensitivity and Range

Two important concepts to understand when analyzing any sensor are *sensitivity* and *range*. Sensitivity is a measure of the degree to which the output signal changes as the measured quantity changes. Let's call the sensor output r and the measured physical quantity x. For example, a photodetector might output a voltage of say, 0.87 volts (r) when it is struck by 2.3×10^{13} photons per second (x). The sensitivity of the sensor is defined by:

$$\frac{\Delta r}{r} = S\frac{\Delta x}{x}$$

Here, a small change in the measured quantity, Δx, is related to a small change in the sensor response, Δr, by the sensitivity, S.

A sensing device reacts to varying levels of some physical stimulus by outputting a characteristic voltage (or current, frequency, etc.). Typically, the circuitry associated with the sensor then amplifies or otherwise transforms this voltage and feeds it into an analog-to-digital converter connected to a microprocessor. The A/D converter is sensitive only to a limited range of voltages, often 0 to 5 V. In the case of the 8-bit A/D converter built into the MC68HC11, this voltage is then converted into one of 256 discrete levels. This, then, is the microprocessor's window on the world. No matter how complex and subtle, all phenomena are collapsed into a number, or set of numbers, with values between 0 and 255.

It is, therefore, important to consider carefully how a physical quantity is transformed into a digital value accessible to the microprocessor. Figures 5.6 and 5.7 illustrate two options—both linear and logarithmic mappings of voltages to numbers.

Suppose the motion of a robot arm is restricted to a well-defined range, 0 to 90 degrees. We wish to know the position of the arm with equal sensitivity over all portions of its range. Under these circumstances, a linear mapping of joint angles to A/D readings, as provided by the simple potentiometer circuit shown in Figure 5.6(a), is appropriate. Figure 5.7(a) shows the mapping.

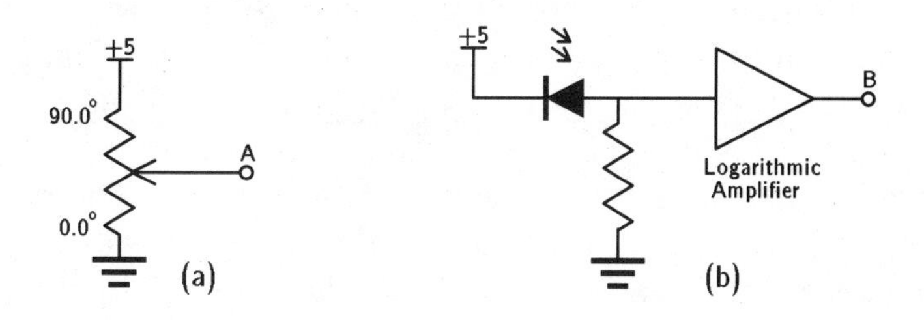

Figure 5.6. The potentiometer in (a) is connected to the joint of a robot arm. The voltage across the network between point A and ground has a linear relationship to the angle to which the joint is set. The photodiode in (b) produces a linear response to a very wide range of illumination levels. After the signal from the diode has been amplified by the logarithmic amplifier however, the voltage at B is proportional to the logarithm of the illumination.

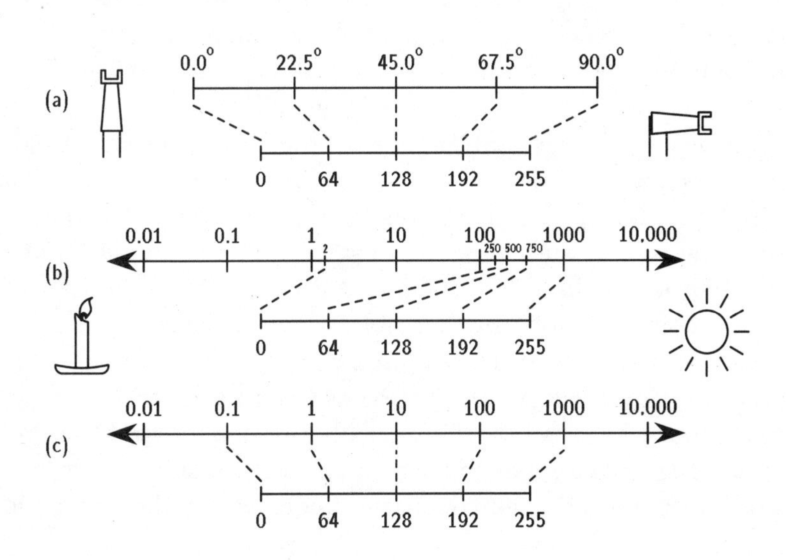

Figure 5.7. It is always necessary to consider how the quantity measured by a sensor will be mapped into the range of digital values available to the microprocessor. (a) The linear mapping illustrated here would map an arm joint angle of 0° from the vertical to the number 0 and an angle of 90° to 255. (b) A linear mapping of illumination units to numbers would map 250 illumination units to the number 64 and 1,000 illumination units to 255. (c) A logarithmic mapping gives a larger dynamic range, from 0.1 illumination units to 1,000 illumination units for an 8-bit (0 to 255) A/D converter.

The situation for the photodiode is more complicated. The level of illumination provided by sunlight is several orders of magnitude greater than that typically produced by artificial lighting. Still, we would like for our robot to be able to sense varying light levels whether it is in a bright room or a dark room. The graph in Figure 5.7(b) illustrates the problem that occurs if we try to use a straightforward linear mapping from photodiode output to A/D levels.

On a scale of arbitrary illumination units, suppose that illumination in a typical bright room varies from, say, 10 up to 1,000 units, while in a dark room, illumination takes on values from 10 down to 0.1 units. If we choose components for our sensor circuit such that illumination levels in the range 0.1 to 1,000 are mapped linearly into A/D values 0 to 255, then the robot has good sensitivity in a bright room, as illustrated in Figure 5.7(b). However, any illumination level below about 2 units is mapped into 0 A/D units. Thus, the robot is practically unable to detect any differences between light levels in a dimly lit room.

One way to fix this problem is with the circuit shown in Figure 5.6(b). Here, a logarithmic amplifier produces a voltage proportional to the logarithm of the photodiode's output. This circuit has the effect of increasing the sensitivity to small changes in light intensity when the robot is in a dark room and decreasing the sensitivity in a bright room. The robot is then able to operate over a much wider range of illumination levels, as sketched in Figure 5.7(c).

In general, the output of a sensor will be neither linear nor logarithmic in any strict sense. This usually presents no problem, however, as long as the robot builder has a clear understanding of the sensor's response and the conditions under which the robot must operate.

5.3 Light Sensors

Visible light sensors and infrared sensors span a broad spectrum of complexity. Photocells are among the easiest of all sensors to interface to a microprocessor, and the interpretation of a photocell's output is straightforward. Video cameras, on the other hand, require

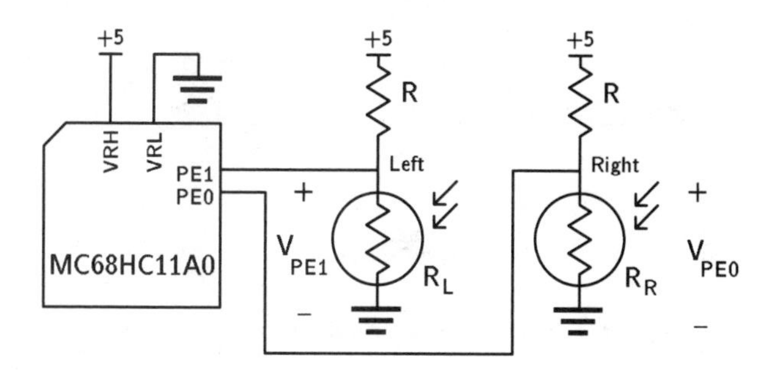

Figure 5.8. Radio Shack 276-1657) are shown in a voltage divider configuration connected to port E, bits 0 and 1. Port E is used here in analog-to-digital converter mode.

a good deal of specialized circuitry to make their outputs compatible with a microprocessor, and the complex images cameras record are anything but easy to interpret.

5.3.1 Photoresistors

Light sensors can enable robot behaviors such as hiding in the dark, playing tag with a flashlight, and moving toward a beacon. Simple light sensors can be purchased as photoresistors, photodiodes, or phototransistors. A photoresistor (or photocell) is easy to interface to a microprocessor. As shown in Figure 5.8, only one external component is needed. Photoresistors are simply variable resistors in many ways similar to potentiometers, except that the resistance change is caused by a change in light level rather than by turning a knob.

Phototransistors provide greater sensitivity to light than do photoresistors. A phototransistor is almost as easy to interface to a microprocessor as a photoresistor. Figure 5.9 illustrates a simple configuration using a phototransistor.

Photodiodes possess great sensitivity, produce a linear signal over a very wide range of light levels, and respond rapidly to changes in illumination. This makes them useful in communication systems for detecting modulated light; the remote control receiver in almost every TV, stereo, and compact disk (CD) player on the market makes

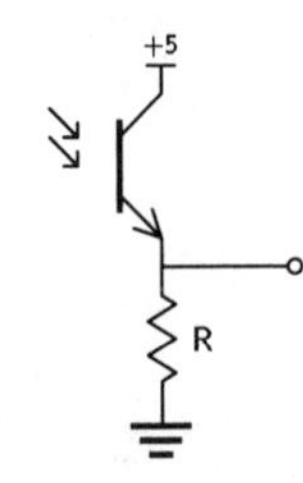

Figure 5.9. A common phototransistor circuit.

use of a photodiode. The output of a photodiode does, however, require amplification before it can be used by a microprocessor.

Because the photoresistor is so useful and easy to incorporate, we will further analyze a practical circuit for connecting one to a microprocessor. Consider the circuit for the left photoresistor in Figure 5.8. Here, two resistances form what is called a *voltage divider*. The total resistance in this circuit, R_T, is the sum of the individual resistances: $R_T = R + R_L$. According to Ohm's law, the current, I, through the circuit is $I = V/R_T$. In order for the A/D converter in the microcontroller to measure a voltage, some current must flow into pin PE1. However, because the MC68HC11 has high-impedance inputs, the amount of current required is negligible compared to the currents in the rest of the circuit. In this case, the connection to PE1 can be ignored while analyzing the voltage divider. Thus, the voltage present on PE1 is:

$$V_{\mathrm{PE1}} = IR_L$$

The resistance of the photoresistor falls as the light level increases. This means that the voltage at PE1 decreases. Substituting for I, we get:

$$V_{\mathrm{PE1}} = \frac{R_L}{R+R_L}V$$

The 8-bit A/D converter in the MC68HC11 maps the variable voltage, V_{PE1}, into the range 0 to 255. Although the mapping provided by the simple voltage-divider circuit is not logarithmic, as was recommended for light sensors in Subsection 5.2.2, a useful output can nevertheless be extracted. A good compromise between sensitivity and range will be achieved if the resistance, R, is set to the same value as the resistance exhibited by the photoresistor when exposed to the light level in the middle of the range of light levels in which the robot must operate.

Typically, photoresistors are made from cadmium sulfide (CdS). Hamamatsu, Clairex, and EG&G manufacture CdS photoresistors; often, photoresistors can be purchased at electronic stores. In addition, most of the semiconductor manufacturers have optoelectronic divisions that fabricate silicon photodiodes and phototransistors. Try Hewlett-Packard, Motorola, Texas Instruments, National Semiconductor, NEC, Siemens and Sharp. Ask for the optoelectronics data book for each company. Texas Instruments sells a TSL250 photodiode with integrated on-chip amplifier. Assemblies of LEDs and photodetectors for encoders or optical switches can be obtained from Omron, Optek, HEI, and Digi-Key. Some companies, such as Hamamatsu and Centronic, also sell photosensor array chips and imagers, although these can be somewhat more expensive. The Texas Instruments TSL214 is a low-cost, 64-element photodiode array.

A Software Driver for Photoresistors

Here, we take a moment to explain in some detail how to configure the analog-to-digital converter and program a software driver for photoresistors. These tasks encompass both the capabilities of the hardware and the responsibilities of the programmer.

As was mentioned in Chapter 3, port E of the MC68HC11 can be configured as either an 8-bit input port or an 8-channel analog-to-digital converter. Internally, there is only one A/D circuit for the entire port and only four registers to store results from the eight channels. Thus, to achieve the full potential of the A/D port, a certain software protocol must be enforced.

First, the voltage reference pins on the MC68HC11 (VRH and VRL) must be set to calibrate the hardware. If these pins are set to +5 V and GND, respectively, then A/D result values of 255 and 0 will correspond to those limits, respectively. Voltages between the limits are proportionately scaled. Two control registers, ADCTL and OPTION, are used to configure the mode of conversion. Reference should be made to the *MC68HC11 Programmer's Manual* to see which bits in these registers should be set to turn on the A/D and to select its various modes. Conversion sequences can be chosen that repeat on a single channel four times or on four channels, once each. In this latter mode, the eight pins of port E can be converted in two banks of four: PE0–PE3 and PE4–PE7. The high bit of the ADCTL mode should be polled periodically because it denotes the conversion complete flag (CCF). Conversions are complete 34-clock cycles after the ADCTL register is written. After each conversion, results are posted in the internal result registers: ADR1, ADR2, ADR3, and ADR4. The converter can also be set up in either mode to convert continuously or just once.

ADCTL	Bit 7							Bit 0
$1030	CCF	–	SCAN	MULT	CD	CC	CB	CA
	0	0	0	1	0	0	0	0

Bits 4 and 5 of register ADCTL are MULT and SCAN, respectively. When SCAN = 0, four conversions are performed, once each, to fill the four result registers. When SCAN = 1, conversions continue in a round-robin fashion. When MULT = 0, four conversions are repeated on a single channel of port E. The selected channel is set by the lower four bits of ADCTL: CD, CC, CB, and CA. When MULT = 1, one bank of four channels is converted. The bank is specified by bits 2 and 3 of ADCTL. If bits 2 and 3 are set to 0, channels PE0–PE3 are converted. If bits 2 and 3 are set to 1, channels PE4–PE7 are converted.

In the following example, written in both assembly language and **C** code, we create a very simple software driver for acquiring a reading from the photocells. The assembly code version might be written:

```
ph-right equ $10   ;Create variable for right photocell
ph-left  equ $11   ;Create variable for left photocell
option   equ $1039 ;Address of OPTION register
adctl    equ $1030 ;Address of ADCTL register
adr1     equ $1031 ;Result register for A/D channel 1
adr2     equ $1032 ;Result register for A/D channel 2

update-photo
  bset option #%10000000            ;Enable A/D system
  bset adctl #%00010000             ;Begin A/D conversion
```

```
check-result
  brclr adctl #%10000000 check-result;Wait in tight loop
  ldaa adr1                          ;Get value from rt photocell
  staa ph-right                      ;Save right value
  ldaa adr2                          ;Get value from lf photocell
  staa ph-left                       ;Save left value
  rts                                ;Return to calling code
```

The **C** version of the photocell code is somewhat simpler:

```
int ph_right = 0;          /* Variable for right photocell data */
int ph_left = 0;           /* Variable for left photocell data */

void update_photo()
{  poke(option,0b10000000);  /* Enable A/D system */
   poke(adctl,0b00010000);   /* Begin conversion */
   while( (peek(adctl) & 0b10000000) == 0 )
   {}                       /* Wait until conversion finished */
   ph_left=peek(adr1);       /* Get and store A/D channel 1 */
   ph_right=peek(adr2);}     /* Get and store A/D channel 2 */
```

In both versions, we first designate locations where the results of the A/D conversions will be stored: `ph-right`, `ph-left` for the assembly version and `ph_right`, `ph_left` for the **C** version. We enable the A/D system by writing the proper value to the OPTION register; then we begin a conversion by writing to the ADCTL register. The next part of both programs polls the conversion-complete bit of the ADCTL register, remaining in a tight loop until the conversion flag is set by the internal hardware of the A/D. Finally, the results of the conversion are moved from the result registers, ADR1 and ADR2, to the designated locations.

To learn the details of which registers and which bits control the various functions of the A/D converter and the microprocessor's other systems, you should really consult the documentation for the MC68HC11A0.

5.3.2 Near-Infrared Proximity Detectors

Following behaviors are easy to implement on a mobile robot. Using a sonar rangefinder to measure range to a person and then staying

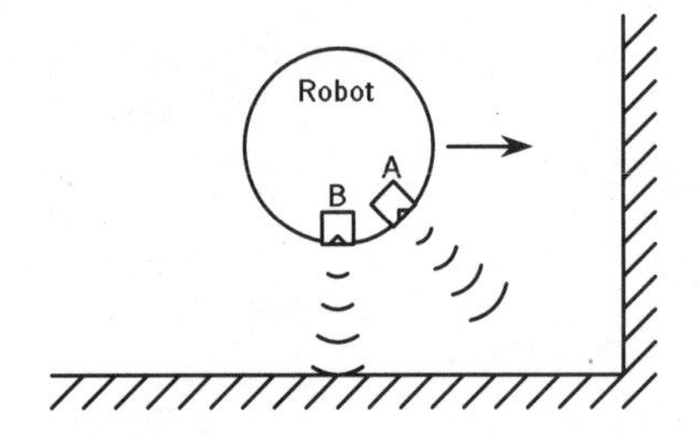

Figure 5.10. The robot can be made to follow a wall using two detect/no-detect infrared sensors, A and B. When neither sensor detects an obstacle, the robot arcs to the right, searching for a wall. When only sensor B detects something, the robot moves forward. When sensor A detects an obstacle, either alone or with sensor B, the robot turns left.

within some tag-along distance is one approach. A simpler strategy is to use a near-infrared proximity detectors. Although these sensors typically do not return actual distance to an object, they do signify whether or not something is present within the cone of detection. These types of sensors usually have much narrower beam widths than sonar rangefinders. Following along walls using two detectors (one pointed directly at the wall and one pointed 45 degrees more forward) is a common strategy, as sketched in Figure 5.10. It is even possible to follow a wall using only one detector by tacking as a sailboat does. In this case, the robot must arc away from the wall when its sensor detects something and arc toward the wall when nothing is detected.

Near-infrared proximity detectors are often called *IRs* for short, but this term can be misleading. These detectors are insensitive to the long infrared wavelengths detected by pyroelectric sensors; rather, they are sensitive in the range just below visible light, often around 880 nanometers (nm) wavelength. In fact, although the human eye cannot see this light, charge coupled device (CCD) imagers are sensitive to it, and if you ever take a video of your robot using a camcorder, it will look lit up like a Christmas tree. Indicator cards are available from Edmund Scientific and Radio Shack that fluoresce when exposed to radiation from an infrared LED. This can facilitate debugging.

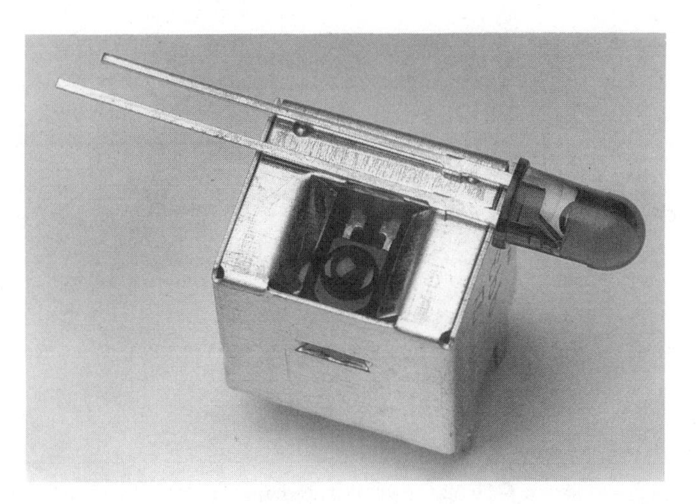

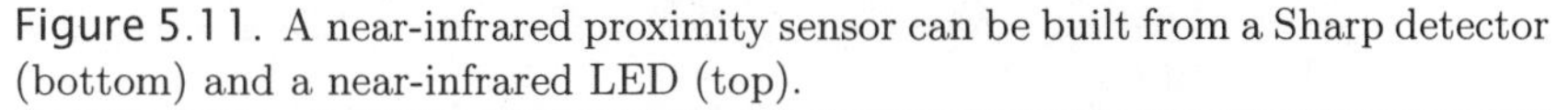

Figure 5.11. A near-infrared proximity sensor can be built from a Sharp detector (bottom) and a near-infrared LED (top).

An infrared emitter and detector pair are illustrated in Figure 5.11. The emitter (top) is an LED made from gallium arsenide, which emits near-infrared energy at 880 nm. Both emitters and near-infrared detectors (photodiodes and phototransistors) can be purchased from nearly any semiconductor company that has an optoelectronics division (Siemens, Motorola, Hewlett-Packard, etc.). Radio Shack also carries both near-infrared LEDs and near-infrared phototransistors. More conveniently, Sharp sells two detector packages the GP1U52X and the newer IS1U60, that contain integrated amplifiers, filters, and a limiter. The GP1U52X unit is distributed by Radio Shack, Sterling Electronics and a number of others.

The Sharp detector responds to a modulated carrier put out by the near-infrared LED. This means that the programmer is responsible for blinking the LED in a certain pattern such that the detector will respond. This modulated carrier protocol increases the signal-to-noise ratio. A minimalist circuit (only one IC is needed, a 74HC04 inverter), which achieves an interface of such a proximity sensor to a MC68HC11, is shown in Figure 5.12.

The Sharp detector responds to a carrier frequency of 40 kHz. A 40 kHz frequency means the LED is blinked on and off with a period of 25 microseconds (μs). According to the device specification, this

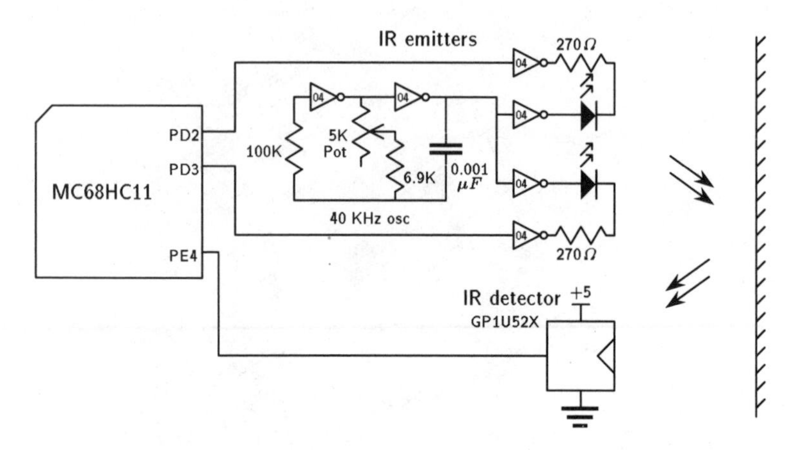

Figure 5.12. A Sharp G1U52X near-infrared proximity detector (Radio Shack 276-137) detects reflected power emitted from near-infrared LEDs, such as a Siemens SFH 484 LED.

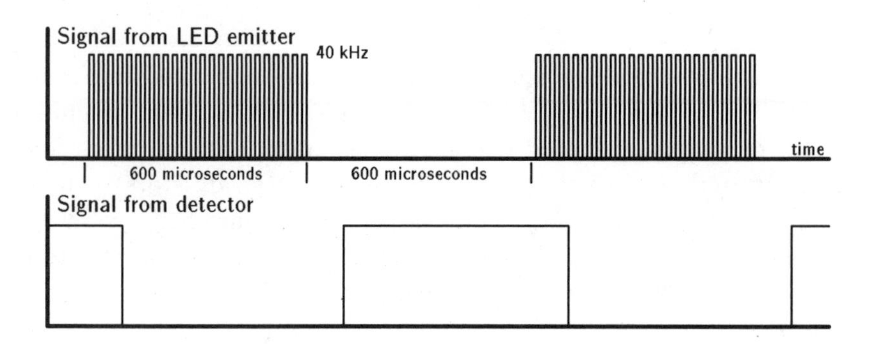

Figure 5.13. The obstacle-detecting infrared beam has a 40 kilohertz (kHz) carrier modulated at 1667 Hertz (Hz). Note that the transmitted signal must be broadcast for several cycles before being acknowledged by the detector. Likewise, when transmission ceases, a few microseconds pass before the detector changes state. Both these delay times can depend on the signal strength.

signal should then be modulated at a lower frequency. The blinking should be on for 600μs and then off for 600μs. Figure 5.13 gives the timing diagram and protocol for this emitter-detector pair.

The 40 kHz oscillator portion of the infrared emitter circuit in Figure 5.12 is implemented using two inverters, a capacitor, a resistor, and a potentiometer. This 40 kHz oscillator runs constantly while Rug Warrior is on, but the LEDs blink only when pins PD2 and PD3 of port D are asserted. Thus, the programmer is responsible for turning these on and off for 600μs each. The Sharp detector outputs a low signal when it detects reflected energy and a high signal when it detects nothing. Figure 5.13 shows the low signal asserted by the Sharp detector when an object reflects energy from the emitter back to the detector. The output of the Sharp detector is a digital signal, either 0 or 5 V. Consequently, pin PE4 of the MC68HC11 can be used in the normal digital input mode. The A/D converter capability is not necessary here.

The circuit that controls the emitters is a rather odd one. It is uncommon to have the outputs of inverters connected together. Normally, an AND gate would be used to allow signals PD2 and PD3 to modulate the oscillator output. (An AND gate outputs a high signal only when both inputs are high.) We chose instead the circuit shown here for practical reasons: It provides the same functionality as an AND gate, and it does not require adding another chip to the circuit.

The geometrical layout of the sensors has the detector mounted at the center-front of the robot and pointed straight ahead. The emitters are set one to each side and aimed slightly outward to the left and right. This saves having two detectors. Rug Warrior can get by with just one and yet still see to both left and right.

An obstacle-detection program can be written very easily in **C** using the `sleep` function, as the following code fragment shows. PD2 is asserted and a sleep period begins. After 600μs, PE4 is polled and its state is saved in the variable `val_on`. Then PD2 is deasserted and the program waits another 600μs. Next, we poll PE4 again and store its value in `val_off`. An obstacle is detected if the detector output is low when the emitter is on and high when the emitter is off. The function `ir_detect()` should be called as often as necessary to keep the variable `ir_status` updated. A similar loop is repeated for the other LED.

```
int ir_status = 0;          /* Global var for IR detection status */

void ir_detect()
{ int val_off, val_on;     /* Intermediate vars for IR detection */
   bit_set(port_D,0b00000100); /* Turn on one emitter */
   sleep(0.000600);           /* Wait for 600μs */
```

```
    val_on = peek(port_E);    /* Get value of detector */
    bit_clear(port_D,0b00000100); /* Turn emitter off */
    sleep(0.000600);          /* Wait for 600μ */
    val_off = peek(port_E); /* Get value of detector */
    if ((val_off & ~val_on & 0b00000100) == 0b00000100 )
        ir_status = 1;        /* Obstacle detected */
    else
        ir_status = 0;        /* No obstacle detected */
}
```

Common fluorescent lights put out a great deal of noise, to which the IR detector is sensitive. Using the turn-on, test, turn-off, test strategy just outlined will help to eliminate spurious obstacle detections due to noise.

Hamamatsu makes some very convenient-to-use optical sensors, ranging from photocells and near-infrared emitters and detectors to position-sensitive devices, photodiode arrays, and triangulation-based near-infrared rangefinders. One very simple implementation of a near-infrared proximity detector uses the Hamamatsu S3599 light-modulation photo IC. This detector contains an on-chip oscillator to drive an accompanying LED and also an integrated correlating receiver. This means the entire system can be built in a very small package. (The discrete-component 40kHz oscillator of the previous example is extraneous here.) Figure 5.14 illustrates a sample circuit.

There is a trick you can play to squeeze a little more information out of an IR proximity sensor. The detector responds to the IR power it receives by activating if the incoming power is high enough and not activating if the power is too low. If the power output by the emitters can be varied then it is possible to determine whether a detected reflection comes from a nearer or more distant object. To estimate range, start by setting the output power at some high level, then check for a reflection. If a reflection is detected, reduce the power and check again. Continue in this way until no reflection is detected. The output power level at which the reflection becomes undetectable is related to the distance of the object.

The effective power seen by the detector can be varied in several ways. The brute force method is to build a digitally-controlled analog circuit where the output power is set by some number of input bits. A second method is to tune the oscillator frequency

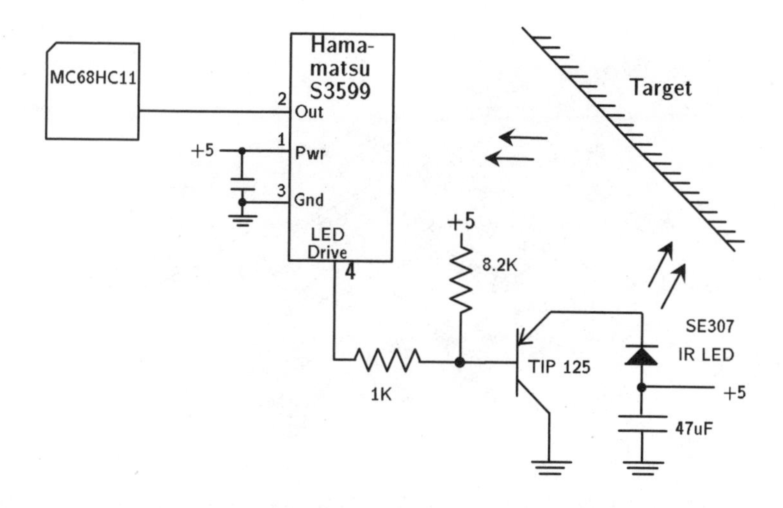

Figure 5.14. A Hamamatsu S3599 near-infrared receiver contains an on-chip frequency generator, which drives a near-infrared LED for correlated detection.

away from the nominal 40 kHz preferred by the detector. The more the frequency differes from 40 kHz, the shorter the range at which the detector will respond to obstacles. A third method is generally most convenient for a fully-digital implementation; simply change the *duty factor* of the oscillator. The detector delivers optimum response when the duty factor, the fraction of the oscillator period when the signal is high, is approximately 50%. (This is the duty factor of the circuit in Figure 5.12.) Building an oscillator circuit with a digitally-controllable duty factor allows estimation of the obstacle's range. This is the scheme used by some commercial research robots.

5.3.3 Near-Infrared Range Sensor

The GP1U52X IR detector discussed in the previous section is a popular, inexpensive, and easy to use proximity sensor. However, a sophisticated new sensor, able to accurately determine the range to a nearby object, has recently been made available by Sharp. The GP2D02 consists of an IR emitter and *position sensitive detector*, PSD, in a single package (see Figure 5.15). Unlike IR proximity detectors, the GP2D02 computes an actual range to an object based on triangulation. This means that (also unlike proximity detectors)

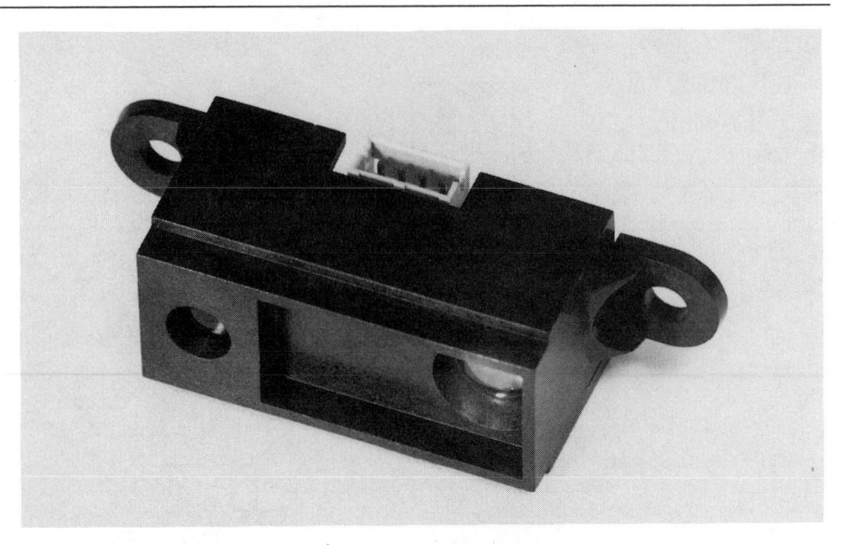

Figure 5.15. The Sharp GP2D02 ranging sensor measures the distance to nearby objects using triangulation.

the GP2D02 is relatively insensitive to the color and texture of the object at which it is pointed.

Figure 5.16 shows how the detector works. The emitter, the lower element in the rectangular package, illuminates a small spot on an obstacle with modulated IR light. A lens forms an image of the spot on the active element at the back of the detector. The output of the detector element is a function of the position on which the image falls. In Figure 5.16(b), the image forms at the center of the active element. When the device is farther from the obstacle as in *a* the image is closer to the bottom. And in Figure 5.16(c), with the device close to the obstacle, the image of the projected spot forms near the top of the active portion of the detector element.

As is suggested by the drawing, when the distance between the detector and the obstacle reaches some minimum, about three inches, the image misses the active portion of the detector element entirely. Thus, the GP2D02 cannot detect obstacles that are too close. Also, at some large distance, the reflected energy is too weak to activate the detector. The maximum distance at which the GP2D02 can detect an obstacle depends on the color and surface properties of the obstacle. From about three to about 15 inches, the output is almost linear with distance, and objects of almost any color can be detected.

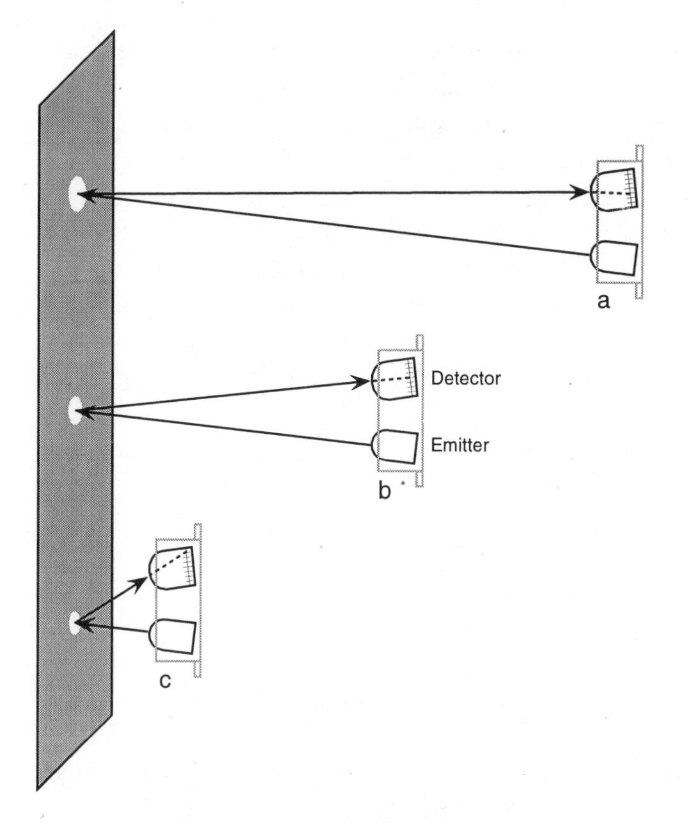

Figure 5.16. The image of a projected spot of modulated IR light forms at different positions on the active portion of the detector element depending on the distance between the GP2D02 and the obstacle.

Unfortunately, the interface of the GP2D02 is not nearly as friendly as that of the IR proximity detectors described earlier. The GP2D02 mates with a miniature 4-position connector that, outside of Japan, is very difficult to acquire. No matter, you can solder wires directly to the pins of the connector.

The next problem is getting data out of the device. The GP2D02 returns range as an eight bit value. Since there is only one output pin, you must actively clock an input pin to get the GP2D02 to output data. Start with the V_{in} high then hold it low for 70 ms or longer. The detector makes its measurement during this period. Then send 8 pulses whose positive half lasts for 200 μs or less. On the positive going part of each pulse, read V_{out}. The most significant bit of the range measurement comes first, the least significant last. Higher numbers in the result correspond to shorter ranges. Consult the manufacturer's literature for more details of the procedure.

There is one last confusing part to the interface. If you simply connect a digital output line from your microprocessor to the V_{in} input line of the GP2D02, it won't work! The GP2D02 uses a curious "open drain" input circuit. Your output line can pull this circuit low, but must never pull it high. That means you should connect a diode between your output line and V_{in}, the anode of the diode goes to V_{in}.

If you can put up with its idiosyncracies, the GP2D02 makes an excellent short-distance range sensor. It even performs well in bright light.

5.3.4 Pyroelectric Sensors

One of the most useful sensors for endowing your robot with a means of interacting with humans is a pyroelectric sensor. A *pyroelectric sensor* is the essential component in certain types of motion-detecting burglar alarms. The output of a pyroelectric sensor changes when small changes in the temperature of the sensor occur over time. The active element in such a sensor is typically a lithium tantalate crystal. Charge is induced as the crystal is heated. Inexpensive pyroelectric sensors are optimized to detect radiation in the 8–10 μm range (the range of infrared energy emitted by humans) and require no cooling to produce a useful signal. This makes them suitable for use in motion sensors and security alarms.

Pyroelectric sensors are sold by a number of companies. Figure 5.17 depicts a dual-element sensor with integrated amplifier, the 442-3, sold by Eltec. The package is shown in the can with the window at the left. To the right is a construction-paper cone for holding a plastic fresnel lens (made by Fresnel Technologies) at the focal distance from the window.

Other companies (Watlow, Mikron Instrument, Detection Systems, Microwatt Applications, Hunter Products, Linear, Spiricon, etc.) make pyroelectric sensors. Nippon Ceramic makes a low-cost version of the pyroelectric sensor shown in Figure 5.17 but without

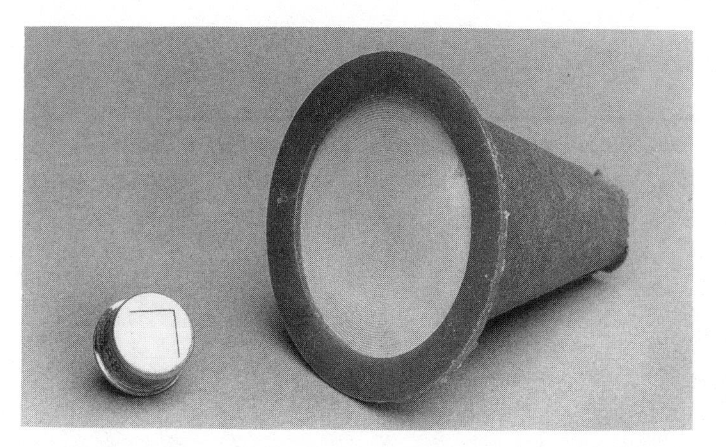

Figure 5.17. Eltec sells a pyroelectric sensor. The 442-3 dual-element sensor is shown at the left. A fresnel lens with a paper-mounting cone that fits over the sensor is shown at the right.

the integrated amplifier. Acroname, Inc. offers a pyrosensor complete with Fresnel lens designed for compatibility with Rug Warrior.

Figure 5.18 illustrates the interface between the MC68HC11 and a pyroelectric sensor. The Eltec 442-3 sensor shown incorporates two lithium tantalate crystals. The amplified difference of the voltage across the crystals is the output of the sensor. In the case that both crystals are at the same temperature, the sensor produces an output signal that remains steady at about 2.5 V (assuming a 5 V power supply). If a person walks in front of the sensor moving from left to right, the signal will rise above 2.5 V by about one volt and then fall below it, finally returning to the steady-state value. Should a person walk in front of the sensor moving from right to left, the reverse will happen. The signal will first fall, then rise, and then settle at 2.5 V. Figure 5.19 illustrates the time-varying output signal of the Eltec sensor.

By taking advantage of the MC68HC11's A/D port, we can implement the interface with a minimum of components. The same "flavor" software driver as used in the photocell routines can gather pyroelectric data. A program to notice when the readings go above or below a preset threshold can trigger some robot behavior. More sophisticated software could look for trends and try to determine

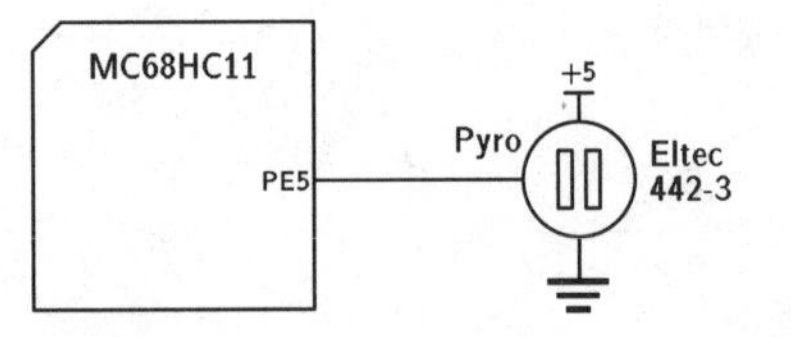

Figure 5.18. An Eltec 442-3 differential pyroelectric sensor with built-in amplifier needs no external components.

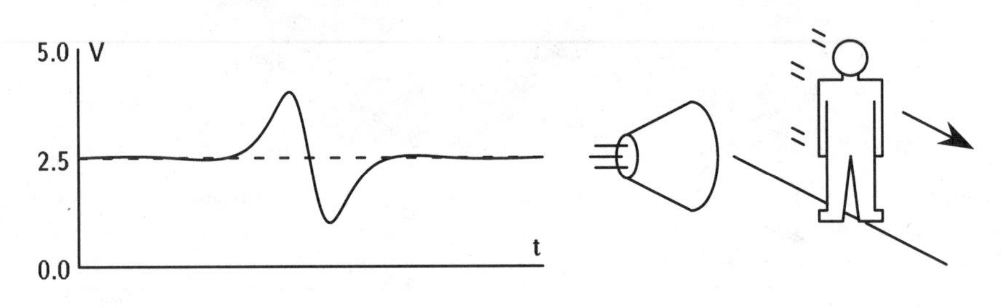

Figure 5.19. A typical signal from a pyroelectric sensor as a heat source passes.

which way the person is moving and attempt to follow.

It is worth pointing out here that most mobile robot building materials are opaque to the long-wavelength infrared radiation that the pyroelectric sensor detects. In particular, if you mount a pyroelectric sensor behind the clear acrylic body shell of your robot, the sensor will remain blissfully ignorant of any passing heat sources you might like it to detect.

5.3.5 Ultraviolet Sensors

On the opposite end of the spectrum from pyroelectric sensors, Hamamatsu offers a line of *ultraviolet sensors* called the UVtron series. These devices are sensitive to radiation in the 185 to 260 nanometer (nm) range but are very insensitive to light in the visible range. In most environments the only source of UV light is a flame. Hamamatsu UVtron sensors have been used with good results by contestants in the Robot Home Firefighting Contest (see Appendix F) held each Spring at Trinity College in Hartford, Connecticut. For information on the UVtron visit Hamamatsu's website at: optics.org/hamamatsu/hps_home.html

Figure 5.20. This small inexpensive camera is sold by Chinon.

5.3.6 Cameras

Video camera technology continues to become more compact and more inexpensive everyday. Small cameras from security systems are a good buy, as illustrated by the Chinon camera in Figure 5.20. Sony also sells small Watchcam cameras.

While onboard vision computations with a MC68HC11 probably are not feasible (especially given all the other sensors connected to Rug Warrior's processor), transmitting to an offboard workstation can be viable. A cable may be used for this application, although a television transmitter is preferable. Some inexpensive and amazingly small (postage stamp sized!) video transmitters are now available. These transmitters operate on the experimental TV (ham radio) frequencies and require a license from the Federal Communications Commission. Contact Supercircuits for information.

5.4 Force Sensors

In general, *force sensors* have proven the most reliable, exhibit the lowest noise, and produce the most easily interpreted signal of all

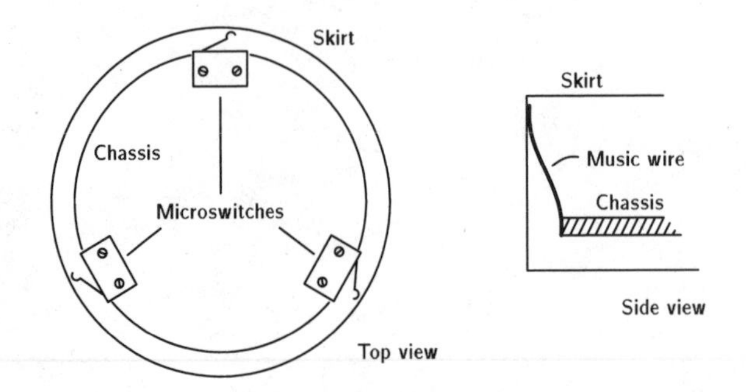

Figure 5.21. How a full-coverage, force-detecting bumper can be implemented on a cylindrical robot. Three microswitches are arranged symmetrically around the perimeter of the chassis so that the activating levers contact the skirt (Top view). The skirt "floats" relative to the chassis, held in place by three or more lengths of stiff steel wire (Side view). This wire, available at hobby shops, is sometimes called *music wire* or *piano wire*.

sensors. Force sensors can be used to determine when the robot is in contact with another object and where that object is in relation to the robot. Such information allows the robot to maneuver away from collisions.

5.4.1 Microswitches

Microswitches, such as the two shown in Figure 5.5 (page 116), are small, momentary switches that can be attached to bumpers to signal when the robot has run into an obstacle. Such switches can be purchased from a number of suppliers, such as Gerber or Digi-Key.

Figure 5.21 illustrates one method for using microswitches to detect collisions between the robot and various obstacles. The switches are mounted in such a way that, when the robot contacts an object, one or two switches will close, thus revealing the relative positions of robot and object.

Figure 5.22 show two ways to interface the bump switches to the microprocessor. The circuit in Figure 5.22(a) is straightforward: One pin of port E is used for each switch. When the robot collides with an obstacle, one or two switches close, changing the state of the corresponding bit(s) from 0 to 1. This approach has the advantage

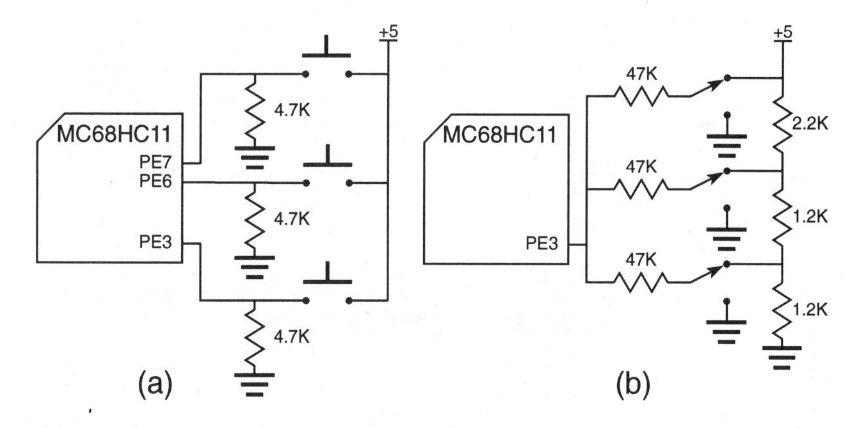

Figure 5.22. Two approaches to force detection. In (a), each switch goes to a separate pin of port E. A digital read of port E reveals the state of the bumper. Circuit (b) channels all bump switches to one pin of port E. Here, we must use the analog-to-digital converter to determine which set of switches is closed.

of being easy to understand and implement, but it uses up three of the MC68HC11's input lines.

There is another way to achieve the same functionality that uses only one MC68HC11 input pin. This second approach is shown in Figure 5.4.1(b), where a network of resistors is used to create different voltages at the MC68HC11 input pin, depending on which switch is closed. (The A/D mode for port E must be used.) The bump switch software driver must read pin PE3, do a conversion, and then set one of eight flags. The correct flag signifies which switch or set of switches is closed; this is determined by in which of eight ranges the measured voltage falls.

A careful analysis will show that the circuit in Figure 5.4.1(b) is essentially a voltage adder. As long as the current flowing from the +5 V supply through the single 2.2 kilohm (KΩ) resistor and two 1.2K resistors to ground is large compared to the current flowing through any other part of the circuit, this approximation will hold. If, as shown, the powering voltage divider has taps at 1/4, 1/2, and 1 times the supply voltage, then the voltage sum will be $1/3 \times (A + B + C)$ (where each of points A, B, and C is connected either to its corresponding tap or to ground). Since the A/D converter produces

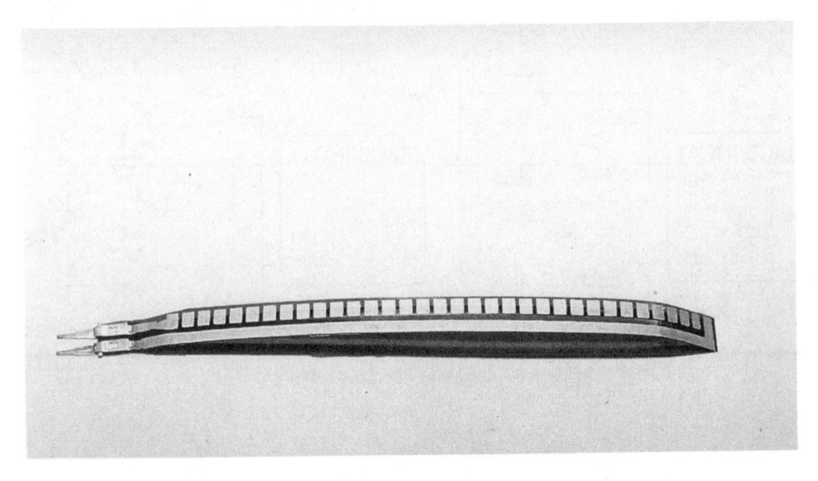

Figure 5.23. A bend sensor is a variable resistor that can be used for bump detection. Depending on the amount of bending, conductive ink between two electrodes creates a larger or smaller resistance.

digital values between 0 and 255, the set of voltages it reads will be 1/3 of 255 times the sum of the voltages from the switches. For example, when only switch A is closed (connected to ground), the A/D output will be $1/3 \times 255 \times (0 + 1/2 + 1/4) \cong 64$. When switches B and C are closed, the output will be 85.

Microswitches can be attached in a number of ways to enhance their applicability. They can be connected to one or more whiskers extending from the robot; deflecting the whisker causes switch closure. For grippers and hands, too, microswitches make very simple but effective touch sensors.

5.4.2 Bend Sensors

Another sensor useful in the domain of contact detection is the *bend sensor*. An example bend sensor, illustrated in Figure 5.23, is distributed by Images Company and by Jameco. This device uses a conductive ink deposited between two electrodes to give a variable resistance, depending upon the degree of bending. This variable resistor can be interfaced to a MC68HC11 in much the same way as a photoresistor, that is, by using a voltage divider with the output signal connected to an A/D channel. Total resistance changes by

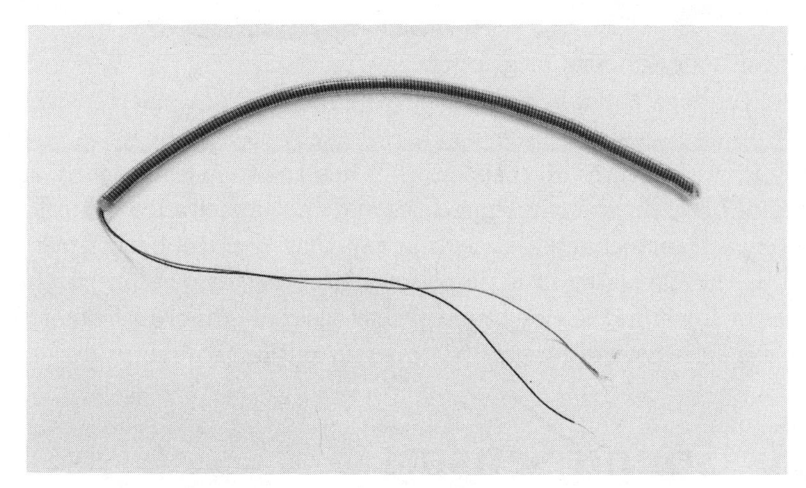

Figure 5.24. The Rubbery Ruler provides a unique way to measure deformations and relative positions.

a factor of about 3 to 5 as the bend sensor goes from straight to maximum bend.

5.4.3 Force-Sensing Resistors

Interlink Electronics manufactures a line of force-sensing resistors that, like bend sensors, are based on conductive ink technology. The resistance of a force-sensing resistor can change by several orders of magnitude as force is applied. (This is a much greater change than the bend sensor exhibits.)

Force-sensing products come in a variety of shapes and sizes, from 0.2 inch diameter circles to strips 24 inches long. A linear potentiometer is also available, which can determine the position of a contact anywhere along its length.

5.4.4 Rubbery Ruler

An interesting, yet quite simple sensor, was recently developed at the University of Melbourne. Called the Rubbery Ruler, this sensor can accurately measures changes in its own length (see Figure 5.24). Such a device might be used in the bumper of a robot to detect deformations or displacement or could be used to determine the position of an arm or gripper.

A Rubbery Ruler is formed by winding two unconnected wires side-by-side in a single layer inside a stretchy tube. The wires effectively form the plates of a capacitor. When the tube is stretched, the distance between the wires increases and the capacitance decreases. The capacitance changes is such a way that if a Rubbery Ruler is used as the capacitor in a simple oscillator circuit, the output frequency is linear in the elongation of the sensor. For more information visit the site: www.ph.unimelb.edu.au/inventions/rubberyruler.

5.5 Sound Sensors

Sensors for sound in the audible range can allow the robot to interact with its operator. Ultrasonic transducers help the robot detect and avoid obstacles.

5.5.1 Microphones

A microphone can easily be interfaced to a microprocessor. Typical behaviors instigated on a robot are: moving toward or away from noise, listening for a specific pattern of sounds, and localizing a sound's position within a room. The microphone shown in Figure 5.5, (page 116) came from Radio Shack, but Digi-Key also sells microphones, as do a number of surplus stores.

The signal from the microphone typically must be amplified before being read by the microprocessor. Figure 5.25 shows one approach, using an LM386 op-amp. Again, the output of the amplifier is connected to an A/D pin of port E and software driver routines similar to the previous examples can be used to read the data.

One significant problem with using a microphone is the need to sample the signal very frequently. Figure 5.26 illustrates the type of signal output from a microphone. If the robot is trying to detect a hand clap or a whistle, for example, it must sample the signal from the microphone often enough so that is does not miss the event. (Instantaneously, the reading from the microphone is just a voltage between 0 V and 5 V.) The signal produced by a hand clap may last only a millisecond or so. This means that the microprocessor

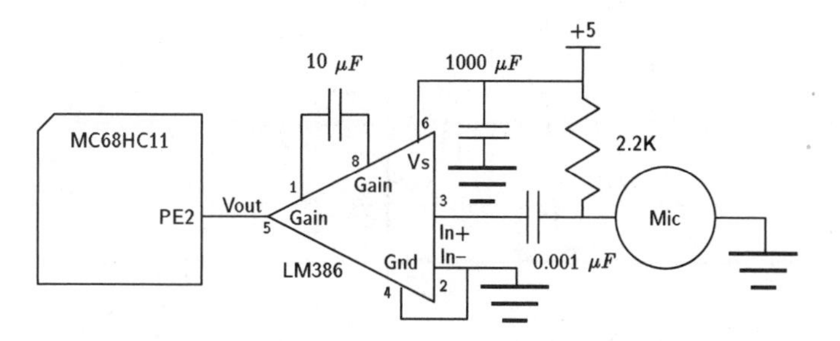

Figure 5.25. A microphone circuit with simple amplifier uses an LM386 op-amp.

must check the output of the microphone at least that often.[1] Thus, looking for very brief or high frequency signals can require all of the microprocessor's time. It may be necessary to dedicate a microprocessor or other custom hardware solely to the task of monitoring the microphone.

Another important problem is that a microphone mounted on a robot is most likely to detect the sound made by the robot's own motors. It will usually be necessary to shield the microphone in some way to guard against this.

More sophisticated acoustic sensors are available that can digitize and record voices for later playback. Other systems do rudimentary (usually speaker-dependent) voice recognition. Still, these systems see continuous improvement and lower prices as time goes on. Speech-synthesis boards are also available from suppliers such as RC Systems. Writing data strings to various registers signals the device to output an assortment of phonemes. The programmer can then create a number of sentences to give the robot simple language facilities.

5.5.2 Piezoelectric Film Sensors

Piezoelectric film is a remarkably versatile and inexpensive sensor material. Properly configured, the same material can be used to detect vibrations, changes in applied force, changes in temperature,

[1]According to the well-known Nyquist theorem, the microprocessor must sample at twice the highest frequency present in the input.

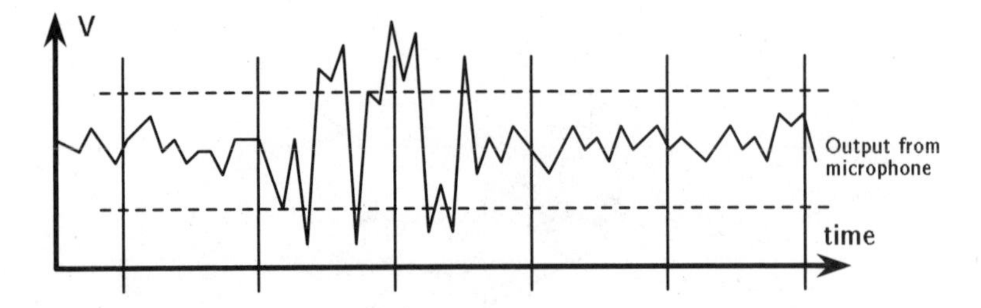

Figure 5.26. The robot is to take some action when it detects a loud sound, that is, when the signal from the microphone goes above the upper dashed line or below the lower dashed line. Each vertical bar represents a sample, the moment when the microprocessor reads the A/D converter channel connected to the microphone. Unless samples are taken at very frequent intervals, the sound of interest can easily be missed.

and even far-infrared radiation. In each case, the sensing operation consists of measuring the voltage imposed on a pair of electrodes on opposite sides of a polyvinylidene fluoride film. Piezoelectric film sensors produce a voltage only when subjected to changes in the sensed quantity. For example, when used as a collision detector, the piezoelectric sensor will generate a voltage spike at the moment the robot bumped into an object but will produce no signal while the robot is pressed against the object. Piezoelectric film allows the robot builder to construct highly customized sensors. Piezoelectric film, evaluation kits, and sensor components are available from AMP (www.amp.com/sensors/sensors.html).

5.5.3 Sonar

While the most common near-infrared detectors deliver only proximity information (something is or is not there), a sonar transducer can actually provide distance information because it is possible to measure the time of flight between the initiation of a ping and the return of its echo. By measuring the time of flight and knowing the speed of sound in air, it is possible to calculate distance covered by the round-trip of the ping.

Figure 5.27 shows the Polaroid sonar rangefinding system, which is one of the most commonly used sensors on mobile robots. These

Figure 5.27. The Polaroid 6500 ranging module mated with a 600 series ultrasonic transducer is a popular combination when building a sonar range measuring sensor for a mobile robot. These two Polaroid units are components of RugBat™, a sonar ranging unit designed to plug into the Rug Warrior Pro™ robot. RugBat™ is available from A K Peters.

rangefinders were developed as autofocus mechanisms for cameras, but the units can be purchased separately. The driver board has a very simple protocol for interfacing to a microprocessor. Figure 5.28 illustrates the necessary interface electronics.

To measure the distance to an object, the ranging board begins by sending a brief 400 volt signal to the transducer. This creates an ultrasonic chirp. After transmitting the chirp, the ranging board monitors the transducer for a returning echo. The board automatically increases its gain with time to better detect the fainter echoes returning from more distant objects. When an echo is detected, the ranging board asserts (sets to high) its output line. By measuring the time between initiation of the chirp and return of the echo, the robot's microprocessor can determine the distance to the object responsible for the echo.

In the example circuit we use two input capture registers to record the time when the sonar ping begins its flight and the time when the echo is detected. The input lines associated with the timers are designated PA1 and PA2. These lines are unassigned and thus available to the sonar module. However, in Rug Warrior's standard configuration, all of the MC68HC11's outputs are dedicated to built-

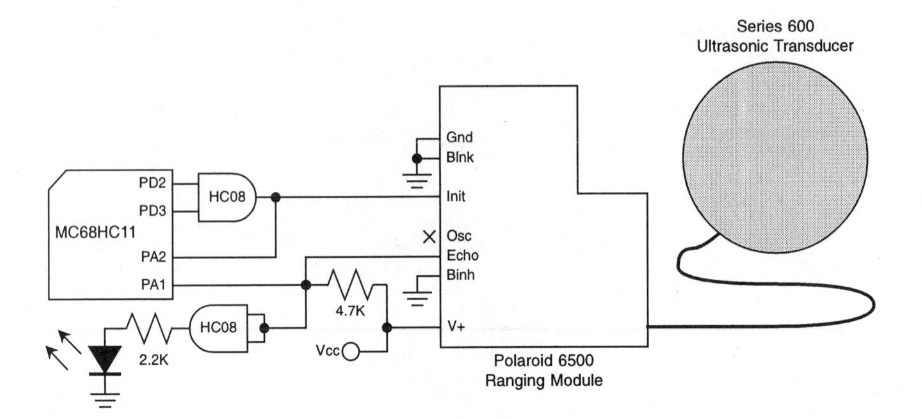

Figure 5.28. Setting the Init line of the Polaroid 6500 ranging module initiates a sonar ping and causes the state of the microprocessor's internal free running timer to be captured in the register associated with line PA2. When an echo is detected the Echo line goes high storing the current value of the same timer to PA1's associated register. The difference between the two registers is the sonar ping's round trip time of flight.

in features.[2] To control a sonar module a creative addition must be made to Rug Warrior's circuit board to provide a new output.

Rug Warrior uses outputs PD2 and PD3 to control its infrared (IR) emitters. Alternatively the left, then the right IR is turned on. When a reflected signal activates the IR detector, the presence of an obstacle on the left or right, respectively, is indicated. In normal operation, Rug Warrior never turns on both emitters at once. We can take advantage of this fact by using the logical AND of the signals from PD2 and PD3 to initiate a sonar ping. We have used a quad AND gate, a 74HC08 chip.

When PD2 and PD3 are asserted, INIT and PA2 are forced high; the former initiates a sonar ping. Upon detection of a returning echo, the ranging board asserts Echo, driving PA1 high. One of the gates of the AND chip is also connected to Echo; this gate acts as a buffer to drive the LED. When Echo goes high, the LED lights, giving a visual indication that an echo has been detected. The 2.2K resistor limits the current to the high efficiency LED. According to the specifications of the sonar ranging board, a 4.7K resistor must be used to pull the Echo line high.

[2]Line PA4 can be used to control the sonar if the LCD screen is eliminated.

When the sonar pulse occurs, the ranging board draws 2 amperes of current for a fraction of a millisecond. Such a large current can be a challenge for the robot's power supply; you will get better results if you install a capacitor of about 500 μF from power to ground near the ranging board. It is also useful to wire a second capacitor of around 1 μF directly to the backside of the ranging module between the V+ and Gnd pins. The ranging board is a sensitive, high-gain device and without this second capacitor, noise on the V+ line can cause the Echo line to go high as soon as the sonar pulse terminates.

Figure 5.28 shows the Blnk and Binh signals connected to ground and no connection to the Osc line. Advanced features can be enabled by using these lines in different ways. Consult the manufacturer's documentation for more information. A technical manual and application notes are supplied with Polaroid's Developers Kit, Designers Kit, and OEM Kit.

A software driver of surprising simplicity can calculate the range, given the circuit in Figure 5.28. The following three functions are all that is needed.

```
/* Enable input capture on rising edge on lines PA1 and PA2 */
void init_sonar()
{ bit_set(tctl2,0b010100);    /* Use bit_set and bit_clear rather */
  bit_clear(tctl2,0b101000); /* than poke to avoid changing */
                             /* other tctl2 bits */
}

/* Initiate a sonar ping */
void ping()
{ poke(tflg1,0b10);  /* Writing a 1 bit clears echo received flag */
  bit_set(portd,0b001100);   /* Turn on PD2 and PD3 => Start ping */
  sleep(0.030);              /* Wait 30 milliseconds for an echo */
  bit_clear(portd,0b001100); /* Clear echo line */
}

/* Determine if an echo was received, if so compute the range */
float range()
{ if ( (peek(tflg1) & 0b10) == 0 )
    return -1.0;              /* IC2 didn't capture echo */
  else
    return     /* Echo detected, compute time and convert to feet */
      ((float) ((peekword(tic2) - peekword(tic1)) >> 1) * 0.000569);
}
```

It is possible to determine the time-of-flight using only one input capture line. To do this, the robot must store to a variable the time when it commanded the ping. After the echo is received this variable is subtracted from TIC1. The only problem with this approach is latency. Because the microprocessor can be interrupted at any instant, there is no top-level way to be sure that the sonar ping was sent immediately after the time was recorded.

As described above, the INIT and ECHO lines of the ranging board are wired to the MC68HC11's PA1 and PA2 inputs. PA1 and PA2 are associated with internal input capture systems IC2 and IC1, respectively, sic. The input capture system allows the current time to be captured (written to an associated register) the moment the signal on an input capture line goes from low to high. The time-storage registers associated with IC1 and IC2 are TIC1 and TIC2, respectively. Time is measured in units of ticks; a tick is 0.5 microseconds long.

Whenever the signal on an input capture line does go high, another internal register (called TFLG1) records the fact by automatically setting a bit corresponding to the particular input capture line that went high. We can use this feature to determine if the sonar board received an echo.

The IC1 and IC2 timers must be initialized so that they behave as described. This is accomplished by the `init_sonar()` function. From the MC68HC11 documentation we know that the register TCTL2 controls IC1 and IC2. Rising edge capture is enabled by writing binary %00010100 to TCTL2. The command `init_sonar()` must be executed before sonar ranging is attempted.

All we need do to initiate a sonar chirp is turn PD2 and PD3 on together. This is accomplished by `ping()`. After starting the sonar pulse, `ping()` waits 30 ms, then turns off PD2 and PD3, turning off the Init line. We need not wait much longer than 30 ms, for an echo. Echoes taking longer than 30 ms to return are generally so faint that they will not be detected by the ranging board. Another reason for the 30 ms cutoff is that the registers that count and capture the time are only 16-bits long—after about 32 ms, the registers overflow and would give meaningless results.

Finally, `range()` is called after `ping()` to compute the time of flight and convert to units of length. First `range()` checks the bit in the TFLG1 register that tells whether IC2 has actually captured a time. If it has not, then no echo was received; `range()` returns -1.0 to indicate the absence of good range data. If an echo has been received `range()` subtracts the time captured by IC1, the ping initiation time, from the time captured by IC2, the time at which the echo returned. This difference is the time of flight in units of 1/2 microseconds. `range()` multiplies this number by a constant that converts to units of feet. The speed of sound at normal temperature and pressure is 1138 feet per second, one tick is 0.5×10^{-6} seconds, so the ratio is: $1138 / 0.5 \times 10^{-6} = 0.000569$ feet per tick. Multiplying this number by the difference computed above would give the total distance the echo travels. However, since the echo travels both out and back, the distance from robot to obstacle is actually half this. We thus divide 0.000569 by 2.0 to get 0.0002845. Multiplying ticks by this number gives the distance to the obstacle in feet.

A careful examination of the `range()` function in the listing above reveals that this is not exactly what is done. As often happens in robotics, there is an additional complication. The **IC** programming language stores signed integers using only 16 bits. This means that integers are restricted to the range -32768 to 32767. Integers larger than 32767 are interpreted as negative numbers. Subtracting the contents of TIC1 from that of TIC2 may result in a number larger than 32767. This is the reason for the shift right bit operation (`>>1`) in the `range()` function. The number produced by this shift (effectively dividing by 2) cannot be larger than 32767. This solves the sign problem but we must correct for dividing by 2 by multiplying the final constant by 2.0, thus 0.0002845 becomes 0.000569 again.

One caveat should be mentioned. A single transducer is used here to both transmit and receive the sonar ping. After transmitting a ping, the transducer continues to oscillate for a brief time. While these oscillations decay, the ranging board must not attempt to detect an echo because it has no way to distinguish a legitimate echo from residual ringing of the transducer. The sonar unit automatically handles this by blanking detection of an echo until 2.38 ms after the ping begins. The effect of this is that in normal operation the sonar unit cannot detect objects closer than about 16 inches.

Sonar ranging is useful for obstacle detection, corridor following, localization, and map building. However sophisticated the final behavior, this underlying primitive operation of calculating the range of a ping is the same in all cases.

5.6 Position and Orientation

For a robot to find its way about in the world, it often needs to make certain measurements. For example, it may be helpful for the robot to know the direction of gravity, the local compass heading, or how far it has moved or turned since it was in some known position. In this section, we review sensors that can provide such information.

5.6.1 Shaft Encoders

A *shaft encoder* is a sensor that measures the position or rotation rate of a shaft. Typically, a shaft encoder is mounted on the output shaft of a drive motor or on an axle. The signal delivered by this sensor can be either a code that corresponds to a particular orientation of the shaft (such shaft encoders are called *absolute encoders*) or it may be a pulse train. Shaft encoders that produce a pulse train are called *incremental encoders*. Each time the shaft turns by a small amount, the state of its output changes from high to low or vice versa. Thus, the rate at which pulses are produced corresponds to the rate at which the shaft turns.

A potentiometer can be used as an absolute position encoder. Each position of the shaft produces a unique resistance. Absolute encoders are commonly used for determining the positions of robot arms.

One way for the robot to get feedback on how far its wheels have turned or on synchronizing two wheels' velocity is to connect an encoder to each motor shaft. Shaft encoders can be purchased as enclosed units or built in as an integral part of a motor. Some incremental shaft encoders contain a spinning disk that has slots cut in it. The disk attaches to the motor shaft and spins with it. A near-infrared LED is placed on one side of the disk's slots and a phototransistor on the other. As the disk spins, the light passing

Figure 5.29. One very simple way of building a shaft encoder. Only two parts are needed: a striped pattern glued to a wheel, which is attached to the motor shaft, and a photoreflector. For Rug Warrior, we use the Hamamatsu P5587 photoreflector, shown taped to the motor and mounted so as to be only a few millimeters from the rotating striped pattern on the wheel when the wheel is mounted on the motor shaft.

through the disk is interrupted by the moving slots, and a signal in the form of a pulse train is produced at the output of the phototransistor. By using a microprocessor to count these pulses, the robot can tell how far its wheels have rotated. The combination of such an infrared LED emitter and a photodetector, packaged for the purpose of being mounted on either side of a shaft encoder's disk, is called a *photointerrupter*.

Another implementation of a shaft encoder is a *photoreflector*, which shines light from an infrared LED onto a striped wheel, which then reflects the light back to a phototransistor. A palette of radially alternating black and white stripes will alternately reflect or not reflect light to the phototransistor, yielding a similar pulse-train output. The photoreflector used by Rug Warrior is packaged with the two devices next to each other in a very compact unit. Figure 5.29 illustrates one of these small devices, attached to the side of a servo motor in such a way as to be within a few millimeters of

Figure 5.30. A close-up of the Hamamatsu P5587 photoreflector, double-sticky taped to the top of a servo motor. The servo motor's shaft is shown at the left, with the small, white pinion gear attached to it.

and facing the striped pattern on the wheel. A closer view of the mounting scheme is shown in Figure 5.30.

Because the near-infrared energy emitted by the LED can penetrate thin, white paper, it is important to take into consideration what is behind the striped paper pattern. Two pieces of plain, white paper discs backing the striped wheel should be enough to make the white segments adequately opaque so that the beam will be reflected back to the detector. Figure 5.31 illustrates 32-, 48-, and 64-count encoder patterns. You can photocopy these patterns and use them to construct your own reflective shaft encoders.

The photoreflectors we have chosen for Rug Warrior are the Hamamatsu P5587s (these devices replace an earlier P306201 part). We have chosen these devices because they have circuitry integrated in the package to amplify and condition the output of the phototransistor. The only interface components required for connecting to the MC68HC11A0 are two resistors: one for pulling up the phototransistor's open-collector output and one for limiting the current through the LED. For reading the shaft-encoder data into Rug Warrior's control system, we have chosen to take advantage of the timer-counter hardware connected to the MC68HC11A0's port A. Port A's 8 pins have various input capture and output compare registers associated

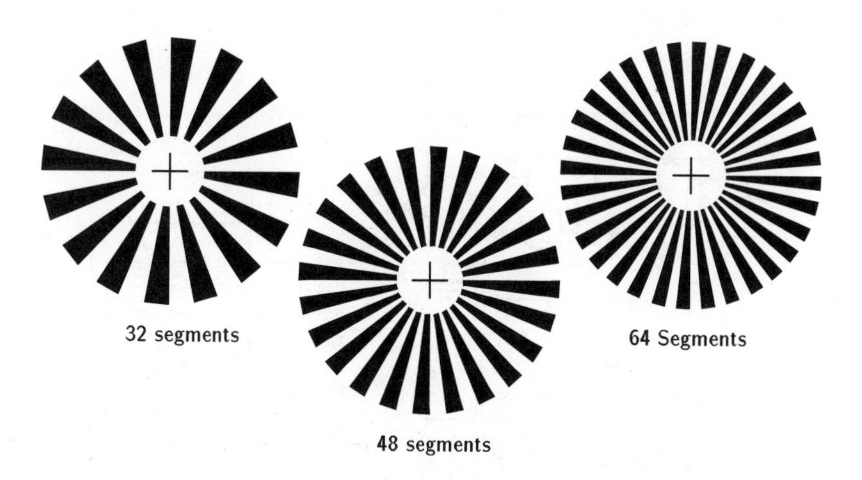

Figure 5.31. Alternating white and black stripes make reflecting and non-reflecting surfaces, respectively, for light emitted from a photoreflector's LED. More stripes give greater resolution to the output measurements, but the stripes cannot be narrower than the field of view of the photoreflector.

with them, which are able either to mark the time that events happen on those pins or to initiate events at preprogrammed times. We use PA7 and PA0 as the port A pins to accept the input from the left and right shaft encoders, respectively, as shown in Figure 5.32.

A pulse accumulator function is associated with PA7 making it easy to count the pulses produced by the left shaft encoder in software. It would have been convenient if the MC68HC11A0 designers had included two of these features on their chip (more advanced versions of the MC68HC11 do have more features for reading shaft encoders and for pulse width modulating motors), but since we do not have that luxury, we connect the right shaft encoder to the PA0 pin and use its input capture function to count the pulses.

Figure 5.33 illustrates a simple open-loop control scheme, where a motor is given a speed command and the shaft encoders are used simply to monitor its velocity. Later, in Chapter 7, we will use other portions of port A's timer system, output pins PA5 and PA6, to drive the motors, and we will also discuss how to use shaft-encoder feedback data to implement in software a velocity controller. In this section, however, we concentrate on describing how to get the shaft-encoder sensor data into the microprocessor.

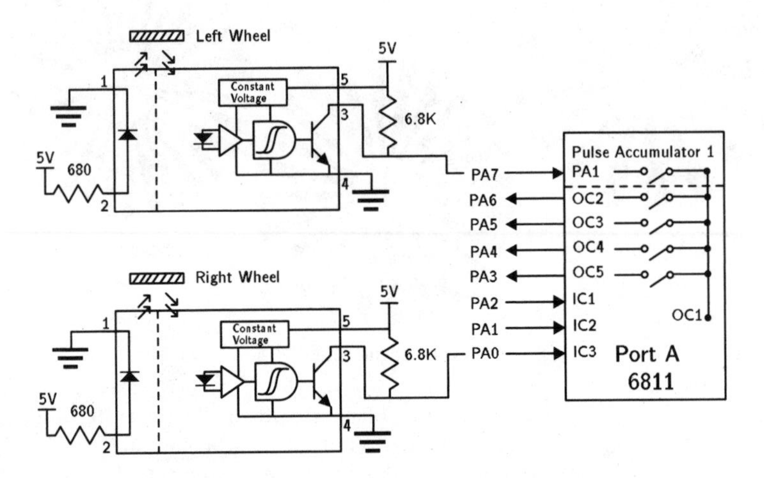

Figure 5.32. The interface between Hamamatsu photoreflectors for Rug Warrior's shaft encoders and the MC68HC11A0 port A pins PA7 and PA0. Shaft-encoder data from the left wheel are counted by the pulse accumulator hardware associated with PA7. For the right wheel, interrupts are triggered by the input capture hardware (IC3) connected to PA0. Shaft-encoder pulses are counted in software in an interrupt-handler routine.

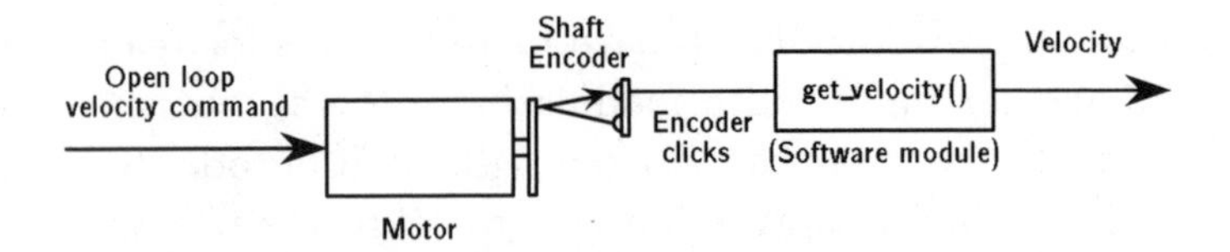

Figure 5.33. In an open loop control scheme, a command is given to the motor to make it turn at a certain speed. Depending on the load, the motor might not actually go at that speed. Shaft encoders can be used to monitor the motor's true speed.

Reading Shaft Encoders

In order to use the shaft-encoder sensors in some sort of velocity control scheme for Rug Warrior, we must first interface the photoreflectors to the microprocessor and store the ensuing counts for each wheel in two variables. One shaft encoder is fed into the pulse accumulator on port A pin PA7, and the other shaft encoder is fed into PA0 with its associated input capture three (IC3) register. Reference should be made to the Motorola MC68HC11 data books for more complete descriptions of the timer-counter system than those undertaken here.

The Pulse Accumulator

The pulse accumulator is an 8-bit counter register, PACNT, associated with port A pin PA7, that makes it very easy to count the number of rising or falling edges input to that pin. This register will overflow after 2^8, or 256, counts:

PACNT	Bit 7							Bit 0
$1027	-	-	-	-	-	-	-	-

In order to configure the system for our needs, we first have to assign pin PA7 as an input. This can be done by setting the data-direction bit for pin PA7 (which is in the pulse-accumulator control register, PACTL) to 0 for configuration as in input pin. Three other bits in the PACTL register must also be assigned. The pulse-accumulator enable bit, PAEN, must be set to 1 to enable the pulse accumulator; the mode-select bit, PAMOD, must be set to 0 for event counting; and the edge-select bit, PEDGE, must be set to 1 or 0, depending on whether it is desired to choose rising (PEDGE = 1) or falling (PEDGE = 0) edges of the shaft encoder's output. We will arbitrarily select to count rising edges and so set the PEDGE bit to 1:

PACTL	Bit 7							Bit 0
$1026	DDRA7	PAEN	PAMOD	PEDGE	0	0	RTR1	RTR0
	0	1	0	1	x	x	x	x

Once the PACTL register has been configured, the pulse accumulator will start counting the number of stripes passing in front of the photoreflector. The main program running on Rug Warrior then simply needs to poll the PACNT register at certain intervals to see how fast the wheel is turning.

Shaft-Encoder Pulse-Accumulator Software Driver

Following is **C** code that initializes the pulse-accumulator system and returns the number of pulses since the last reading. To activate the pulse-counter system, call `init_velocity()` during system initialization. Velocity of the left wheel can be found by calling `get_left_vel()` at regular intervals. Velocity is in units of encoder clicks per time interval (where the time interval is the time between two successive calls to `get_left_vel()`).

```
int PACTL = 0x1026; /* Pulse accumulator control, 8-bit reg */
int PACNT = 0x1027; /* Pulse accumulator counter, 8-bit reg */

void init_velocity() /* Initialize hardware for vel monitoring */
{ poke(PACTL, 0b01010000); /* PA7 input, enable pulse acc, */
                           /* rising edge */
  poke(PACNT,0); }  /* Start with 0 measured velocity */

float get_left_vel() /* Left vel from PA7 using pulse counter */
{ float vel;
  vel = (float) peek(PACNT);
  poke(PACNT,0);      /* Reset for next call */
  return(vel); }
```

Once the pulse accumulator hardware has been initialized, it will run in the background, automatically incrementing the count every time a stripe on the encoder wheel moves past the photoreflector. The robot's main program does not have to keep track of this activity but is free to attend to other sensors and actuators. When it needs to know the encoder count, the main program calls the function `get_left_vel()`.

Although an assembly language routine to start the pulse accumulator would also be very simple, we use an example of **C** code here for a particular reason: Namely, later, in Chapter 7, we will describe how to use shaft-encoder data as the feedback in a velocity controller for Rug Warrior's two motors. As that algorithm will require some multiplication and a fair amount of bookkeeping, it is easier to describe control algorithms by sticking solely to **C**.

Input Capture Registers

For the encoder wheel connected to port A pin PA0, more software complexity is in store. Because the MC68HC11A0 has only one pulse accumulator, we must use an interrupt to count encoder clicks from the right wheel. We will use the IC3 register associated with PA0 to generate an interrupt on every rising edge. The interrupt-handler routine, which automatically runs whenever a rising edge is detected, must increment a counter, clear the interrupt flag, and return from the interrupt.

To configure IC3 for this operation, a few associated registers must be initialized in a way similar to setting up the pulse accumulator. In this case, we will be generating interrupts and writing an assembly language interrupt-handler routine that keeps track of the count.

The TMSK1 register contains the bits that must be set to enable interrupts associated with events on any input capture pin. We will set the bit associated with IC3I, enabling interrupts:

TMSK1	Bit 7							Bit 0
$1022	OC1I	OC2I	OC3I	OC4I	OC5I	IC1I	IC2I	IC3I
	x	x	x	x	x	x	x	1

The TFLG1 register contains a flag bit, IC3F, which is set whenever the interrupt condition is met. If IC3F is set while global interrupts are enabled (the I bit of the condition code register is clear), then the hardware will automatically initiate an interrupt—the user's interrupt-service routine is called. Code in the interrupt-service routine must clear the IC3F flag; otherwise, when an attempt is made to return from the interrupt, the hardware will think the IC3 interrupt is pending and immediately service it again. Clearing the interrupt flag is accomplished by writing a 1 to the bit in the TFLG1 register that corresponds to that interrupt's flag. We will write the binary number %00000001 to TFLG1 to clear the IC3F flag.

TFLG1	Bit 7							Bit 0
$1023	OC1F	OC2F	OC3F	OC4F	OC5F	IC1F	IC2F	IC3F
	x	x	x	x	x	x	x	1

EDGxB	EDGxA	Configuration
0	0	Capture Disabled
0	1	Capture on Rising Edge
1	0	Capture on Falling Edge
1	1	Capture on Any Edge

Figure 5.34. The four actions possible by any input capture pin are to never capture, to capture on rising edges, to capture on falling edges, or to capture on any edges. Two bits in the TCTL2 register (the most significant bit, EDGxB, and the least significant bit, EDGxA) set the desired response for any successful input-event detection.

Another matter to take care of is assigning on which type of edge the input capture interrupt will trigger. Figure 5.34 gives the possibilities and the associated 2-bit code for assigning the desired trigger. We will trigger on rising edges, since that was the arbitrary choice made for the encoder connected to PA7.

These bits must be written to the TCTL2 register to configure it for rising edge-triggered interrupts. Storing %00000001 to TCTL2 will assign this properly:

TCTL2 \$1021	Bit 7							Bit 0
	0	0	EDG1B	EDG1A	EDG2B	EDG2A	EDG3B	EDG3A
	x	x	x	x	x	x	0	1

After these interrupts are configured, the main program loop must enable interrupts globally with the `CLI` instruction. Until this instruction is executed no interrupts can occur. Once this is done, any rising edge arriving on pin PA0 will trigger an interrupt. The vector address for the IC3 interrupt is \$FFEA. The two byte address stored at this location is the address at which the user's interrupt handler code must begin.

Shaft-Encoder Input Capture Software Driver

The Interactive **C** compiler used on Rug Warrior, **IC**, has a means of interfacing to MC68HC11A0 assembly language routines. (It does this by following certain naming conventions for routines and variables and by using certain file-loading protocols.) We use these features here to write an interrupt-handler routine for input capture register IC3, which counts the shaft-encoder pulses and stores the running sum in `right_clicks`, a global variable accessible by the main **C** program.

```
TFLG1 EQU $1023                    ;Timer Flag 1, 8-bit reg
ORG MAIN_START                     ;Origin for assembly module

subroutine_initialize_module:      ;This module runs on reset
  ldd #IC3_interrupt_handler       ;16-bit addr of intrpt handler
  std $FFEA                ;Store in IC3 intrpt vector
  cli                      ;Enable interrupts generally
  rts                      ;Return from subroutine

variable_right_clicks:     ;Create a C variable, right_clicks
  fdb 0                    ;Fill double byte, 16 bits.right_clicks = 0

IC3_interrupt_handler:
  ldd variable_right_clicks
  addd #1                  ;Add one more encoder count
  std variable_right_clicks
  ldaa #%00000001          ;Clear the IC3 flag by writing a one
  staa TFLG1               ;Store in TFLG1 to clear IC3 flag
  rti                      ;Return from interrupt
```

These code fragments accomplish several goals. A code initializer module, `subroutine_initialize_module`, is created, whose purpose is to store the address of the interrupt handler in the correct location. The **IC** system calls `subroutine_initialize_module` each time the reset button is pushed. A variable, `variable_right_clicks`, for storing the encoder counts from the right shaft encoder is also created. (**C** routines will reference this variable using the variable name `right_clicks`.) Finally, `IC3_interrupt_handler`, an interrupt-handler, is written, which increments the right-encoder counts variable each time the reflective photosensor sees the stripe it is looking at change from black to white.

If we compare this example with the code for the other shaft encoder connected to PA7, the contrast is clear. The pulse accumulator provided us with special purpose hardware to relieve the main program of the duty of incrementing a counter every time an event occurred. Here, the programmer must specifically set up an interrupt-handler routine to attend to this chore.

Now we add a function, get_right_vel(), to our existing **C** code, which returns the value of right_clicks (then resets it) whenever it is called. Our supervising **C** program must now also include the commands to initialize the appropriate registers for using the IC3 input capture interrupt.

For instance, our **C** program might look like the following:

```
int TCTL2 = 0x1021; /* Timer Control 2,8-bit reg,interrupt edge */
int TMSK1 = 0x1022; /* Timer Interrupt Masks, 8-bit reg */
int TFLG1 = 0x1023; /* Timer Flags, 8-bit reg */
int PACTL = 0x1026; /* Pulse accumulator control, 8-bit reg */
int PACNT = 0x1027; /* Pulse accumulator counter, 8-bit reg */

void init_velocity()                /* Call to begin vel monitoring */
{ poke(PACTL, 0b01010000);         /* PA7 in, ena pls acc, rising edg */
  poke(PACNT,0);                    /* Start with 0 measured velocity */
  bit_clear(TCTL2,0b00000010);/* IC3 interrupts on rising edges */
  bit_set(TCTL2,0b00000001);  /* IC3 interrupts on rising edges */
  bit_set(TMSK1,0b00000001); } /* Enable only IC3 interrupts */

float get_left_vel()                /* Left vel from PA7, pulse ctr */
{ float vel;
  vel = (float) peek(PACNT);
  poke(PACNT,0);                    /* Reset for next call */
  return(vel); }

float get_right_vel()               /* Right vel PA0 using interrupt */
{ float vel;
  vel = (float) right_clicks;
  right_clicks = 0;                 /* Reset for next call */
  return (vel); }
```

The functions get_left_vel() and get_right_vel() provide a uniform way to acquire each motor's shaft encoder data. This is the essence of an abstraction barrier. Even though the hardware interface to each shaft encoder is implemented differently, the programmer simply relies on the function get_left_vel() and the function get_right_vel(). The programmer need not worry about how these functions interface to the hardware.

Later, we will use these primitive operators to create a higher-level program, a velocity controller, which will cause the two motors

Figure 5.35. Futaba makes a small, rate gyro for model airplanes. The input is a pulse-width-modulated signal, and the output is an increased or decreased pulse width, depending on the rate of rotation.

to always go at the same speed, enabling the robot to maintain a constant heading.

5.6.2 Gyros

Another sensor that is useful in monitoring how the robot moves is a *rate gyroscope.* Mechanical gyroscopes use the principle of conservation of angular momentum to keep one or more internal axes pointed in the same direction as the exterior of the gyroscope, the gyroscope case, translates and rotates. Thus, a gyrosocpe attached to a robot makes it possible to determine either how rapidly the robot is rotating or how far it has rotated, relative to a fixed coordinated system.

Humphrey, Columbia, and Murata sell small gyroscopes, as does Futaba. The inexpensive model from Futaba, shown in Figure 5.35, is a single-axis rate gyro made for model helicopters. A rate gyro produces a signal proportional to the rate of rotation about an axis perpendicular to the axis of the gyro, but it does not provide absolute orientation information. The Futaba gyro takes a pulse-width-modulated signal provided by the MC68HC11 and modifies it (increasing or decreasing the pulse width) based on the rate of rotation of the gyroscope case.

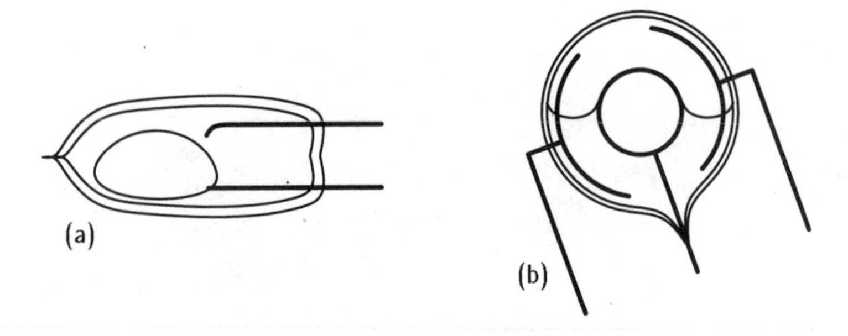

Figure 5.36. Sensor (a) is a mercury switch. When this sensor is tilted, the drop of mercury closes the contact between the two electrodes. In an electrolytic-tilt sensor (b), the amount of conduction between the center electrode and each of the outer electrodes is determined by the degree to which the outer electrode is immersed in the electrolytic fluid.

5.6.3 Tilt Sensors

Determining whether your robot is level or tilted can mean the difference between negotiating rough train smoothly or tumbling over. Many types of sensors can provide information about the relative angle between the robot body and the gravity vector. The simplest and generally least expensive tilt sensor is the *mercury switch*, such as the one illustrated in Figure 5.36(a). This sensor consists of a small, glass bulb containing two or more contacts and a drop of mercury. Depending on which way the bulb is tilted, the bead of mercury will close or open the circuit.

Such a sensor is easy to interface in a microcontroller. When mounted properly, it provides a digital signal, alerting the microprocessor that the robot has tilted too far in one direction. Several mercury switches fixed at different orientations can provide information about the degree and direction of tilt. Software conditioning of the signal from a mercury switch is almost always required, however. As the robot starts, stops, and bounces about, the bead of mercury frequently makes contact, even when the robot is not dangerously tilted.

The electrolytic-tilt sensor, a type of inclinometer, offers an improvement over the mercury switch in many applications. Figure

5.36(b) diagrams an inclinometer. This sensor has two or more electrodes immersed in a conductive fluid. The conduction between the electrodes is a function of the orientation of the sensor relative to gravity. The electrolytic-tilt sensor produces an analog signal proportional to the degree of tilt. Such sensors are typically much more expensive than mercury switches. Spectron offers a full line of electrolytic-tilt sensors.

An exciting recent development in sensor technology is the micromachined accelerometer. This device is a chip with a tiny suspended mass machined into the silicon. Piezoresistors embedded in the structure are used to sense minute changes in position of the mass as the chip undergoes acceleration. Such devices can also be used to detect the direction of gravity. Micromachined accelerometers offer an accurate, rugged, and reliable means for determining the direction of tilt of a mobile robot. IC Sensors and Lucas Novasensor are good sources for these sensors.

5.6.4 Compasses

A *compass* provides a way for your robot to acquire absolute information about its orientation. This can be very helpful when writing a navigation algorithm. In open areas, compasses are very reliable, and once calibrated to local magnetic north, they are also accurate. If your robot is to be used indoors, however, the serviceability of a compass becomes more problematic. Magnetic fields from electrical wiring, structural steel in buildings, and even the metal components of the robot itself can all produce large errors in the compass reading. As long as errors of, say, ± 45 degrees can be tolerated, the compass is a viable option. Certain electronic compasses intended for use in automobiles can, with sufficient modification, be employed by your robot. ETAK manufactures digital compasses. Precision Navigation, Inc. whose compasses are distributed by Jameco, has several useful models. Some models from Precision Navigation contain automatic tilt compensation. The Fetch robot described in Section 11.3.2 uses such a compass.

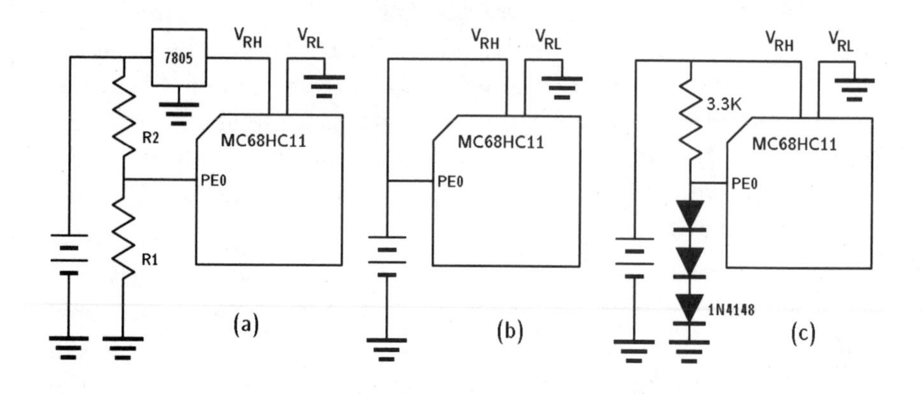

Figure 5.37. Circuit (a) shows one way to construct a battery level monitor if the microprocessor is operated from a regulated supply. When using an unregulated supply, circuit (b), although tempting, will not work. Circuit (c) corrects the deficiency of (b) by using the diode voltage drop to provide a reference voltage.

5.7 Proprioceptive Sensors

A *proprioceptive sensor* is any sensor used to measure the internal state of the robot. Monitoring these sorts of sensors can tell the robot when it is time to recharge its batteries, when a motor is overheating, or when a component has malfunctioned.

5.7.1 Battery-Level Sensing

By sensing its battery voltage, a robot can determine when it is time to return to the charging station or curtail power-draining operations. Designing a battery-level indicator is a simple matter when the microprocessor operates from a regulated supply, as in Figure 5.37(a). As shown, only a voltage divider is needed.

In the circuit of Figure 5.37(a), when VRH has been connected to the regulated output voltage from an LM7805, VRL will go to ground. We wish to determine V_B, the battery voltage. The voltage supplied by the batteries must always be higher than the regulated voltage in order to achieve good regulation. In this case, suppose that the batteries are effectively exhausted when their voltage reaches 7.0 V.

If we simply connected one of the A/D channels, say, PE0, to the positive battery terminal, it would not be possible to determine the battery voltage. Since the voltage at PE0 would always be greater than that at VRL, the A/D converter would always report a value of 255 to the ADR1 result register.

We must engineer a circuit that will deliver a maximum of 5.0 V to PE0 when the batteries are fully charged and a smaller voltage as the batteries discharge. This is the purpose of the voltage divider. We will choose resistors $R1$ and $R2$, such that the voltage at PE0 begins at 5.0 V and decreases as the batteries discharge. Suppose that, when fresh, the batteries supply a maximum voltage of $V_{B,\mathrm{max}}$. With the voltage divider connected as shown in Figure 5.37(a), the maximum voltage that can be present at PE0, $V_0 = \frac{R1}{R1+R2} V_{B,\mathrm{max}}$. To compute $R1$ and $R2$ we choose V_0 to be 5.0 V, since higher voltages cannot be measured. Given that we also know $V_{B,\mathrm{max}}$, we can now solve for $R1$ and $R2$ if we arbitrarily choose the sum $R1 + R2$. This sum should be high enough so that the drain on the battery due to the voltage divider is insignificant compared to that of the rest of the electronics; at the same time, the sum should be small compared to the internal impedance of the A/D converter.

To complete the example, assume that $R1+R2 = 4700\Omega$ and that power is supplied by eight NiCd cells whose fresh voltage is 9.6 V. Now we have $R1 = \frac{5.0}{9.6} \times 4700 = 2447\Omega$, $R2 = 2252\Omega$. By measuring the voltage at PE0, we can determine V_B: $V_B = \frac{4700}{2447} \times V_0$.

There is a complication if the microprocessor supply does not include a regulator, as in Figure 5.37(b) and (c). As we have seen, the A/D converter works by comparing the voltage at PE0 with the reference voltages at VRH and VRL. If connected as shown in (b), the ratio of these voltages remains constant as battery voltage declines. Thus, the A/D converter always reports that the battery voltage equals VRH, and the result of the conversion is always 255.

In Figure 5.37(c), we make use of the *diode voltage drop* to produce a reference to which we can compare the battery voltage. Whenever current through a diode exceeds a certain minimum, a characteristic voltage (usually about 0.6 V) develops across the diode. In the circuit in Figure 5.37(c), the A/D converter will compare the constant $3 \times 0.6 = 1.8$ V at pin PE0 with the changing voltage at VRH. If the battery pack is fully charged at, say, 7.0 V and depleted at 4.5 V, then the result from the A/D converter will be $255 \times \frac{1.8}{7.0} = 66$ and $255 \times \frac{1.8}{4.5} = 102$, respectively.

5.7.2 Stall Current Sensing

One reliable way to determine if a robot is stuck is to monitor the current being used to drive the motors. If all other sensors fail to detect an imminent collision, the robot will, in short order, come to rest against the obstacle. In this situation, the wheels will stop rotating while current to the motors will go to a maximum. Thus, motor current serves as a collision detector of last resort. One way to detect motor current is to put a small resistance in series with the motor (typically, a fraction of an ohm), amplify the voltage across the resistor, and measure the voltage with one of the A/D channels. Some motor-driver chips have built-in circuitry to simplify this measurement. The L293E and IR8200 motor-driver chips have such features.

The software that monitors motor current in order to detect a collision should not respond too quickly. Each time the robot accelerates from a dead stop, motor current will typically go to a maximum, then decrease as the robot speeds up.

5.7.3 Temperature

It is often a good idea to monitor certain temperatures within the robot. If the electronics get too warm, the microprocessor may crash. High temperatures can also shorten the lives of motors, and NiCd batteries may be damaged by heat if high current charging continues after the batteries are already fully charged. Certain motor-driver chips, the IR8200 for example, have built-in, over-temperature sensors. For other applications, many companies manufacture discrete temperature sensors including Murata, EDO Corporation, and RCD Components.

5.8 Exercise

To this point, we have seen how to take a large number of simple sensors and interface them to a microprocessor. In Chapter 9, we will

Figure 5.38. A photograph of Rug Warrior about to fall over the edge of a step. What kinds of sensors could be used to detect a drop-off? Whiskers? Microswitches? Bend sensors? Two near-infrared beams separated a few inches and aimed to cross at the level of the floor? Sonar? Invent your own!

see how to arrange higher-level programs, using behavior control, to enable the robot to act in response to its sensor readings to create seemingly intelligent behaviors. As we have seen, sensors merely deliver voltages to the microprocessor. What the robot manages to achieve with these signals depends upon how clever the programmer can be with software.

Many times, however, the programmer just does not have enough variables in her or his environment to juggle. The problem often dictates going back to hardware and inventing a new sensor for the job. For instance, in Figure 5.38, Rug Warrior is about to tumble off the edge of a step. All its sensors point upward and all its code implicitly assumes that it will always travel on level surfaces. Try to invent a sensor that will detect a step. Mount it on your Rug Warrior's chassis, and interface it using connectors we discussed earlier in some spare prototyping space you left open on your board for expansion features. Try programming a software driver, and see how it works!

5.9 References

Whole volumes have been written about sensors for mobile robots, but here we have had the opportunity to touch only briefly on a few simple sensors that can be incorporated inexpensively into Rug Warrior. For the definitive reference on sensors for mobile robots consult (Everett 95). This book gives in depth coverage of a wide variety of sensors.

More sophisticated robots, such as Robart II, from the Naval Ocean Systems Center shown in Figure 5.1 (Everett, Gilbreath, and Tran 1990), and Attila from the MIT Mobile Robot Lab, shown in Figure 5.2 (Angle and Brooks 1990) take advantage of redundant sensors to endow themselves with increased awareness of their surroundings.

Robart II predated and influenced much of the hardware design later undertaken at the Mobile Robot Lab, especially in the realm of sensors. Everett and Stitz (1992) gives a complete exposition on the workings and wonders of a wide variety of sensors applicable to mobile robots.

Angle (1991) describes how the six-legged Attila was designed to use its legs as sensors as the robot moved through its environment, and how various sensors of increasing reliability were situated to trigger the lowest-level behaviors in a layered control system. Ferrell (1992) expands on that theme and discusses the notion of creating virtual sensors from combinations of concrete physical sensors to make Attila more reliable.

For books on sensors and interfacing electronics, Beckwith and Marangoni (1990) detail making mechanical measurements from position sensors, force sensors, accelerometers, and the like, while Jung (1986) presents a "cookbook" of useful op-amp designs for amplifying and conditioning small sensor signals. Seipple (1983) is another useful sensor text.

Mechanics

6.1 Locomotion

From slithering to hopping, there are a great variety of ways to move across a solid surface. Among robots, the three most common systems use wheels, tracks, and legs.

Wheeled vehicles are by far the most popular for several practical reasons. Wheeled robots are mechanically simple and easy to construct. The payload weight-to-mechanism ratio is also favorable. Both legged and tracked systems generally require more complex and heavier hardware than wheeled systems designed for carrying the same payload. Additionally, a wide variety of wheeled devices, such as toys, can be modified for robot use.

The principal disadvantage of wheels is that, on uneven terrain, they may perform poorly. As a rule, a wheeled vehicle has trouble if the height of the object it must surmount approaches the radius of the wheels. One solution is simply to use wheels that are large compared to all likely obstructions. In many instances, however, this is impractical.

For robots that must operate in a natural environment, tracks are an appealing option because tank treads allow the robot to nego-

Figure 6.1. Some clever arrangements of wheels can provide functionality similar to that of tracks. By mounting the wheels on pivoting outriggers, Sojourner, developed by NASA's Jet Propulsion Laboratory, was able to climb Martian rocks three wheel radii high. (Sojourner™, Mars Rover™ and spacecraft design and images, copyright 1996-97, California Institute of Technology. All rights reserved. Further reproduction prohibited.)

tiate relatively larger obstacles and are less susceptible than wheels to environmental hazards, such as loose soil and rocks. The major disadvantage of tracks, however, is inefficiency. Friction within the tracks themselves dissipates power, and energy is wasted whenever the vehicle turns because the treads must slip against the ground. The dead-reckoning ability of tracked vehicles suffers for the same reason. If the robot computes its position by counting the number of times the track-driving wheels have rotated, then the error in the robot's estimate of where it is grows whenever the vehicle turns. In fact, to a greater or lesser degree, the dead-reckoning ability of all robots suffers from this problem of wheel, track, or leg slippage. One wheeled robot that dramatically demonstrated its ability to maneuver through rough terrain is the Sojourner robot. Sojourner, pictured in Figure 6.1 successfully explored a small portion of Mars in 1997.[1] For more information on Sojourner please refer to Section 11.3.1.

[1]Rocky III, a predecessor of Sojourner, was pictured in the first edition of *Mobile Robots.*

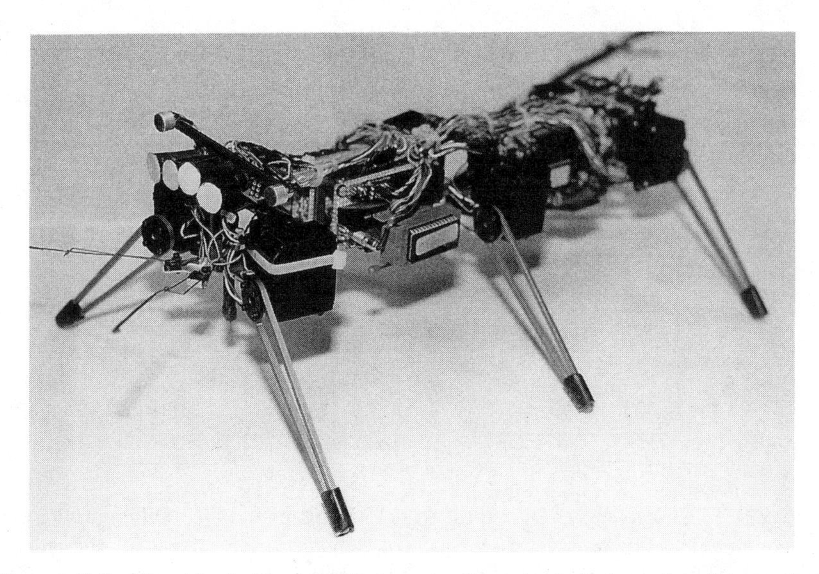

Figure 6.2. Genghis, built at MIT, is a shoebox-sized six-legged walking robot. Genghis is now on display at the Smithsonian Air and Space Museum, in Washington, D.C.

Walking robots can potentially overcome more of the problems of rugged terrain than either wheeled or tracked robots. Figure 6.2 shows Genghis, a six-legged robot built at the MIT Mobile Robot Lab. While there is great interest in the development of practical systems, legged robots face a number of challenges. Many of these challenges stem from the large number of degrees of freedom required by legged systems. Since each leg must have at least two motors, the cost of building the robot is higher relative to those with wheels or tracks; the walking mechanism is also more complex and thus more prone to failure. Furthermore, control algorithms become more involved, as there are more motions to coordinate. Optimal control of walking and running machines is still an active area of research.

6.1.1 Wheel Arrangements

For a wheeled robot, the designer may choose among several significantly different arrangements of driven and steerable wheels. Among these arrangements, as illustrated in Figure 6.3, are differential drive, synchro drive, tricycle drive, and car drive (also known as *Ackerman steering*).

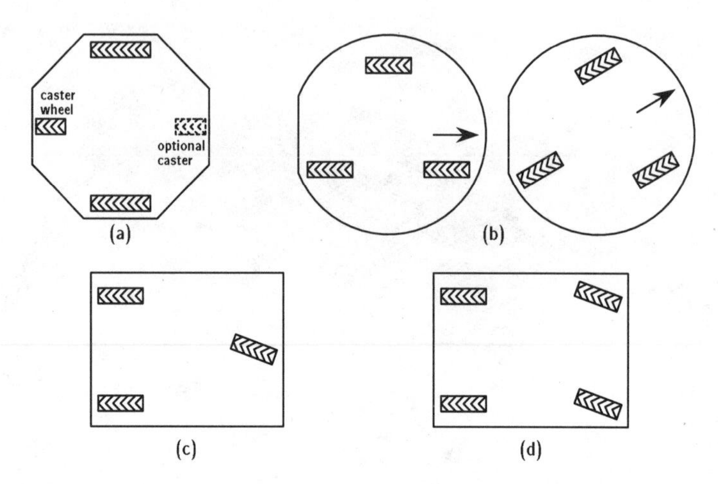

Figure 6.3. Bottom views of several wheel arrangements. (a) Differential drive uses one or possibly two caster wheels. (b) Synchro drive rotates all the wheels together. The drive/steer wheels are shown in two different orientations. (c) Tricycle drive has the steering motor on one wheel and the driving motor on the back pair of wheels. (d) Car-type drive rotates the front two wheels together.

Differential Drive

From both programming and construction standpoints, differential drive can be one of the least complicated locomotion systems. The TuteBot employs this type of drive, as does the robot illustrated in Figure 6.4. The differential scheme consists of two wheels on a common axis, each wheel driven independently. Such an arrangement gives the robot the ability to drive straight, to turn in place, and to move in an arc.

An important design problem for a differential drive robot is how to ensure balance. Some additional support, besides the two drive wheels, must be provided to prevent the robot from tipping over. Usually, this is done by adding one or two caster wheels, arranged in a triangle or diamond pattern. Depending on the robot's weight distribution and the strength of its motors, a triangle pattern may still leave the robot vulnerable to tipping. If the robot shown in Figure 6.3(a) without the optional caster, moves forward (to the right) rapidly and then suddenly stops, it will tip in the direction of motion unless its center of gravity is well to the left.

Figure 6.4. A differential drive robot (such as this floor-cleaning prototype, from the Robot Talent Show) can pivot about its center.

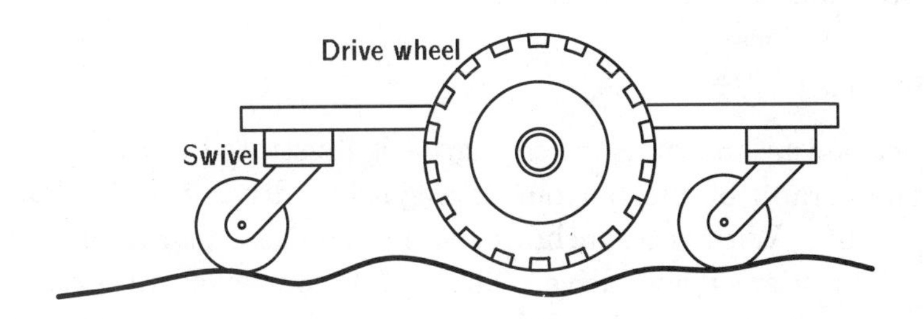

Figure 6.5. This side view of the chassis of a differential drive robot with two rigidly mounted caster wheels illustrates how undulations in the terrain can break the contact between drive wheels and ground, thus leaving the robot unable to move.

A diamond pattern solves this problem but may introduce another problem, as is illustrated in Figure 6.5. If the caster wheels are attached rigidly to the robot body, then small undulations in terrain can leave the robot supported only by the casters. The drive wheels lose contact with the surface and become unable to move the robot. Mounting the caster wheels in such a diamond pattern thus requires some sort of suspension system so that the casters can move up and down relative to the drive wheels.

Another design consideration for differentially driven robots is how to make the robot go straight. As we saw with the TuteBot, even when the same voltage is applied to the two motors, they will turn at different speeds and the robot will veer to one side or the other. To make the robot go straight, we must ensure that the wheels turn at the same velocity.

When the motors encounter different loads (e.g., one wheel is on carpet and the other, on a hard floor) motor speeds will vary and the robot will turn even if it was initially adjusted to go straight. This means that motor velocity must be controlled *dynamically*—there must be a means to monitor and change motor speed while the robot is underway. One type of control scheme is discussed later in section 7.8.2 (page 257). The simplicity of differential drive is thus somewhat offset by the increased complexity of the system required to control it. However, decreasing mechanical complexity in favor of increasing electronic and software complexity is often the most reliable and cost-effective trade-off.

Synchro Drive

A mechanism known as *synchro drive* is illustrated in Figure 6.3(b). A photograph of the bottom of a synchro drive base is shown in Figure 6.6. With the synchro drive mechanism, all wheels (usually three) both steer and drive. The wheels are linked in such a way that all point in the same direction at all times. In order to change direction, the robot simultaneously rotates all wheels about a vertical axis, as shown in Figure 6.3(b). Thus, the robot's direction of motion changes but the chassis continues to point in the same direction. If the robot is to have a front (presumably where the sensors are concentrated), additional linkages must be provided to keep its body pointed in the same direction as its wheels. The synchro scheme overcomes many of the problems of differential, tricycle, and car-type drives at a cost of greater mechanical complexity.

Car and Tricycle Drives

Car-type drive (Ackerman steering), with its four points of suspension, provides good stability. Tricycle drive has a similar feature, with the advantage of being mechanically simpler, since car drive

Figure 6.6. This Real World Interface base uses a synchro drive, which steers all three wheels together at the same time.

requires some sort of link between the two steerable wheels. In general, for both tricycle drive and car drive, the two fixed wheels will be connected to a drive motor and the steerable wheel(s) will not be driven. On some robots, however, the steering wheels are also driven. With car and tricycle drive, it is not necessary to monitor wheel velocity in order to make the robot go straight. Simply positioning the steerable wheel at its neutral position is sufficient. This simplicity, however, comes at a price, as we will see in the next section.

6.1.2 Robot Kinematics

Robot *kinematics* addresses how robots move. Given that steering is set to such and such an angle and that each wheel turns so many times, where will the robot end up and which way will it be pointed?[2]

Differential and synchro drive robots have a subtle advantage over car and tricycle drive types. The difference is their kinematics. Consider the robot shown in Figure 6.7, which has three degrees of freedom when moving on a flat surface. Precisely what we mean

[2]In robotics, the inverse problem is usually more interesting (and more difficult). Given that we want the robot to arrive at some position, pointed in some particular direction, the problem of *inverse kinematics* is to compute the set of robot operations which will achieve the goal.

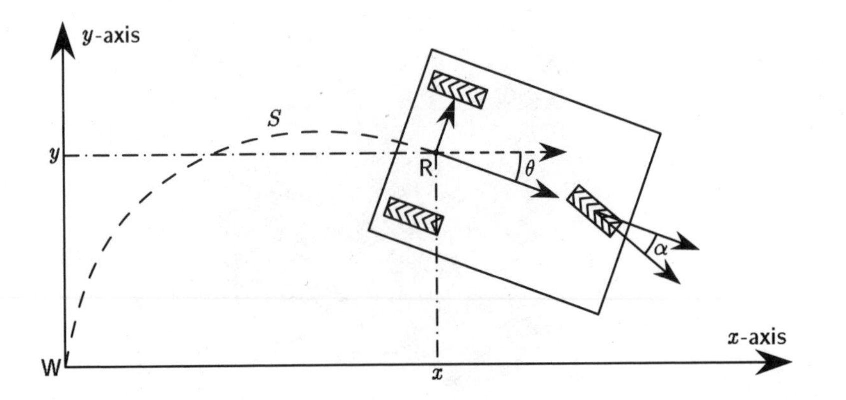

Figure 6.7. The kinematics of a tricycle drive mobile robot. A wheeled robot has three degrees of freedom in the plane but only two controllable parameters.

by this is the following: Relative to some global coordinate system (labeled W in the figure), the robot can be at any position specified by two coordinates, x and y, and pointed in any direction specified by a third coordinate, angle θ. These three degrees of freedom (x, y, θ) give us the distance to and the angle between the global frame, W, and a local reference frame, R, on the robot. (We could have put frame R anywhere on the robot but because the robot's center of rotation is the point midway between its two drive wheels, we chose that point.)

We would like the ability to position and orient our robot anywhere on the plane. That is, regardless of where it starts out, if we give it x, y, and θ coordinates, the robot should be able to move to that location. There is a problem, however. To achieve these three degrees of freedom the robot has only two parameters that it can control: the steering angle, α, and the total distance it travels, S. This means that the robot's orientation and its position are coupled: In order to turn, it must move forward or backward. The robot cannot go directly from one position and/or orientation to another, even if nothing is in the way. In order to achieve a desired position and orientation simultaneously, the robot must follow some path, possibly complex. The details of such a path are greatly complicated by the presence of obstacles. This is the reason parallel parking is

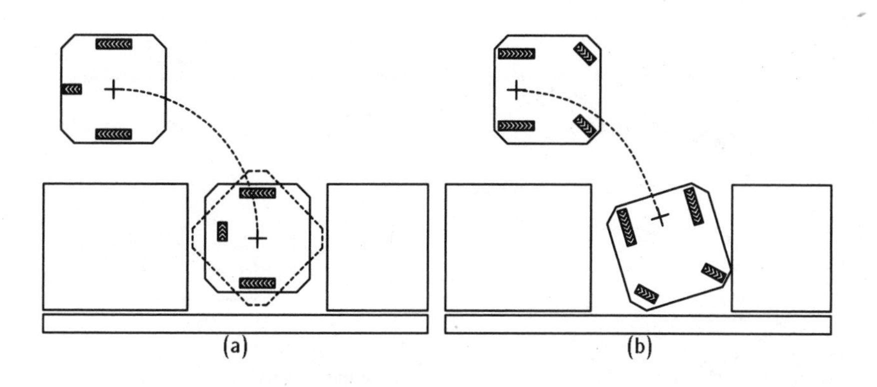

Figure 6.8. A comparison of differential drive and car drive kinematics is made vivid by this parallel parking example. To achieve the desired goal of being positioned between the two rectangles and facing to the right, the differential drive robot (a) moves to position and rotates in place. The position of the caster wheel reveals the rotation that has just occurred. No similarly simple path will achieve this goal for the car drive robot (b).

difficult. However, a robot based on differential or synchro drive can, by turning in place, effectively decouple its position from its orientation.

These ideas are illustrated in Figure 6.8. In both parts of the figure, the goal is the same: for the robot to position itself between the two rectangles and to point to the right. The differential drive robot in (a) achieves this easily; it drives to position and then turns in place to attain the desired orientation. But for the car drive robot in (b), no simple procedure will yield the same results. Although it likely can achieve the goal, a long series of turns and forward and back motions will be required. Deciding exactly which motions are to take place can be a difficult problem.

6.1.3 Robot Shape

A robot's shape can have a strong impact on how robustly it performs. A noncylindrical robot runs a greater risk of being trapped by an unfavorable arrangement of obstacles or of failing to find its way through a narrow or cluttered space.

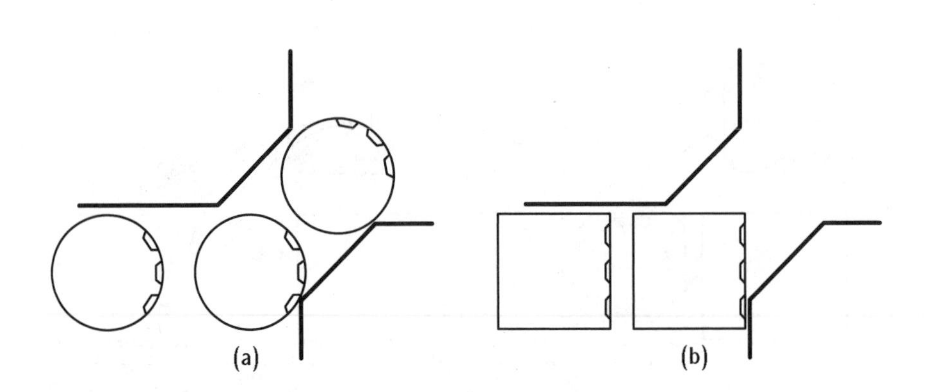

Figure 6.9. Robots of the same width but different shapes encounter a narrow passage. A simple turn-while-in-contact algorithm allows the round robot (a) to negotiate the passage. Success for the square robot (b) is problematic.

In Figure 6.9, a cylindrical robot (a) and a square robot (b) of identical width encounter a narrow passage while moving to the right. A simple algorithm will allow the cylindrical robot to find its way through the passage. The robot will drive until its bumper detects a collision; then it will stop. Since the collision is on its right side, the robot will turn to the left until it is able to go forward again. It will then proceed through the passage. This scheme is simple because the robot is able to rotate *while in contact with an obstacle.*

The square robot, by contrast, must both back up and rotate if it wishes to use the same tactic. However, it is not clear how far the robot should backup and what it should do if it suffers another collision while escaping from the first. Thus, an algorithm designed to navigate a square robot through a narrow passage requires more complexity than one for a cylindrical robot. To understand the reason for this, we must appeal to an advanced concept in robotics called *configuration space* (see Section 6.7). Configuration space analysis allows us to find a path for a robot of arbitrary shape in an arbitrary environment. The configuration space for the robot in Figure 6.9(a) collapses to a two-dimensional channel. A path through this channel can be easily found using only local methods. The configuration space of the situation shown in Figure 6.9(b), however, is a complex, three-dimensional mathematical construct. Such an arrangement is necessarily more difficult to search.

6.2 Adapting Mobile Platforms

An abundance of inexpensive and readily available mobile platforms are adaptable for use as mobile robot bases. These include radio-controlled cars, wire-guided (tethered) vehicles, and other battery-powered toys. Most drive types except synchro are well represented in the toy store.

A number of strong reasons recommend choosing the drive and suspension system of a toy as the base of a mobile robot. Less design and construction are required, as a major portion of the robot has already been built and the problems of mechanical power transmission and component placement have largely been solved by the manufacturer. Also, it is often much less expensive to adapt a mass-produced toy than to purchase similar component parts separately.

The robot designer, however, should be aware of some typical problems with this approach. Such a base is usually optimized for use as a particular toy, not as a robot. The motors in toys typically require high current and provide low efficiency, which means that the design of the drive electronics will be more complicated and robot running time will be reduced.

In general, the motors and gearing used by toys are designed to make the toys fast. Thus, control problems are often encountered when the robot is required to move slowly in order to respond to sensors. Also, shaft encoders for measuring distance and implementing a velocity-control system are usually not present and can be difficult to add. Figure 6.10 illustrates one type of drive train that can be acquired from toys used as radio-controlled cars and sold at stores such as Radio Shack. This particular drive train came from an old model toy, no longer sold, called a Red Fox Racer. The interesting feature of the Red Fox Racer drive train was that it came equipped with separate drive motors for left and right wheels, which meant it could be fairly easily adapted for the locomotion system for a mobile robot.

6.2.1 Identifying the Drive Type

The least expensive mobile toys have only one motor and maneuver using a sequence of forward and back-and-turn motions. When the

Figure 6.10. The differential steering mechanism from an inexpensive Radio Shack wire guided car. Two motors connected to gear trains drive the left and right wheels separately.

motor spins in one direction, the toy moves straight forward. When the motor reverses, a simple clutch built into the back axle causes one wheel to slip and the robot to turn. That is, the toy turns only when backing up. It *is* possible to design a robot that operates in this simple manner (such as Squirt, illustrated in Figure 6.11), but it may become stuck in situations where backing up is not possible. It is easy to recognize a toy with a back-and-turn mechanism, as its remote control will usually have only one button. When the toy is switched on it begins moving forward. When the remote control button is pressed, the toy backs up and turns.

More generally useful toys have either differentially driven wheels or tracks or a separate drive motor and steering motor. In the latter case, the steering motor may often be a simple solenoid that allows the toy to steer in only a small number of preselected directions.

One way to determine which type of drive mechanism a toy possesses is to switch it on and observe its behavior. If the drive wheels change velocity relative to each other as the remote steering mechanism is manipulated, then the toy is probably a differential drive type. If the toy has steerable wheels that flip between only two or

Figure 6.11. Squirt (right) built at MIT, is slightly larger than 1 cubic inch and goes forward or backs-and-turns using one motor and a clutch in the rear axle. Goliath (left), another MIT robot, which once claimed the title of the world's smallest autonomous robot, with two motors, six sensors, two batteries, and an onboard computer in just over 1 cubic inch of volume. Goliath uses tank-drive differential steering.

three different positions, it most likely uses a steering solenoid. If the steerable wheels change direction smoothly as the remote steering mechanism is moved, then steering is most probably accomplished with a servo motor.

6.2.2 Electrical Modifications

The point of modifying a toy is to make microprocessor control possible. Often motors, servos, and gear trains can be used in situ while the toy's original electromechanical controls must be discarded. Thus, it will be necessary to design new drive circuitry to replace the old manually controlled system. Before this can be done, however, information must be obtained about the characteristics of the motors. Some of this information can be acquired most easily by temporarily leaving the toy's motors and servos connected to the original circuitry while measurements are made.

The first step, then, is to disassemble the toy to the point that the motors and steering actuators are exposed. Identify the drive motor or motors, which will be connected to the toy's drive wheels via a gear train. The voltage the motors and servos are designed to accept is most probably equal to the voltage supplied by the toy's batteries.

For example, if the toy is powered by four 1.5 volt (V) alkaline batteries, then the motors probably are designed to run on 6.0 V. Sometimes, however, a split power supply is employed, which directs half the battery voltage to each motor. This split power supply setup is often seen because it is simpler to design a motor-reversing circuit if different power supplies are used for forward and reverse. If the supply type cannot be determined from an analysis of the original wiring, turn on the toy and measure the voltage across each motor while it is running.

The drive motors in virtually all toys are connected to the rest of the circuit by only two wires. But often a capacitor will be soldered directly across the leads of the drive motors.

The capacitor suppresses voltage spikes produced by the motors and should be left in place. Disconnect the motors from the toy's drive electronics, and attach a wire to each motor lead. Ultimately, these two wires will be connected to the microprocessor-controlled motor-drive circuitry you will design.

Measure the resistance across the terminals of the motor with the rotor in several different positions. Often, the measured resistance will change as the brushes contact different parts of the commutator. The maximum current that the motor-driving circuitry must provide is the supply voltage divided by the average resistance. See Chapter 7 for a thorough description of how to design motor-driver circuitry.

Solenoids are generally two-state devices with two electrical terminals. When voltage is applied, the movable core of the solenoid moves to its activated position. When voltage is removed, the core returns to its normal position. The core is attached via a linkage to the steerable wheels of the toy. Some solenoids can assume either of two activated positions, depending on the polarity of the applied voltage.

Servo motors used for steering adhere to some more or less general control standards. See Section 7.6.4 (page 225) for a description of how servo motors are used in velocity-control feedback systems.

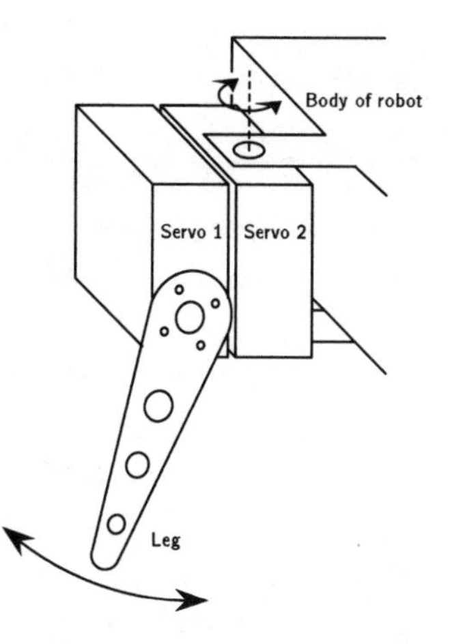

Figure 6.12. A simple two-degree-of-freedom leg can be constructed using a pair of model airplane servo motors. The servo motors are attached to each other, with their axes of rotation 90 degrees apart.

6.3 Legged Locomotion

In general, legged locomotion systems are quite complicated. There are however, a few simple variations. An insect-like leg can be constructed using only two model airplane servos, as shown in Figure 6.12. This is the same construction used on the Genghis robot pictured in Figure 6.2 (page 171).

To take a step, servo 1 first swings the leg outward, away from the body. This is designed to raise the leg over any obstruction. Next, servo 2 rotates the servo pair so as to move the leg forward. Servo 1 then rotates the leg downward until it makes contact with the ground. Finally, servo 2 rotates back, pushing the robot forward. A coordinated motion of six such legs allows the robot to move forward or backward or to turn.

6.4 Construction Systems

There are also a number of readily available construction systems that may be adapted to mobile robots. Such systems have interlocking motors, gear trains, and other mechanical parts; some even include simple sensors and switches. LEGO, Fischer-Technik, Meccano, Capsella, Erector Set, and others offer products of this sort. LEGO, in particular, is the construction medium used in a popular mobile robot design course at MIT. These building sets make robot construction simple and quick because all mechanical components are available from a single source and all are guaranteed to interface easily with each other.

The primary disadvantage of such systems is the constraint on component placement: You must put things where they will fit rather than where you want them. Another problem is the unfavorable strength-to-weight ratio typical of plastic components. This can make such systems unusable for the construction of large robots or robots that must carry heavy loads. Nevertheless, Fischer-Technik and the other sets are good choices for prototyping new robots. Rug Warrior II, the tank, used LEGO parts for the mechanical structure of its base.

6.5 Custom Construction

If the requirements of a proposed robot cannot be met by adapting an existing toy vehicle or by using a building set, it may be necessary to construct the robot base from scratch. Rug Warrior I, the cylindrical version of Rug Warrior shown in Figure 6.13, uses a differential drive mechanism that was constructed from scratch, using tools and materials from a workshop.

6.5.1 Wheel Mounting

When building your own robot base from scratch, one thing to consider is the attachment of the wheels to the motors. Rug Warrior I has wheels mounted directly on the shaft coming from the gearbox. This configuration is diagrammed in Figure 6.14(a). Although simple and straightforward, there are potential problems with this design. The gearbox of the motor is required to support the entire weight of the robot. If the robot weighs too much or bounces too violently as it moves over uneven terrain, the acceptable side load (force perpendicular to the output shaft) of the gearbox can easily be exceeded. The manufacturer typically specifies the acceptable

Figure 6.13. Rug Warrior I uses two motors to drive the left and right wheels in a differential manner. A nylon caster mounted on a fixed axle slips on the ground when the robot turns in place.

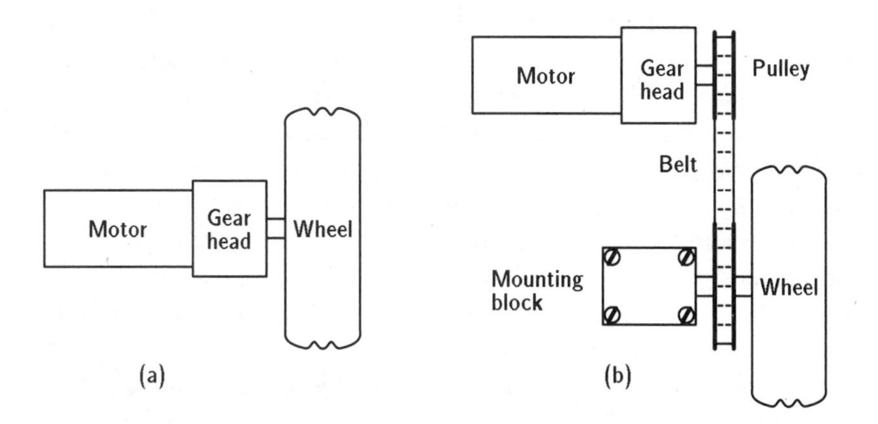

Figure 6.14. Two common wheel-mounting systems are direct attachment and belt-and-pulley systems. (a) A wheel can be directly attached to a gearbox shaft. (b) Gears can also be isolated from shock and wheel load by a belt.

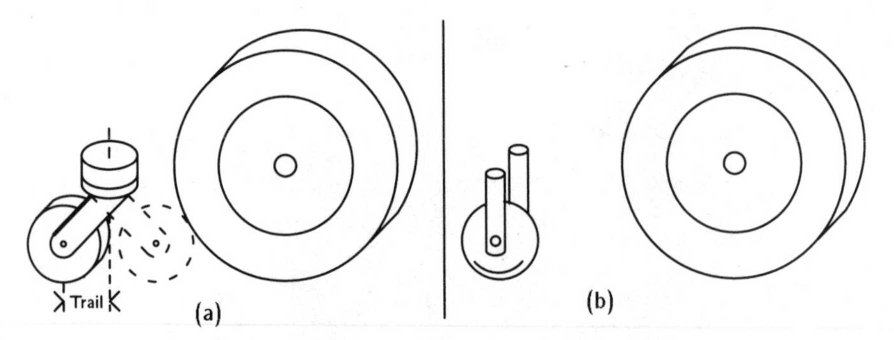

Figure 6.15. A caster wheel invokes a number of design considerations. (a) It should spin freely, so that the robot can turn easily, and be large enough to surmount obstacles. (b) A simpler solution is the caster used by Rug Warrior I, which is simply a nylon ball mounted on an axle. When the robot turns, the ball slides.

side load for a motor or gear output shaft in the motor data sheet. You should check carefully the gear specifications before making this design choice. If the side load is exceeded, the life of the motor or gearbox will be shortened.

There are several ways to avoid this problem. A rugged, more expensive gearbox can be used, or the wheel can be supported at two points by running the gearshaft through the wheel and attaching the shaft to a mount on the other side. Another alternative is shown in Figure 6.14(b), where the motor and gearbox can be isolated from the side loads and shocks using a belt-and-pulley system.

As mentioned earlier, in order to be balanced, a differentially driven robot must have at least one supporting wheel in addition to its two drive wheels. Ideally this would be a caster—a wheel free to rotate and to swing. But there are several conflicting constraints on the design of this wheel, which is depicted in Figure 6.15. The caster must have a large diameter so that it can ride over obstacles as large as the drive wheels can surmount. It also must have a large trail so that it can swing freely when a side force is applied, and it must fit entirely beneath the force-sensing skirt so that it does not collide undetectably with obstacles.

In order to simplify the mechanics, a compromise was made on Rug Warrior I in the design of the supporting wheel. The wheel is not a true caster at all; it is a ball with a fixed axle running through

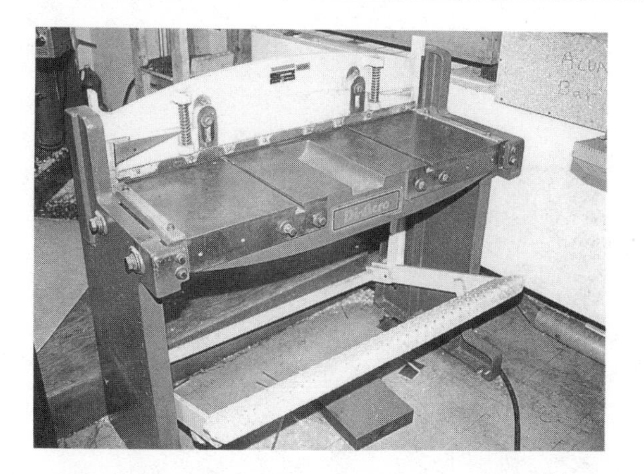

Figure 6.16. A shear machine is used for cutting sheet metal. This machine is operated with a foot pedal; when the pedal is pressed, a large knife edge moves down and slices the piece of aluminum placed underneath.

the center. As such, the wheel must slide sideways when the robot turns in place. This is easily accomplished, however, because the wheel is spherical and made of nylon so that it slides easily. Also, the robot is balanced in such a way as to minimize weight on the rear wheel.

6.5.2 Sheet Metal

One of the simplest yet most effective ways to build a robot is to design a body made of formed sheet metal, in particular, aluminum. There are different kinds of aluminum. Some are designed to be bent, while others are hard and brittle and will break rather than bend. Aluminum is easy to work with and can produce a lightweight, rugged chassis. Metal-working tools found in a machine shop, such as a shear and a brake, are convenient for forming aluminum pieces. (A *shear* is a tool that slices off strips of metal; see Figure 6.16. A *brake* is a machine for bending metal; see Figure 6.17.)

An effective way to work with sheet metal is to lay out cuts, bends, and holes on a piece of paper (perhaps using a computer drawing program) and then tack this template directly to the aluminum sheet with rubber cement. This trick, illustrated in Figure 6.18, will save a great deal of work over transferring your markings

Figure 6.17. Sheet metal can be bent into a wide variety of shapes using a brake. The brake has one arm that can clamp the aluminum onto the base table of the brake. The table can then be rotated up, which makes the aluminum fold.

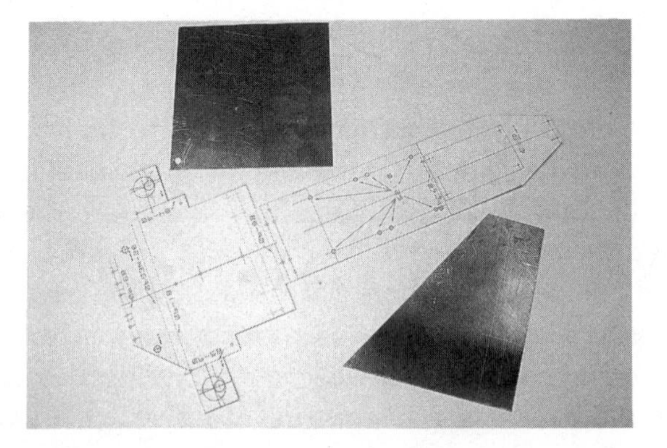

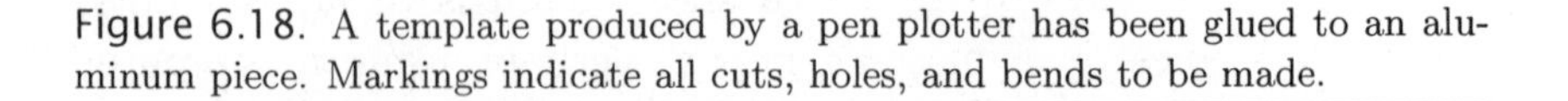

Figure 6.18. A template produced by a pen plotter has been glued to an aluminum piece. Markings indicate all cuts, holes, and bends to be made.

to the metal. One caveat is in order here. Except for small parts, templates should be made with pen plotters rather than laser or ink jet printers. Because of the uncertainties in the way paper feeds through such printers, the aspect ratio cannot be guaranteed and so, in general, a vertical inch will not equal a horizontal inch.

A punch is the fastest, most effective, and safest way to cut holes or other shapes in sheet metal. Although drills are commonly used for this purpose, they can be quite dangerous unless the metal is clamped or otherwise held in place. If the drill bit binds while a hole is being cut, the entire work piece may begin to spin. This can be almost as dangerous as using a circular saw from which the guard has been removed.

6.5.3 Acrylic

In addition to aluminum, another popular choice for robot body material is acrylic. Like aluminum, acrylic forms a strong lightweight body and can be worked with readily available tools; it also bends easily with the application of heat. Rug Warrior I was built from an acrylic chassis.

The body of Rug Warrior I was constructed by first using a band saw to cut out the acrylic chassis and skirt pieces. A band saw is shown in Figure 6.19. The chassis was drawn using a computer drawing program, and the drawing was printed out and then attached to a piece of acrylic. The template of the mechanical base of Rug Warrior I is shown in Figure 6.20. Mounting holes were then drilled in the 1/16-inch thick acrylic sheet of the skirt and the 1/8-inch thick chassis pieces. To form the skirt, we heated the acrylic sheet in an oven at approximately 300 degrees Fahrenheit for several minutes. Then, using oven mitts, we wrapped the sheet around a cylindrical object with a diameter close to that of the robot and held the acrylic in place until it cooled.

Readers who attempt a similar procedure should use great caution. Acrylic may catch fire if it comes into contact with hot electrical elements or flame. Touching the hot acrylic can also cause painful burns. As always, it is best to practice a new procedure with some scrap material.

Figure 6.19. A band saw is a useful tool for forming acrylic. Cuts can be made with accuracy and speed.

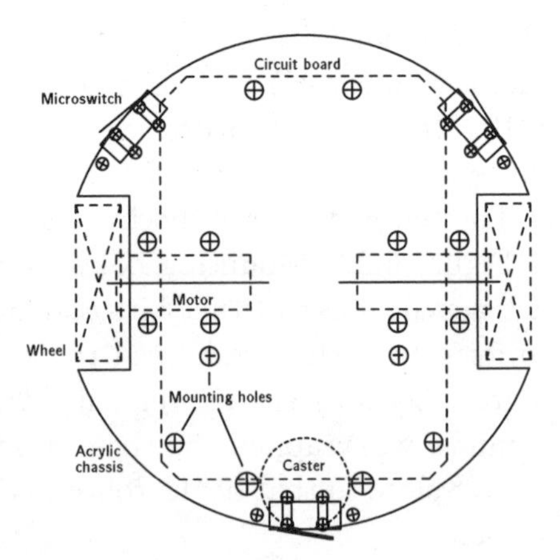

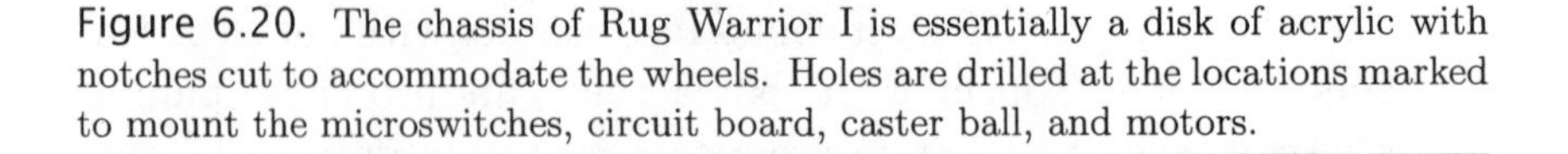
Figure 6.20. The chassis of Rug Warrior I is essentially a disk of acrylic with notches cut to accommodate the wheels. Holes are drilled at the locations marked to mount the microswitches, circuit board, caster ball, and motors.

6.6 Exercise

Try building your own platform for a Rug Warrior. Fischer-Technik or a toy car might be an easy starting point. Perhaps if you have access to a machine shop, you can create a more sophisticated base than the ones shown here. Because the Rug Warrior board is so small, your base will not have to carry much payload weight. Lots of interesting mechanisms can be created from Fischer-Technik and other construction kits. Don't hesitate to use double-sticky tape, glue, and Velcro. Try lots of ideas. Build a walking machine!

6.7 References

Many of the photographs illustrating the variety of possible mechanical platforms in this chapter were robots built at the MIT Mobile Robot Lab. A few were built at other research laboratories.

Sojourner, the Mars rover, was built at the Jet Propulsion Laboratory. Visit the web site mpFwww.jpl.nasa.gov for more information. Experimental results from Rocky III, one of Sojourner's predecessors, are reported in (Miller et al. 1992). Genghis, the six-legged robot of Figure 6.2 was designed in a bachelor's thesis (Angle 1989) at the Mobile Robot Lab by Colin Angle and programmed by Rodney Brooks (1989). The floor-cleaning robot in Figure 6.4 was a product of the Robot Talent Show and built by Joe Jones. The synchro drive base pictured in Figure 6.6 was a product of Real World Interface.

Once the world's smallest robot, Goliath (left in Figure 6.11) was an undergraduate project at the Mobile Robot Lab designed by James McLurkin. Squirt (right in Figure 6.11), which previously held the "tiny" title, was built in 1988 by Anita Flynn, Rodney Brooks, William Wells, and David Barrett (Flynn et. al. 1989).

The section on robot kinematics and configuration space stemmed largely from the work of Tomás Lozano-Pérez. More in-depth discussion can be found in Lozano-Pérez, Jones, Mazer, and O'Donnell (1992).

7

Motors

7.1 Variety Abounds

A few years ago, a computer was the largest and most expensive component of a robot, while motors and batteries consumed only small percentages of the budget. These days, while motors and batteries have changed little, the relationship has flipped. Microelectronics have shrunk in size and cost so drastically that, for the types of mobile robots we describe in this book, the motors and gears will typically be the most costly items.

An electric motor converts electrical energy to mechanical energy. Motors come in all manner of shapes and sizes. There are electromagnetic direct current (DC) motors and electromagnetic alternating current (AC) motors and a number of variations of each. AC motors are typically used for large machinery (such as machine tools, washers, dryers, and the like) and are powered from an AC power line. You might run across AC motors with titles such as single-phase, split-phase, capacitor start, permanent split-capacitor, shaded-pole and three-phase motors. AC motors are seldom used in mobile robots because a mobile robot's power supply is typically a DC battery.

We will focus on DC motors in this book. DC motors are commonly used for smaller jobs and will suit our purposes well. They also appear in a large variety of shapes and sizes: permanent magnet iron core, permanent magnet ironless rotor, permanent magnet brushless, wound field series connected, wound field shunt connected, wound field compound connected, variable reluctance stepper, permanent magnet stepper, and hybrid stepper motors.

For a robot's needs, a DC motor usually runs at too high a speed and too low a torque. In order to swap these characteristics, a DC motor must be geared down. Connecting the shaft of a motor to a geartrain causes the output shaft from the geartrain to rotate much more slowly and to deliver significantly more torque than the input shaft. A geartrain can be assembled discretely and attached to the motor shaft, or a DC motor can be purchased with the geartrain already prepackaged inside the motor housing.

These compact motors are termed *DC gearhead motors* and will be most useful in putting together a small robot. DC gearhead motors are normally based on permanent magnet ironless rotor motors in order to be as lightweight as possible. They can also be purchased with position encoders integrally connected. Figure 7.1 illustrates two conventional DC gearhead motors.

Most DC motors have two electrical terminals. Applying a voltage across these two terminals will cause the motor to spin in one direction, while a reverse polarity voltage will cause the motor to spin in the other direction. The polarity of the voltage determines motor direction, while the amplitude of the voltage determines motor speed.

However, some DC motors, such as stepper motors, have more than two electrical terminals, often up to six or eight. Signals are applied to these wires, which energize different coils inside the motor sequentially. The rotor is subsequently attracted to each portion and "stepped around" in a continuous fashion. Thus, the timing of these signals determines the motor speed, the phase between the signals determines the motor direction, and the number of commands determines the motor position.

Another type of DC motor with more than two electrical terminals is an assembly known as a *servo motor*. Although the term *servo motor* is used in a variety of contexts, what we are talking about here

Figure 7.1. These DC gearhead motors manufactured by Escap are permanent magnet ironless rotor models with 54:1 and 27:1 geartrain ratios. The motor on the left has an attached printed circuit board, which interfaces to a position encoder encapsulated in the motor housing.

is the three-wire DC servo motor that is often used for a control surface on a model airplane or for a steering motor on a radio-controlled car. This type of assembly incorporates a DC motor, a geartrain, limit stops beyond which the shaft cannot turn, a potentiometer for position feedback, and an integrated circuit for position control. Of the three wires protruding from the motor casing, one is for power, one is for ground, and one is a control input where a pulse-width signals to what position the motor should servo. When we speak about a motor servoing to a position, we mean that an electrical circuit directs the motor to rotate to the commanded position and keeps it there. If you try to grab the motor shaft while the servo loop is running, and forcibly rotate the shaft to a different position, the electrical circuit will read the angle of the potentiometer, realize that the shaft is no longer at its commanded position, and increase the current to the motor. This will increase the torque the motor puts out and the motor will push back against the torque you are applying with your hand. The servo motor will continue to do this until the shaft has rotated back to its commanded position. A servo motor then is an assembly which consists of a DC gearhead motor,

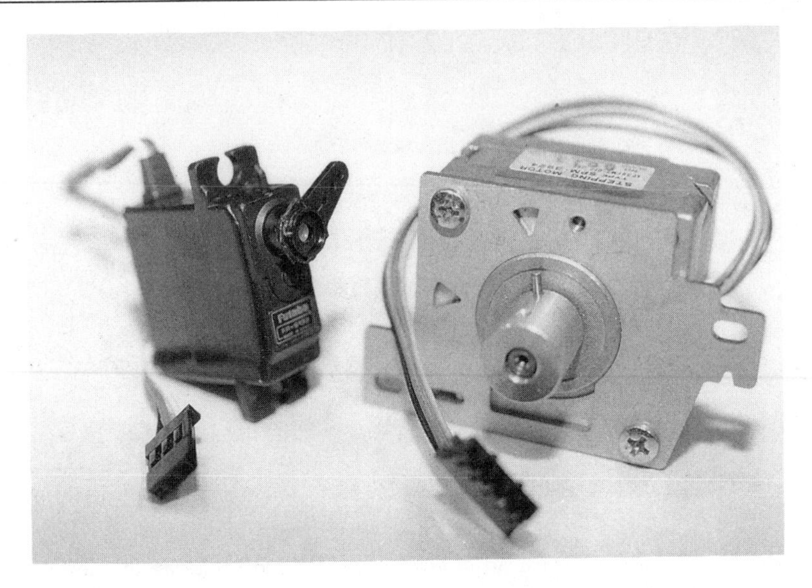

Figure 7.2. Shown on the left is a Futaba servo motor and on the right, a stepper motor. Note the three-wire lead on the servo motor and the six wires protruding from the stepper motor.

a position sensor on the shaft, and an integrated circuit for control, all packaged into the casing of the servo motor.

The flaps and control surfaces on model airplanes do not have to rotate continuously, so limit stops are added to these motors and a single-turn potentiometer then suffices to provide position information back to the integrated circuit that controls the motor position. Servo motors can be extremely compact and easy to control, and because they are mass produced for the toy industry, they are often cheaper than other DC gearhead motors. Although they rotate less than 360 degrees and hence are not suitable for wheeled robot propulsion, these model airplane servo motors often find their way into robot grippers, arms, and legs. Figure 7.2 shows both a servo motor and a stepper motor.

If you want to skip ahead to building Rug Warrior's locomotion system, we will tell you right now that our choice was to take Royal Titan Maxi Servos, available from Tower Hobbies, strip out the controller chips and potentiometers and remove the limit stops, and use these motors as continuously revolvable DC gearhead motors to drive Rug Warrior's wheels. This is the cheapest, simplest solution we could find for this book's example robot.

DC motors are also characterized another way: as either brush-type or brushless motors. These designations refer to the manner of commutation used that converts direct current from the battery into the alternating current required to generate motor action. If the DC current is commutated mechanically with brushes, the commutator segments at the ends of the rotating rotor coil physically slide against the stationary brushes that are connected to the motor's terminals on the outside of the case. If the DC current is converted into AC current in the rotor electronically, with position sensors and a microprocessor controller, then no brushes are needed. Brush-type motors are more common and cheaper. Brushless DC motors have an advantage over brush-type motors in that friction is reduced, leading to longer life and finer control for the motor. Also, brushless motors can produce less radio frequency interference. The trade-off is that brushless DC motors require more extensive control circuitry in order to do the commutation electronically.

In addition to electromagnetic DC and AC motors, there are a few other types of motors that are not electromagnetic. Piezoelectric ultrasonic motors, which can be found in autofocus lenses in some Japanese cameras, work on the principle of mechanical bending of a piezoelectric ceramic, using frictional coupling to a rotor. The Japanese have also introduced these motors into headrest actuators in new luxury cars, paper pushing mechanisms in copiers, and in tinier versions in wristwatches for use as silent (vibrating) alarms. Ultrasonic motors, in contrast to conventional electromagnetic motors, spin at lower speeds and with higher torques, alleviating the need for geardown. This means they can be compact and lightweight, but the frictional coupling between rotor and stator results in problems of wear. A small piezoelectric ultrasonic motor is shown in Figure 7.3.

Also, in research labs around the world, *electrostatic* motors are being micromachined out of silicon in dimensions on the scale of a human hair. Electrostatic motors work on the principle of charge attraction, where a force is created as two charged plates slide past each other. At small scales, electrostatic forces can be relatively strong, but for large motors, electromagnetic forces are more effec-

Figure 7.3. This 8 millimeter (mm) diameter piezoelectric ultrasonic motor, built at the MIT Mobile Robot Lab, is composed of two pieces: the stator and the rotor. The *stator*, shown on the left, is a steel ring with piezoceramics bonded onto the bottom that causes a wave to travel around the ring. The top piece, the *rotor*, is made of brass and, when pressed against the stator, is dragged along and spins. The stator with a rotor on top is illustrated on the right.

tive. Although micromotors have not reached the stage of practical use, they are intriguing.

Shape memory alloys can also be used for robot actuation. A shape memory metal such as Nitinol changes shape reversibly on being heated and cooled. Mondo-tronics, Inc., sells a small, six-legged robot (shown in Figure 7.4) that is actuated by these materials. When the wire is heated by passing current through it, the wire changes shape and shrinks, causing a leg to lift. When the wire is cooled (i.e., when no current is passing through it) the wire changes back to its original longer shape and the leg goes back down. The wires are attached to the legs in such a way that, while three legs lift the others push backward. Alternating this pattern between the two sets of three legs causes the robot to propel itself forward. Plans and instructions for building a similar microrobot called *Stiquito* are available to the public. If you have access to the Internet, you may acquire this information via anonymous FTP. Connect to site cs.indiana.edu, and look in the pub directory.

Even more esoteric is a new class of actuators that are starting to a ppear in research laboratories around the world. These are cotton-like fibers that act similarly to artificial muscles. With the alternating addition of acidic and basic solutions, these actuators can shrink and expand up to 1,000 times their original volume with strength and speed equal to those of human muscle. While still a lab-

Figure 7.4. This 10 centimeter (cm) robot from Mondo-tronics weighs 50 grams (g) and is actuated by shape memory wires which are wrapped around various screws mounted on the legs and body. Passing 200 milliamperes (mA) of current through a sequence of wires causes alternating legs to lift up and move forward.

oratory curiosity, these polymer gels may prove to be the technology of the future.

7.2 How a DC Motor Works

For the project at hand, let us focus on how permanent magnet DC gearhead motors work. Understanding the mechanism behind the production of torque is helpful when trying to read a motor specification sheet for choosing the correct-sized motor. Such understanding will be helpful again later, when designing the power electronics for controlling the motor from a microprocessor.

Electromagnetic forces in DC motors come about when current-carrying conductors are placed in magnetic fields, as illustrated in Figure 7.5. Magnetic fields can be generated by permanent magnets. Flux lines across an air gap flow from one magnet's north pole to another magnet's south pole. The Lorentz force law states that current-carrying conductors placed in magnetic fields create forces. The force, F, created is perpendicular to both the direction of the

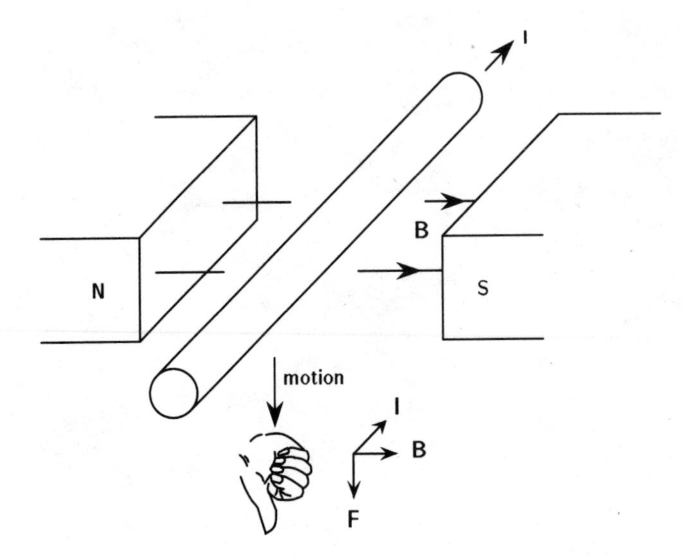

Figure 7.5. A magnetic flux field, B, is set up by the permanent magnets in the direction from north pole to south pole. A current-carrying conductor placed in such a field experiences a force acting on it. The resultant force is directed downward.

current, I, and the direction of the flux field, B. The direction of F is determined by the right-hand rule, where the fingers curl from the direction of the current toward the direction of the flux field and the thumb points in the direction in which the resultant force is created. In the case of Figure 7.5, the force produced is in the downward direction.

Rotary motion requires a loop of wire. Figure 7.6(a) shows a loop of wire mounted on an axis of rotation and situated in the flux field set up by the permanent magnets. Figure 7.6(b) illustrates the resulting forces. Because forces are created in a direction perpendicular to both the current's direction and the magnetic field's direction, current going into the loop along the top generates, according to the right-hand rule, a force acting downward. Current coming out along the bottom portion of the loop creates a force acting upward. The force disparity, acting at a distance from the center of rotation, causes the loop to experience a torque. The loop will rotate until a force disparity no longer exists. That point would be reached when the plane of the loop is vertical and the forces on the top and bottom

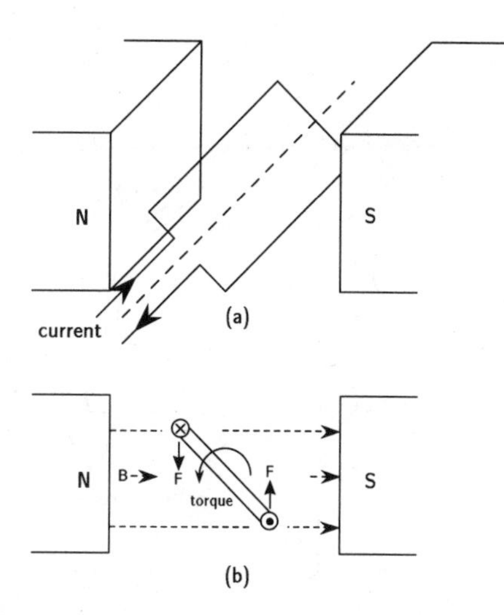

Figure 7.6. (a) This loop of wire has current flowing into the page on the left side and out of the page on the right. (b) The resulting oppositely directed forces, acting at a distance from the center of rotation, cause the loop to rotate until it is vertical.

portions of the loop would both act through the center of rotation, resulting in zero torque.

Continuous rotary motion can be achieved by reversing the direction of the current just as this point is about to be reached. The process of deriving this necessary alternating current from a DC battery is called *commutation*. Mechanical commutation requires a set of brushes that allow the ends of the loop of wire to slip across the contacts of the battery. The commutator setup is shown in Figure 7.7.

A disassembled DC gearhead motor is shown in Figure 7.8. A large number of loops of wire are usually incorporated in order to increase the torque of the motor. These loops are wrapped around an armature that can contain an iron core for increased flux or be ironless for lighter weight. Two half cylindrically shaped permanent magnets are housed along the inside of a steel casing, which provides a flux return path. The wound armature is fitted between the magnets, leaving a small air gap. As the current through the armature alternates, a force is created, causing the armature and the shaft to rotate.

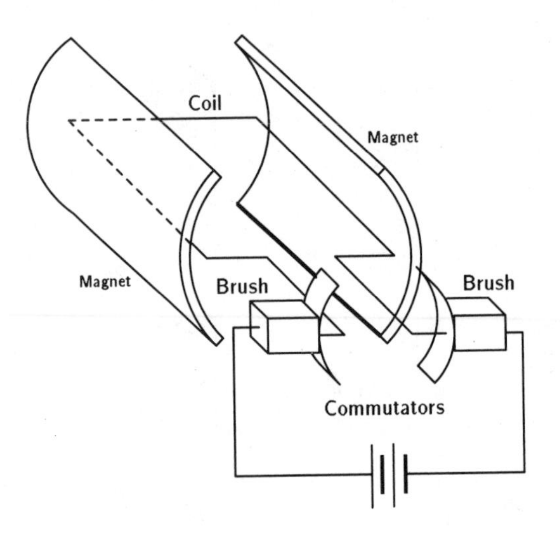

Figure 7.7. A commutation system using brushes is one way to make a DC motor. The commutator segments are attached to the loop of wire and rotate with it, while the brushes remain stationary as the commutator segments slide past.

Figure 7.8. A permanent magnet DC gearhead motor shown here has been removed from its housing. Windings of the armature around a central core with ends connected to commutator segments can be seen at the right, while the geartrain is mounted on the shaft at the left. A cylindrical housing (not shown) fits around the armature and holds two permanent magnets along its inner shell.

7.3 Sizing a DC Motor

Selecting an appropriate motor for your robot involves both understanding the loads that the robot will impose on the motor and the performance that the motor can deliver, as detailed in the manufacturer's data sheets. Some manufacturers present the pertinent characteristics in the form of a graph, while others list the specifications in table format. Sometimes, if the motor is obtained from a surplus dealer or extracted from a toy, it is not possible to obtain data sheets, in which case simple experiments can be performed to measure the pertinent characteristics. Whatever the case may be, it is useful to have a clear understanding of the motor language and to brush up on the conversions between various units of measurement.

7.3.1 Torque, Speed, Power, and Energy

Torque is the angular force that a motor can deliver at a certain distance from the shaft. For instance, 5 oz.-in. of torque means that, at a distance of 1 inch away from the shaft of a motor, the motor is strong enough to pull up a weight of 5 ounces through a pulley (see Figure 7.9). In metric units, motor torques are often specified in Newton-meters (Nm). (When you try to imagine how much force a Newton is, think of the weight of an apple. A force of 1 Newton is about equal to the force that gravity exerts on one apple's mass.) Alternatively, metric units for torque can also be found specified in terms of *gram-force-centimeters* (gf-cm), where a gram-force is meant to signify the force that gravity exerts on 1 gram of mass. We will stick to metric units in this book, but some conversions to keep handy are:

$$1\,\mathrm{N} = 1\,\frac{\mathrm{kg\text{-}m}}{\mathrm{sec}^2} = 0.225\,\mathrm{lb}$$

$$1\,\mathrm{kg} = 2.21\,\mathrm{lb}\,(\mathrm{mass}) \qquad \text{and} \qquad 1\,\mathrm{in} = 2.54\,\mathrm{cm}$$

Also, when we begin to talk about electrical power being converted to mechanical power in a motor, it is useful to keep straight the relationships involving *power* (in watts) and *energy* (in joules). *Power* is the rate at which you are using up *energy*. The relationship between power and energy is expressed as:

$$1\,\text{Watt}{=}1\,\frac{\text{Joule}}{\text{sec}}$$

Figure 7.9 illustrates the electrical to mechanical power conversion of a DC motor. The electrical power supplied to the motor, P_e, equals the voltage, V, across the motor's terminals times the current, I, through the motor. The current, measured in units of amperes, is the amount of charge, in coulombs, passing through any cross-section of a conductor per second:

$$P_e = VI$$

$$1\,\text{Ampere} = 1\,\frac{\text{Coulomb}}{\text{sec}}$$

$$1\,\text{Watt} = 1\,\text{Volt}\cdot\text{Ampere} = 1\,\text{Volt}\cdot\frac{\text{Coulomb}}{\text{sec}}$$

Mechanical power, P_m, equals the torque, T, output by the shaft times its angular speed, ω, where the torque is taken in Newton-meters and the angular speed is measured in units of radians per second:

$$P_m = T\omega$$

$$\frac{2\pi\,\text{rad}}{\text{sec}} = 1\,\frac{\text{rev}}{\text{sec}}$$

$$1\,\text{Watt} = 1\,\frac{\text{Nm}}{\text{sec}}$$

Since power is energy per unit time, this tells us that one joule of energy can be expressed in two ways: either as 1 Newton-meter or as 1 coulomb-volt:

$$1\,\text{J}{=}1\,\text{Nm} \qquad \text{and} \qquad 1\,\text{J}{=}1\,\text{CV}$$

This is just reaffirming the fact that energy is energy, whether it comes from a mechanical origin or an electrical origin. A motor is just a transducer transforming energy from one form to another.

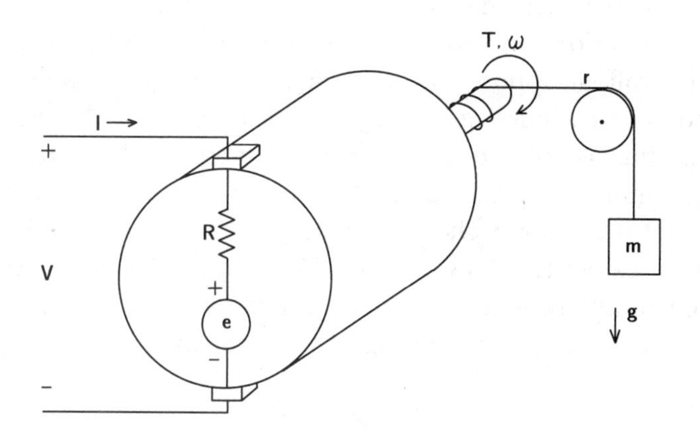

Figure 7.9. A simple model of a DC motor is an equivalent circuit that models the motor windings as having a resistance, R, and generating (when running) a back-emf voltage, e. The electrical power input to the motor is the product $P_e = VI$, and the mechanical power output is the product of torque and rotational speed, $P_m = T\omega$.

7.3.2 A Motor Model

These relationships, describing the conversion of electrical power to mechanical power in a permanent magnet DC motor, can be clearly seen by the equivalent circuit model shown in Figure 7.9. The mechanical output power (due to losses from friction, windage, heating in the coils, and so on) will be some fraction of the electrical input power. This percentage is given as the efficiency, η, where:

$$P_m = \eta P_e$$

The rotor coil that we saw in Figure 7.6 is essentially an inductor with a resistance R. When the rotor is turning, the commutator segments sliding past the brushes create an alternating current in the armature windings. A changing current, $\frac{di}{dt}$, through an inductor induces a voltage across it:

$$v = L\frac{di}{dt}$$

where L is the proportionality constant called the inductance. As the motor turns, this voltage is induced and opposes the applied driving voltage. The faster the motor turns, the more often the current switches direction, and so the larger the induced voltage becomes. Since this voltage opposes the applied drive voltage, as it increases, it tends to limit the current through the resistance, R. As the current falls, less flux is created around the conductor and the torque also falls. In summary, as the speed goes up, the torque goes down.

The rotating motor then can simply be modeled by the induced voltage, e, called the *back-emf* (*emf* stands for electromotive force) and the winding resistance, R. The applied voltage is related to the back-emf and the current through the motor by:

$$V = IR + e$$

Note that, when the motor is not rotating, e is 0 V and the current through the motor is just equal to the applied drive voltage divided by the resistance. This is the current required to start the motor from zero speed, called the *starting current* or *stall current*, I_S:

$$I_S = \frac{V}{R}$$

When the rotor is rotating, e increases proportionally with the speed of the armature:

$$e = k_e \omega$$

where k_e is called the back-emf constant. The applied voltage is then related to the current and the armature speed by:

$$V = IR + k_e \omega$$

The negative feedback provided by the back-emf causes the motor to settle to a steady-state operating point of speed and torque, as determined by the load and the applied voltage. The torque that the motor produces is dependent on the flux field around the loop of the conductor, and that flux is controlled only by the current. The torque increases linearly with current with a proportionality constant k_t, known as the *torque constant*:

$$T = k_t I$$

Solving for I and plugging it into the equation above, we get:

$$V = \frac{TR}{k_t} + k_e\omega$$

It turns out that k_t is actually equal to k_e. We can see this from the fact that the mechanical power output by the shaft will be the electrical power input, minus the I^2R losses due to heating in the resistor:

$$P_m = P_e - I^2R$$

$$T\omega = VI - I^2R$$

Plugging in for T and V,

$$k_tI\omega = (IR + k_e\omega)I - I^2R$$

gives

$$k_t = k_e = k$$

The applied voltage is then related to the torque and speed by the constant k:

$$V = \frac{TR}{k} + k\omega$$

Rearranging, we find that the speed-torque relationship is linear with a negative slope:

$$\omega = -\frac{R}{k^2}T + \frac{V}{k}$$

These relationships can be more clearly seen when plotted along with the motor performance curves.

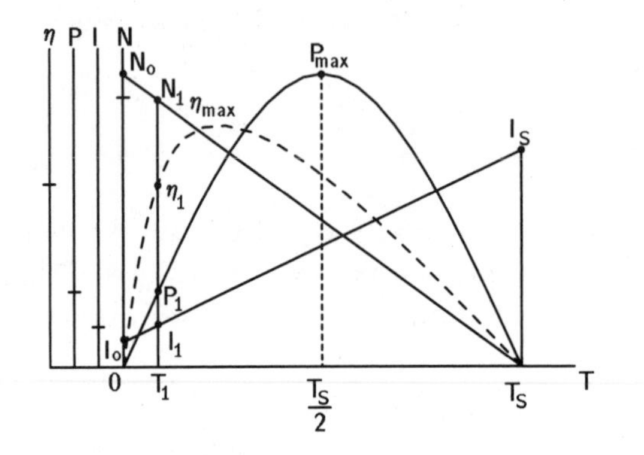

Figure 7.10. For a given voltage, a DC motor has the typical drooping characteristics of speed, N, decreasing linearly torque, T. As the current, I, is increased, the torque is increased, also linearly. Power output, P, is the product of torque and speed and has a quadratic characteristic. Maximum efficiency, η_{max}, occurs at a lower torque than the maximum power output torque.

7.3.3 Speed-Torque Curves

The speed and torque characteristics for a DC motor depend on a variety of parameters that have to do with the geometry of the motor, the materials involved, the number of windings, and the voltage at which the motor is driven. Typically, a manufacturer provides a data sheet showing the pertinent characteristics. These are usually illustrated in a speed-torque graph for a given applied drive voltage. Efficiency, current, and power output are often plotted along with speed on the vertical axis against torque on the horizontal axis, as shown in Figure 7.10.

We can see in Figure 7.10 that the speed-torque curve is linear, with a negative slope as we derived, and has a y-intercept dependent upon the applied voltage. Also, the current increases linearly with torque and is independent of applied voltage, as we showed earlier. The power curve has a negative quadratic form which is understood by remembering that:

$$P_m = T\omega$$

and plugging in our equation for ω:

$$P_m = -(\tfrac{R}{k^2})T^2 + \tfrac{V}{k}T$$

where we see the negative quadratic dependence of power on torque.

You will find it useful to check a few points of interest on a motor data sheet in choosing the most appropriate motor for your robot. The *no-load speed*, marked N_o in Figure 7.10, is the speed, at a given voltage, at which the torque is 0. (N usually refers to the angular speed in units of rpm. Remember to convert to $\frac{\text{rads}}{\text{sec}}$ when plugging into these equations for ω.) This is the speed of the motor with nothing attached to the shaft. That is, the no-load speed, the value of ω for $T = 0$, is just

$$\omega_{max} = \tfrac{V}{k}$$

The current in this no-load condition, I_o, is called the *no-load current* and is that required to overcome motor friction and windage.

At the other end of the scale, the torque that the motor can deliver just as it stalls and can no longer rotate is known as the *stall torque*, T_s. The current at this condition, I_S, is the stall current. Since the motor is not moving when stalled, the back-emf is 0 and the maximum current, I_S, is just the applied voltage divided by the coil resistance, as mentioned earlier. Torque being proportional to I, the maximum torque is:

$$T_S = \tfrac{kV}{R}$$

At any given point of operation of torque and speed, the mechanical power output is the product of the two. The torque at which the maximum power occurs can be found by taking the derivative of the power with respect to the torque, setting the result equal to 0, and solving for T:

$$\tfrac{dP_m}{dT} = 0 = -\tfrac{2RT}{k^2} + \tfrac{V}{k}$$

$$T = \tfrac{kV}{2R}$$

or

$$T = \tfrac{1}{2}T_{max}$$

Thus, the point of maximum power output is attained at half the stall torque. The corresponding speed at this operating point is then found to be:

$$\omega = -\frac{R}{k^2}\frac{kV}{2R} + \frac{V}{k} = \frac{V}{2k}$$

or

$$\omega = \frac{1}{2}\omega_{max}$$

The maximum power then is simply:

$$P_m = \frac{1}{4}\omega_{max}T_{max}$$

The ratio of mechanical power output to electrical power input is the *efficiency*, η. Note that maximum efficiency cannot be achieved at maximum power output. In fact, the point of maximum efficiency, where you would like to drive your motor, is a low-torque, high-speed operating point. Consequently, we typically select an oversized motor so that it can run at an efficient operating point while supplying enough torque.

It turns out that the maximum efficiency, for reasons we will not go into here, can be calculated from the measurements of the no-load current, I_o, and the stall current, I_S:

$$\eta_{max} = (1 - \sqrt{\frac{I_o}{I_S}})^2$$

This can be useful for characterizing a motor for which you do not have data sheets.

The data shown in Figure 7.10 are for one given value of applied voltage. If the motor is run at a lower voltage, the speed-torque line shifts downward as shown in Figure 7.11(a). As the voltage is decreased, the speed and the torque are both decreased. Changing the voltage changes the speed of the motor. Another way to change the speed without having such an adverse effect on the torque (in fact, a method that has an advantageous effect on the torque) is to use a geardown. As shown in Figure 7.11(b), gearing down the motor by, say, a factor of 2, cuts the no-load speed in half while doubling the stall torque. Thus, power is maintained constant through

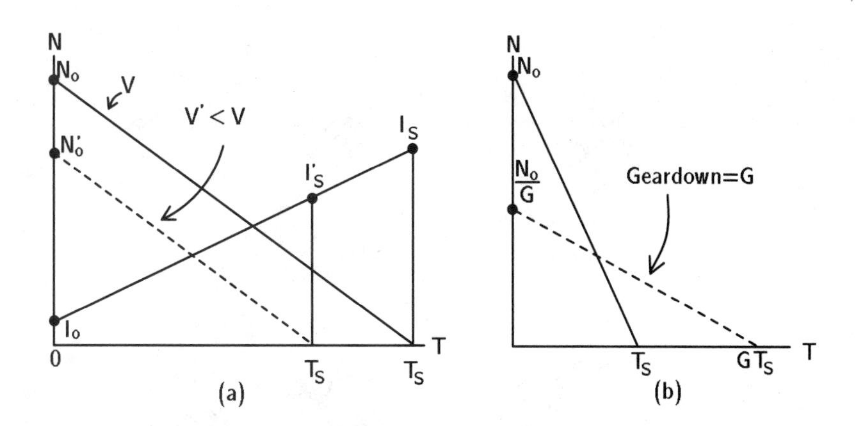

Figure 7.11. (a) Running the motor at a lower voltage causes it to slow down for all values of torque output. (b) Gearing down the motor reduces the speed by the gear ratio, G, and increases the torque by the same factor, G.

a lossless (frictionless) geartrain. Typically though, there are losses both through the motor and again through the geartrain. Good motors these days might have efficiencies of 90% or better, but cheap toy motors (like the ones we will use in the Rug Warrior prototype) might be only 50% efficient, or less. Adding these losses to those through the geartrain and then taking into account the losses between wheels and the ground (from friction, slippage, etc.) results in a system that is not very efficient. For Rug Warrior, most of the energy from its battery pack goes into the propulsion system. Powering Rug Warrior's computers and sensors will be practically insignificant in comparison.

7.4 Gears

Geartrains and transmission systems come in a variety of forms, such as spur gears, planetary gears, rack-and-pinion systems, worm gears, lead screws and belt-and-pulley drives. Figure 7.12 illustrates a number of these mechanisms. High-quality geartrains are usually metal, but plastic gears are often found in toys.

The DC gearhead motor shown earlier in Figure 7.8 had a *spur gear* set on the output shaft. Spur gears are the most common forms of gears found in DC gearhead motors. A schematic of a two-level

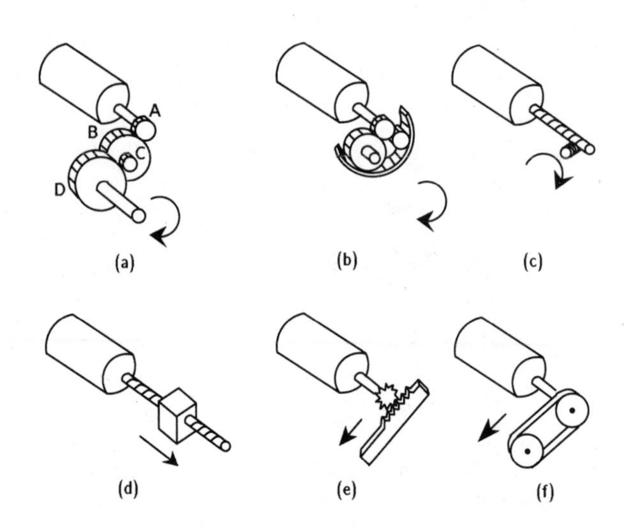

Figure 7.12. (a) Spur gears mesh pairs of gears with different numbers of teeth to achieve speed reduction. (b) Planetary gears have several gears meshed in an outer ring for large reduction. (c) Worm gears produce rotary motion at right angles to the shaft. (d) A lead screw and nut can create linear motion as can (e) rack-and-pinion systems and (f) belt-and-pulley drives.

spur geartrain is shown in Figure 7.12(a). The small gear, mounted directly on the motor shaft, is called a *pinion* and has to rotate many times to turn the gear it is meshed to once. Thus, even though the pinion may spin very quickly, the gear it is attached to spins very slowly. If A, B, C, and D denote the number of teeth on each corresponding gear in the figure, then the speed of the output shaft is related to the speed of the input shaft by:

$$\omega_{out} = \frac{A}{B}\frac{C}{D}\omega_{in}$$

where the final speed has been decreased by the geardown ratio.

Planetary gears are similar to spur gears but are less common in low-end gearhead motors and are slightly more expensive. The difference between planetary gears and spur gears is that planetary gears, as shown in Figure 7.12(b), fit a number of gears concentrically inside a toothed ring. This configuration produces greater efficiency and higher output torques in a smaller package. Planetary geartrains are sometimes found in portable battery-powered screwdrivers and drills.

Worm gears are another means for achieving large geardown in a small space. Worm gears, shown in Figure 7.12(c), instead of having teeth, are threaded and match to a lead screw attached to the shaft of the motor. In this way, the output motion is turned to right angles from the motor shaft.

For linear motion, a *lead screw and threaded nut* can be used. Figure 7.12(d) illustrates how the motor shaft turns the lead screw and a threaded nut moves linearly down the shaft, depending on the number of threads per inch on the lead screw. Lead screws can give very large geardown but are not very efficient.

Rack-and-pinion systems, Figure 7.12(e), also turn rotary motion into linear motion. In this case, a small pinion gear on the motor shaft rotates against a straight length of rack having matching teeth, propelling the rack linearly back and forth.

Another linear motion mechanism is the belt-and-pulley system, shown in Figure 7.12(f). This is the mechanism used to drive a tank-treaded vehicle, such as we will describe later for Rug Warrior II.

7.5 Motor Data Sheets

While it is possible to buy a plain DC motor and attach any number of gear-reduction mechanisms for propelling your robot, we will focus on DC gearhead motors here for Rug Warrior (typically with spur gears) because it makes life easier when geartrain and motor are packaged together. There is no need to find a machine shop and spend time making gearboxes.

Picking an appropriate motor involves understanding a manufacturer's data sheets. A data sheet is usually given for the motor alone, and then another data sheet is supplied for the type of gearbox (with an assortment of reduction ratios) that will fit that motor. The gearbox specification can place constraints on the motor, such as for maximum allowable input speed or maximum deliverable output torque.

Actual data sheets for the small Escap motors (shown in Figure 7.1 of this chapter) are given in Figure 7.13 and Figure 7.14. The data are given here in tabular form instead of graph form, but the reader can reconstruct the graphs that were discussed earlier, (as

D.C. motor
escap® 16 M 11

Standard types available from stock			-210	-208	-207	-205
Measuring voltage		V	6	7.5	9	15
No-load speed		rpm	8400	7800	8300	8200
Stall torque		mNm	3	2.5	2.3	2.4
		oz-in	0.42	0.35	0.33	0.34
Power output		W	0.7	0.5	0.5	0.5
Av. no-load current		mA	7	5	4	2.5
Typical starting voltage		V	0.06	0.1	0.1	0.2
Max. continuous current		A	0.4	0.28	0.24	0.14
Max. recommended speed		rpm	12000	12000	12000	12000
Max. angular acceleration		10^3 rad/s^2	96	114	120	102
Back-EMF constant		V/1000 rpm	0.7	0.94	1.1	1.8
Rotor inductance		mH	0.5	0.8	1	3
Motor regulation R/k^2		10^3/Nms	300	330	380	350
Terminal resistance		ohm	13.4	27	39.5	105
Torque constant		mNm/A	6.7	9	10.2	17
		oz-in/A	0.949	1.28	1.44	2.41
Rotor inertia		kgm^2 · 10^{-7}	0.7	0.56	0.5	0.6
Mechanical time constant		ms	20	18	19	21
Thermal time constant	rotor	s	6	5	4	4
	stator	s	380	380	380	380
Thermal resistance	rotor-body	°C/W	10	10	10	10
	body-ambient	°C/W	35	35	35	35

Figure 7.13. The Escap model 16M11-210 DC motor is a 6 volt (V) motor with a no-load speed of 8,400 revolutions per minute (rpm) and a stall torque of 3 milli-Newton-meters (mNm). (By courtesy Portescap, Inc.)

illustrated in Figure 7.10), since the major features, such as no-load speed, stall current and stall torque are given in these tables. The torque constant given in the table can be used to find the slope of the I-T curve, and the back-emf constant can be used to determine the slope of the ω-T curve. (Note that, if these constants are converted to the same units, they are equal.)

For instance, Figure 7.13 describes the performance of the motor by itself without a gearhead. Four models of this motor are available, each with a different winding and therefore intended to be run at a different voltage. The voltage for which the specifications are given is called the *measuring voltage* or sometimes the *rated voltage*. Thus, the 16M11-210 motor, when run at 6 V, will have a no-load speed of 8,400 rpm, a stall torque of 3×10^{-3} Nm, and a maximum possible output power of 0.7 W.

If the 16M11 motor is purchased with an attached gearhead, the part number for the gearmotor is M1616M11; its specifications, as shown in Figure 7.14, recommend that the -210 winding version be run at 5 V so that the no-load speed of the motor stays within the

D.C. gearmotor
escap® M1616 M11

Standard types available from stock		M 1616 M 11					
Max. recom. dynamic output torque	mNm (oz-in)	50 (7.1) at 20 rpm					
	mNm (oz-in)	30 (4.2) at 150 rpm					
Max. recom. static output torque	mNm (oz-in)	250 (35.4)					
Max. recommended input speed	rpm	7500					
Available reduction ratios		9	27	54	243	486	2190
				81		729	
Average efficiency		0.8	0.7	0.65	0.6	0.55	0.5
Nr. of geartrains / direction of rotation		2/=	3/≠	4/=	5/≠	6/=	7/≠
Length L	mm	37.1	38.6	40.1	41.6	43.1	44.6
Mass	g	28	29	29	30	31	32
Motor specifications		**-210**			**-207**		
Measuring voltage[1]	V	5			8		
No-load speed	rpm	7000			7200		
Stall torque	mNm (oz-in)	2.5 (0.354)			2.1 (0.297)		
Terminal resistance	ohm	13.4			39.5		
Torque constant	mNm/A (oz-in/A)	6.7 (0.949)			10.2 (1.44)		
Other motor characteristics	see page	17			17		
Gearmotor specifications							
Av. no-load current	mA	10			8		
Typical starting voltage	V	0.1			0.3		
Mechanical time constant	ms	21			19		

Figure 7.14. The Escap model M1616M11-210 gearhead motor should be driven at 5 V (instead of 6 V) in order to keep the no-load speed of the motor within the maximum allowable input speed of the gearbox. (By courtesy of Portescap, Inc.)

allowable input speed of the gearbox. The gearmotor with the 54:1 reduction will weigh 29 g, be 40 mm long, be 16 mm in diameter, and have an efficiency of 65%. The no-load speed will be:

$$N_o = \frac{7000\,\text{rpm}}{54} = 130\,\text{rpm}$$

and the stall torque will be:

$$T_S = (54)(2.5\,\text{mNm})(0.65) = 88\,\text{mNm}$$

Earlier, we showed that the maximum possible output power was:

$$P_{m,max} = \tfrac{1}{4}\omega_{max}T_{max}$$

We can calculate this maximum power by converting the no-load speed and the stall torque to the appropriate units. If we want to know how many radians per second are equal to 130 revolutions per minute, the easiest way to keep all the conversions straight is to set up the question this way:

$$? \frac{\text{rads}}{\text{sec}} = 130 \frac{\text{rev}}{\text{min}}$$

Since multiplying the righthand side by 1 does not change the equality, we can multiply 130 rpm by identity relationships, converting revolutions to radians and minutes to seconds in such a way that the old units cancel out:

$$? \frac{\text{rads}}{\text{sec}} = 130 \frac{\text{rev}}{\text{min}} \cdot \frac{2\pi \, \text{rads}}{\text{rev}} \cdot \frac{1 \, \text{min}}{60 \, \text{sec}} = 13.6 \frac{\text{rads}}{\text{sec}}$$

This gives:

$$P_{m,max} = \tfrac{1}{4} T_{max} \omega_{max} = \tfrac{1}{4}(88 \times 10^{-3} \, \text{Nm})(13.6 \tfrac{\text{rads}}{\text{sec}}) = 0.3 \, \text{W}$$

Escap motors are fairly high quality, and like many DC gearhead motors, can cost over $100 each. Escap (actually, Portescap is the name of the company) sells old-inventory motors (catalog motors but ones that have sat on the shelves for too long to be sold as new) for a fraction of their original cost. Although the selection is limited, this source can be useful for hobbyists. Maxon, Micro Mo, Pittman, Inland, Globe, Canon, Copal, and Namiki are a few of the other numerous manufacturers that sell DC motors and have readily available catalogs with specification sheets. Surplus dealers often buy out remains of original equipment manufacturers' (OEMs) unused motors and sell them at significantly reduced costs. Dealers such as Burden's Surplus Center, Herbach and Rademan, America's Hobby Center, Edmund Scientific, Sheldon's Hobbies, Stock Drive Products, and Tower Hobbies sell wide assortments of smaller, cheaper DC gearhead motors.

Most of the low-cost permanent magnet DC motors, such as those found in toys, are made by one company–Mabuchi. Mabuchi produces over 3 million motors a day and sells them in lots of 5,000 or more. They make strictly stand-alone motors, not gearhead motors, but sell them to OEM manufacturers who then incorporate motors into toys, model airplanes, and the like. Typically, a toy manufacturer will use the molding of the toy itself to be the gearbox for the plastic geartrain they add to the motor so it is not always convenient to extract the motor and build it into your robot.

Model airplane servo motors, on the other hand, are very modular and convenient for this purpose. While most model airplane servos continue to be high-priced, mass production of the most common models has led to lower prices for servo motors. Futaba, Royal Products Corporation and Airtronics are a few of the manufacturers of these servo motors. Catalogs from hobby stores, such as Tower Hobbies and Sheldon's Hobbies, list a wide range of models. Higher-quality servos with metal gears and ball bearings are available, also.

Servo motor data sheets (which are typically printed on the backs of the packages) look different from the data sheets for DC gearhead motors. Servo motors usually run from a 5 V supply. For instance, for the Royal Titan Maxi Servo, the specifications are described this way:

Royal Titan Maxi Servo

Weight	3.7 oz.
Output Torque	112 oz.-in.
Current Drain	8 mA
Transit Time	$\frac{0.22\text{sec}}{60^o}$

A *transit time* (in $\frac{\text{sec}}{\text{deg}}$) is given instead of a no-load speed (in rpm) because the integrated circuit servos the motor to a specified position and it never spins all the way around. However, if the servo is stripped down to being just a DC gearhead motor (potentiometer, limit stops, and integrated circuit removed), this transit time is equivalent to the no-load speed. The output torque listed above is simply the stall torque. Converting to proper units to find power output:

$$?\frac{\text{rads}}{\text{sec}} = \frac{60^o}{0.22\text{sec}} \cdot \frac{2\pi\,\text{rads}}{360^o} = 4.8\frac{\text{rads}}{\text{sec}} = 46\,\text{rpm}$$

$$?\text{Nm} = 112\,\text{oz.-in.} \cdot \frac{\text{lb.}}{16\,\text{oz.}} \cdot \frac{1\,N}{0.225\,1\,\text{lb.}} \cdot \frac{2.54\,\text{cm}}{1\,\text{in.}} \cdot \frac{1\,\text{m}}{100\,\text{cm}} = 0.79\,\text{Nm}$$

The maximum possible power then is:

$$P_{m,max} = \frac{1}{4}T_{max}\omega_{max} = \frac{1}{4}(0.79\,\text{Nm})(4.8\frac{\text{rads}}{\text{sec}}) = 0.95\,\text{W}$$

which is 3.2 times as large as the earlier Escap motor — but then, this is a larger motor. To compare weights, we convert to grams:

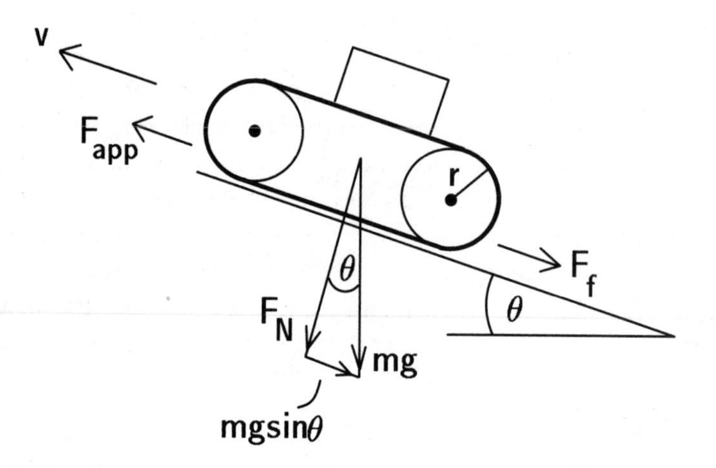

Figure 7.15. This free-body diagram of a tracked-drive Rug Warrior illustrates the forces acting on the vehicle as it climbs a hill. Use of this diagram helps to determine the maximum torques that the robot's motors will be required to deliver.

$$?\mathrm{g}{=}3.7\,\mathrm{oz.}\cdot\frac{28\,\mathrm{g}}{1\,\mathrm{oz.}} = 104\,\mathrm{g}$$

The Royal Titan servo motor then is about 3.6 times heavier than the 29 g Escap DC gearhead motor discussed previously. It turns out, however, that the Royal Titan with the potentiometer and circuit board removed, leaving essentially a comparable DC gearhead motor, weighs only 78 g. This seems to make sense, as the Royal Titan gears are plastic and the Escap gears are metal.

7.6 Motors for Rug Warrior

7.6.1 A Vehicle Model

In order to get a rough idea of how much power the motors for Rug Warrior must be able to deliver, we can sketch the scenario illustrated in Figure 7.15. Assume that Rug Warrior uses a differential drive mechanism (two motors) and needs to climb a ramp of angle θ at constant velocity, v. The free-body diagram makes explicit the forces acting on the vehicle.

Because the vehicle moves at constant velocity, there must be no net force on the car. That is, since:

$$F = ma$$

and the acceleration, a, is 0 (the car moves at constant velocity), the net force F must be 0. This means that the applied force, F_{app}, from the wheels acting in the direction up the hill must balance the forces down the hill resisting that force. These resisting forces are the friction force and the force that is the component of the vehicle's weight acting in the direction down the hill. Thus:

$$F_{app} = F_f + F_W$$

where F_f is equal to the coefficient of friction, μ, times the normal force, F_N:

$$F_f = \mu F_N = \mu mg \cos\theta$$

and F_W is $mg \sin\theta$ (where mg, mass times the acceleration due to gravity, is just the weight of the robot). This leaves:

$$F_{app} = \mu mg \cos\theta + mg \sin\theta$$

The power required from the motors is the product of the force that needs to be applied by the wheels times the velocity, v, the robot travels up the hill:

$$P_m = F_{app} v$$

where each motor must supply half that power, as Rug Warrior has two motors.

The torque and speed requirements of each motor can be calculated from:

$$\frac{P}{2} = T\omega \quad \text{and} \quad \omega = \frac{v}{r}$$

where r is the radius of the wheel.

The range and the running time of the robot are dependent upon the battery pack, since power is the rate of energy usage. If the battery has E joules of energy, then the battery lifetime, t, will be:

$$t = \frac{E}{P}$$

The range of distance, D, the robot can travel will be: constrained by

$$D = vt$$

Typically, battery capacity is not given in joules but in units of ampere-hours. To find the energy contained in a battery pack, we must multiply the capacity rating in ampere-hours by the nominal voltage rating of the battery. (Recall that 1 joule equals 1 coulomb-volt and 1 ampere equals 1 coulomb per second.) For instance, suppose a 3 V battery has a 1,300 milli-ampere-hours (mAh) capacity. How many joules does it contain?

$$?\mathrm{J} = 3\,\mathrm{V} \cdot 1300 \times 10^{-3}\,\mathrm{Ah} \cdot \frac{\frac{\mathrm{C}}{\mathrm{sec}}}{\mathrm{A}} \cdot \frac{3600\mathrm{sec}}{\mathrm{h}} = 14,040\,\mathrm{CV} = 14,040\,\mathrm{J}$$

7.6.2 Selecting a Motor

The model we just described for Rug Warrior is hardly realistic. We certainly do not expect that our robot will be climbing up a ramp forever. Rather, because reality is so complicated (e.g., uneven terrain, stop-and-go crises, unknown coefficients of friction, accidents with chair legs, etc.), we use this model simply to attempt to size the peak power requirements.

Let's say that our goal is for Rug Warrior to weigh under 1.5 pounds, which is roughly 650 g. Furthermore, assume that we would like our robot to climb a 30 degree grade at a steady half foot per second, which is $\frac{0.15\,\mathrm{m}}{\mathrm{sec}}$. We will use two motors and a tank-drive locomotion system. Picking a value for μ is a way of trying to account for slippage and friction from the treads and the like. It is not clear what this coefficient of friction will be, but we can make some assumption and pad our result by oversizing the motors at the end. Let's pick μ to be 0.3. The power required then is:

$$P_m = F_{app}v = mg(\mu\cos\theta + \sin\theta)v$$

$$P_m = (0.65\,\mathrm{kg} \cdot 9.8\frac{\mathrm{m}}{\mathrm{sec}^2})(0.3\cos 30^o + \sin 30^o)(0.15\frac{\mathrm{m}}{\mathrm{sec}}) = 0.73\,\mathrm{W}$$

Figure 7.16. The easiest way to build a Rug Warrior is to start with model airplane servo motors; add LEGO parts for bearings, axles, and treads; and then place the batteries, electronics, and sensors on top.

We want to oversize our motors quite a bit, both because there are so many unknowns and because the maximum efficiency point is at a much lower torque than the maximum power point. If we multiply our power requirement by a whopping factor of 3, that would give:

$$P_m = 2.1\,\text{W} \qquad \text{or} \qquad \frac{P_m}{2} \simeq 1\,\text{W}$$

7.6.3 Converting Servo Motors

What we have chosen, as we mentioned earlier, is to use model airplane servo motors. We recommend these motors for the Rug Warrior project of this book because they are fairly inexpensive and easy enough to modify. Although servo motors are not as cheap as toy motors, the fact that they come with gearboxes already built in means that we need not bother with machining a custom gearbox.

Figure 7.16 illustrates the tank-tread version of Rug Warrior that we built using two Royal Titan Maxi Servos, which cost $25 each, LEGO gears for wheels, LEGO tracks for tank treads, and LEGO axles and blocks for bearings and chassis. The PC board on top is

Figure 7.17. The underside of Rug Warrior contains two servo motors taped to the chassis, and LEGO gears mounted on the motor shafts for wheels. LEGO tracks are then used to make tank treads.

3.4"×4.5" and contains an MC68HC11A0, with the 10 sensors and the accompanying control electronics.

The tank drive is made up of two motors, connected to the back wheels in a differential fashion. The front wheels are passive, each having its own axle and bearing. The tank treads are wrapped around from back wheels to front wheels, so the robot can pivot in place.

Figure 7.17 is a view of Rug Warrior from the underside. The two black boxes are the servos. Attached to each is a LEGO gear for a wheel. The gear acts to mesh easily with tracks also supplied by LEGO. It is possible to build a sturdier and lighter-weight chassis, perhaps something made from aluminum sheet metal using a sheet metal bender and a punch for forming sides and placing holes. Real ball bearings and ground shafts could be used for the front wheels (obtainable from suppliers such as Berg, Small Parts Inc., etc.), but it turns out that ball bearings can cost as much as the MC68HC11A0 computer chip! Instead, we elected to use the LEGO building system, not just for gears and tracks but to continue with it for front wheel axles and bearings, as the axles that come with LEGO are made from hard plastic and spin nicely in the holes in the LEGO

Figure 7.18. Some servo motors are easier to convert to continuous rotation than others. The gearhead of a Royal Titan Maxi Servo is shown here. The leftmost gear is above the potentiometer, and the ball bearing ring is mounted on its top for support of the output shaft. A plastic limit stop is molded onto the gear just below and to the left of the ball bearing.

bricks that we use for the chassis. We used double-sticky tape or black electrical tape to hold the chassis together.

To build this propulsion system for Rug Warrior, first modify the servos so that they can spin all the way around. Figure 7.18 shows the gearhead portion of the Maxi Servo; it has four stages of reduction for a 143:1 geardown. The motor shaft is at the right, the potentiometer shaft is at the left (the motor and potentiometer are below, inside the case), and the third shaft is in the middle. The output power is taken off at the potentiometer shaft. A plastic nib, molded onto the gear there, prevents the shaft from turning multiple revolutions. Above that nib is a metal ring, which is the ball bearing that supports the load.

Next, cut that plastic nib off. A pair of dikes (i.e., diagonal wire cutters) will work fine for the job. Then take that gear off and remove a plastic inset from its underside, which the potentiometer shaft's flat is held against. Not all servo motors have this feature of the removable inset. Some have the inset molded into the gear and have the gear turn directly on the potentiometer's shaft, which means it is not possible to easily make it continuously revolvable.

Figure 7.19. A bottom view of the servo in the previous picture shows a Mabuchi motor is in the righthand portion of the casing. The lefthand portion holds the potentiometer and a small circuit board containing an integrated circuit for servo control.

The Royal Titan servos have the removable inset and also have the gear resting on a bushing around the pot's shaft, which means you can actually remove the potentiometer completely. This brings us to the next step; removing the potentiometer.

Figure 7.19 is a view from the underside of the servo motor, with the cover removed and the potentiometer and servo circuit pulled out. Clip the wires for your motor, removing the circuit board. Take out the potentiometer by removing the screw holding it in place. Note the motor on the right. It is a Mabuchi motor and comes equipped with a capacitor across its leads and two resistors to ground to suppress noise spikes from the motor. Desolder the remains of the wires from the servo circuit, and solder on two new wires to the two terminals of the motor. Replace the cover over the gears, making sure the shafts sit properly in their holes. Try hooking a power supply or a battery pack up to two motor leads. The motor should spin continuously. Reversing the polarity of the applied voltage should reverse the direction of spin.

Adding wheels to a servo motor is convenient because servo motors come with an assortment of attachments (plates, levers, star-shaped mounting brackets, etc.) that are designed to fit snuggly onto the output shaft. Figure 7.20 illustrates a servo motor with the lever attachment. A simple way to mount the LEGO gear is to use the circular plate attachment (instead of the lever attachment), which is roughly the same size as the gear; sand off any small ridges on the plate and/or the gear and glue them together.

It may seem odd to actually throw away a few components from a servo motor and still wind up with the lowest-cost route to a DC gearhead motor. Such are the benefits of mass markets. We will use a MC68HC11A0 and some power electronics (in a form known as an H-bridge) to drive the motors for steering Rug Warrior's treads. However, first let us digress a moment to explain how and where an unmodified servo motor would normally be used.

7.6.4 Unmodified Servo Motors

Typically, a radio-controlled model airplane servo motor is used to adjust a control surface on a wing of a model airplane to a certain position. The integrated circuit and potentiometer are used to implement a closed-loop position control system. The radio sends what is known as a *pulse-code modulated signal* to a receiver on the model plane. As stated earlier, of the three wires emanating from the servo motor, one is for power, one is for ground, and one is connected to this pulse-code modulated signal. Figure 7.20 illustrates the protocol for commanding the servo to a given position.

Basically, a servo motor expects a train of pulses of varying widths. These pulses are repeated at a given period, typically set to 20 ms. The width of the pulse is the code that signifies to what position the shaft should turn. The center position is usually attained with 1.3 ms wide pulses, while pulse widths varying from 0.7 milliseconds (ms) to 1.7 ms will command positions all the way to the right and all the way to the left, respectively.

These position servo motors can be very useful for robot accessories (such as fingers, grippers, legs, and squirt guns) where the range of motion does not require continuous revolution. For continuous motion, we described how to modify the servo and reduce it to a simple DC gearhead motor by throwing away the control circuit and power electronics that come with it and adding our own. However, there is a way to use these motors as continuous revolution DC

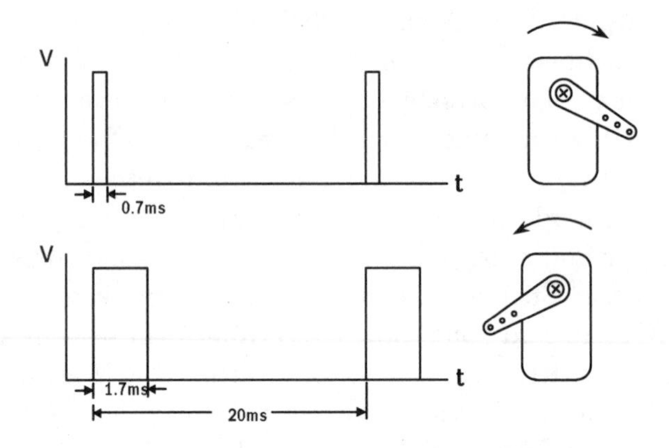

Figure 7.20. An unmodified servo is a three-wire device that takes power, ground, and a pulse-code modulated signal, such as the one shown above. Wider or thinner pulses tell the servo to move to a designated position, either clockwise or counterclockwise from center.

gearhead motors without having to add our own H-bridges and control electronics. The trick is to remove the inset in the plastic gear as before, which affixes itself to the flat of the potentiometer's shaft, but do not actually remove the potentiometer. Set the potentiometer to its central position. Now the gears will turn continuously but the potentiometer will never move. With this configuration, if we send the motor a pulse-code modulated signal to move all the way to the right, the motor will try to comply, never get any feedback, and never stop. Similarly, a pulse-code modulated signal to move to a position to the left will cause continuous rotation all the way to the left. This is an elegant trick (hack, to use the proper term) but we do not pursue it any further for Rug Warrior, because we want to explain how to attack the more common problem of driving a regular DC motor in general, and how to implement a servo loop.

7.7 Interfacing Motors

A microprocessor cannot drive a motor directly, since it cannot supply enough current. Instead, there must be some interface circuitry so that the motor power is supplied from another power source and

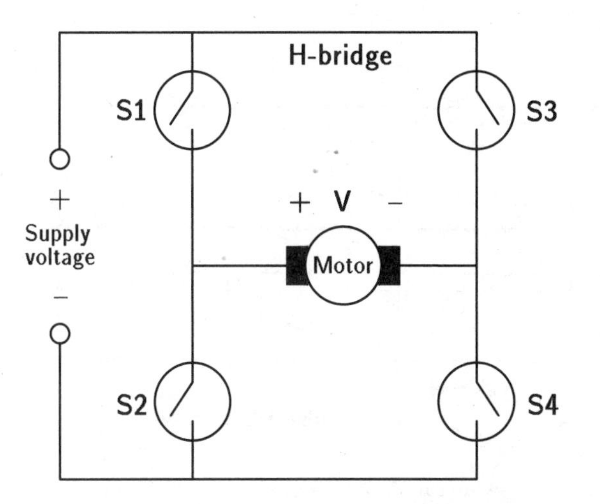

Figure 7.21. A circuit topology known as an H-bridge is used to control a motor. Four switches are controlled by a microprocessor and determine the direction in which current is allowed to pass through the motor. Changing the direction of the current changes the direction of the motor rotation.

only the control signals derive from the microprocessor. This interface circuitry can be implemented in a variety of technologies, such as relays, bipolar transistors, power MOSFETs (metal oxide semiconductor field effect transistors), and motor-driver integrated circuits. In all technologies, however, the basic topology of the circuit is usually the same. This circuit is known as an *H-bridge* and merely consists of four switches connected in the topology of an H, where the motor terminals form the crossbar of the H, as shown in Figure 7.21. You can imagine the abstraction of each switch as being implemented by either relays or transistors, where the power is supplied by the battery and the control signals by the microprocessor.

7.7.1 H-Bridges

In an H-bridge, the switches are opened and closed in a manner so as to put a voltage of one polarity across the motor for current to flow through it in one direction (setting up magnetic fields and causing it to turn) or a voltage of the opposite polarity, causing current to flow through the motor in the opposite direction for reverse rotation.

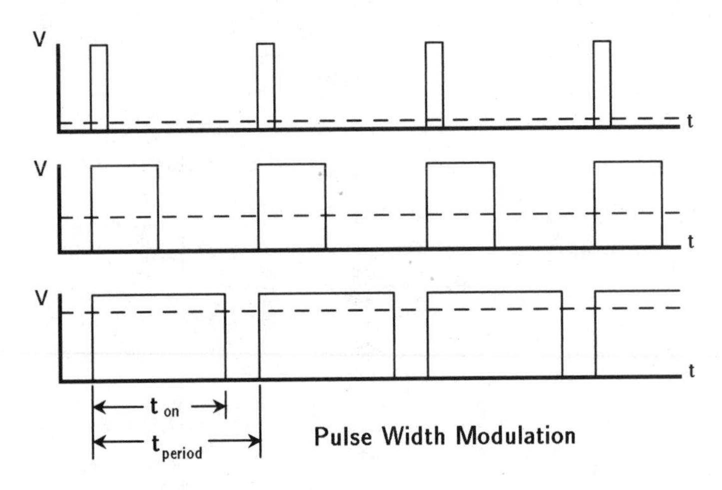

Figure 7.22. Pulse-width modulation of the voltage, by turning switches in the H-bridges on and off for various lengths of time, creates a different average voltage across the motor. Solid lines represent voltages applied when the switches are closed. Dotted lines represent the resulting average voltage applied across the motor.

For example, if switches S1 and S4 in Figure 7.21 are closed while switches S2 and S3 are open, current will flow from left to right in the motor. When switches S2 and S3 are closed and switches S1 and S4 are open, current will flow from right to left, reversing the motor. If the terminals are floating, the motor will freewheel, and if the terminals are shorted, the motor will brake.

To control the speed of the motor, the switches are opened and closed at different rates in order to apply different average voltages across the motor. This technique, called *pulse-width modulation*, is illustrated in Figure 7.22, where V is the voltage across the motor and t is time. For instance, if switches S1 and S4 are used for pulse-width modulation while switches S2 and S3 are left open, the voltage across the motor (as defined in Figure 7.21) will be equal to and of the same polarity as the supply voltage when S1 and S4 are closed and 0 V when they are open. The speed of a DC motor can be adjusted by changing the pulse-width ratio:

$$\text{Pulse-Width Ratio} = \frac{t_{on}}{t_{period}}.$$

of the voltage applied across its terminals.

Note that what we are describing here is different from pulse-code modulation for servo motors, discussed earlier. There, some "intelligence" was added so that the pulse width was a code signifying to what *position* the servo should move. Here, we are merely using varying pulse widths to create different average voltages across the motor to change its *speed*.

We mentioned before that the abstractions of switches in Figure 7.21 can be implemented in a number of ways. Relays can be used to turn motors on and off and reverse their directions as we saw in the TuteBot example, but relays are seldom used in pulse-width modulation speed controllers because they typically cannot switch quickly enough. Relays also tend to wear out. Solid-state switches, such as power bipolar transistors and power MOSFETs, are more convenient for pulse-width modulation schemes, and we will concentrate on these implementations here.

It is possible to design your own solid-state H-bridge controller, but there are also a number of single-chip solutions on the market. We chose one of these for Rug Warrior, and the anxious reader can skip ahead to the section on motor-driver power integrated circuits (see Section 7.7.4). However, if your particular project has requirements not available in a commercial H-bridge chip or if you are simply curious, the following sections give a bit of background on what is inside a motor-driver integrated circuit.

7.7.2 Switching Inductive Loads

Whether using solid-state switches or relays, problems arise when switching inductive loads such as motors, as illustrated in Figure 7.23. We know that the voltage induced across an inductor is proportional to the rate of change of current through it:

$$v = L\frac{di}{dt}$$

If the current through an inductor has reached a steady state and is not changing, the voltage across it is 0 V and the inductor behaves like a straight piece of wire. Figure 7.23(a) shows what happens if that steady-state current is upset by the opening of a switch. Namely, the current cannot instantaneously go to 0 A so a voltage, $v = L\frac{di}{dt}$, is induced in a direction opposing the flow of

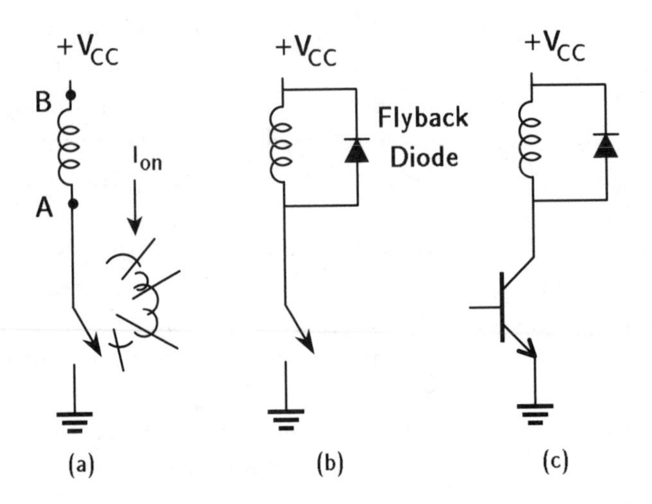

Figure 7.23. (a) The steady-state current through an inductor, I_{on}, cannot immediately go to 0 A when the switch is opened. The changing current induces a voltage across the inductor, making the potential at A greater than at B, causing the switch or relay to arc over. (b) Flyback diodes protect switches from blowing up. (c) Transistor switches must be protected in the same manner.

current. That is, the point marked A will be at a potential positive with respect to point B (which is at V_{cc}). Although the current does not change instantaneously when the switch is opened, it does change very quickly, and the faster the rate of change, the larger the induced voltage spike. Depending on the size of the inductor, the magnitude of the current, and how quickly the switch is opened, these voltage spikes can temporarily reach several hundred volts or more, enough to cause the switch to arc over and blow up.

The solution to this problem is to put what is known as a *flyback diode* in the reverse direction across the inductive load (Figure 7.23[b]) so that the voltage spike will forward bias the diode, creating a return path for the current. In this way, the power will "fly back" to the power supply.

Solid-state switches are just as susceptible to voltage spike destruction as mechanical switches, which is why transistor circuits switching inductive loads are usually shown with appropriate flyback diodes, as illustrated in Figure 7.23(c).

7.7.3 Power Electronics

As we discuss controlling motors from a microprocessor and the power electronics needed for the interface, we will talk about transistors used as switches. In Chapter 5 on sensors, we saw transistors, or collections of transistors in the form of op-amps, used as linear amplifiers to add gain to a circuit for amplifying small signals from sensors into larger signals understood by a microprocessor. The microphone circuit and the sonar circuit were examples. In addition to transistors used as linear amplifiers, we have also seen transistors used in another way: as CMOS (complementary metal oxide semiconductor) logic-gate switches. All the circuitry making up the internals of the 6811, its associated RAM and various discrete NAND gates and inverters, are simply composed of low-power, n-channel and p-channel MOSFET transistors used as switches. MOSFETs are similar to bipolar junction transistors in some sense, yet different in many ways. We will give some comparisons and contrasts between MOSFETs and transistors in a moment.

First, however, transistors can be classified another way, either as signal-level devices or as power devices. Transistors used for linear amplification of sensor signals or for logical manipulation of bits are concerned with processing information and are generally low-power devices. Power transistors, on the other hand, are capable of handling larger currents and voltages. They might be used as linear amplifiers in output stages of high-fidelity audio systems to drive speakers or they might be used as switches in H-bridges to pulse-width modulate motors requiring large currents. Power devices are typically larger than signal-level devices, as they require more silicon area for higher current-handling capability and larger packages for heat dissipation.

Semiconductors and Charge Carriers

Solid-state switches and power electronics are semiconductor devices. What is a *semiconductor* exactly, and why is silicon the material of choice for the semiconductor industry?

In a normal *conductor*, for instance, a metal such as aluminum, free electrons act as charge carriers and move in a direction toward a

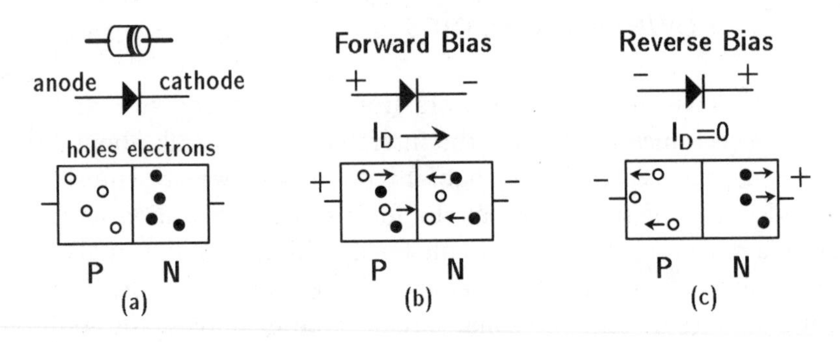

Figure 7.24. (a) A diode is simply a PN junction, where the *p*-type region is the anode and the *n*-type region is the cathode. (b) When forward biased, holes and electrons cross the junction, causing current to flow. (c) When reverse biased, no current flows.

positive potential. (Recall that *positive current* flows in the direction opposite to that of electron flow–so positive current moves away from a positive potential.) An *insulator* such as glass is the complement of a conductor, has no free charge carriers, and does not conduct current. A *semiconductor* on the other hand, lies somewhere in between. It is neither a perfect insulator nor a perfect conductor.

Because silicon has four valence electrons in its outer ring, it loves to bond covalently with other silicon atoms and create a perfect crystal lattice, much like diamond. Silicon is a semiconductor, and by adding various levels of impurity atoms, such as phosphorus or boron, silicon can become increasingly conductive. The reason that silicon is the material of choice for the semiconductor industry is that it is the only semiconductor that grows a native oxide layer. That is, when exposed to air, the silicon at the surface combines with oxygen to form a thin layer of silicon dioxide, essentially, a glass. Thus, in silicon processing, it is very convenient to create both conductors and insulators, a feature useful for patterning devices.

Another important characteristic of a semiconductor such as silicon is that two types of charge carriers are available to conduct current. Not only are electrons available to conduct current, but charge carriers called *holes* can also be developed. When impurity atoms of phosphorus are implanted in silicon, the five valence electrons in phosphorus's outer ring cause phosphorus atoms to bond into the crystal silicon lattice, giving up one free electron as a charge carrier. Since electrons carry negative charges, regions of silicon doped with phosphorus are called *n*-type regions.

When impurity atoms of boron are added to single-crystal silicon, the three valence electrons of boron's outer ring cause boron atoms to bond into the silicon lattice, leaving a vacancy or hole. If electrons from other covalent bonds leave and fill these holes, the holes have essentially moved, creating a passage of positive charge carriers. Regions of silicon doped with boron are then termed *p-type regions.*

All the interesting behavior in silicon devices comes about at junctions of *n*-type and *p*-type regions. In fact, a diode is nothing more than a single PN junction, a junction of *p*-type and *n*-type material. Figure 7.24 illustrates how a diode works. When forward biased, holes and electrons each cross over the PN junction, attracted to the far terminals. They then mix and recombine, becoming neutral. New charge carriers are supplied by the terminals, and a continuous flow of both types of charge carriers is maintained, resulting in a steady-state current. When the PN junction is reverse biased, holes and electrons are each attracted to their nearby terminals and absorbed by them. The charge carriers move away from the junction, and the device becomes depleted of charge carriers. Thus, no current flows. This ability to allow current to flow or not flow, depending on the polarity of applied voltage, is the essential characteristic of a diode.

Bipolar Transistors

We saw that a diode is a single PN junction. A *bipolar junction transistor* is simply two PN junctions, back to back. There are two possible combinations of two PN junctions, *npn* or *pnp*, as shown in Figure 7.25.

Although simply having two PN junctions instead of one would seem only a minor addition at first glance, the realization and implementation of this technology has changed the world, for the third terminal on this dual-charge-carrier device allows the current to be *controlled.* The current can be either amplified when used in an analog fashion or switched when used in a digital manner.

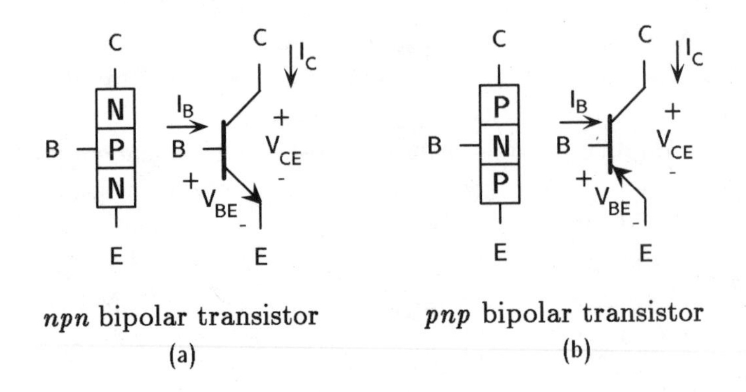

Figure 7.25. Bipolar junction transistors are made up of two PN junctions, back to back. (a) In an *npn* bipolar transistor, the collector and emitter are *n*-type while the base is *p*-type. (b) In a *pnp* bipolar transistor, the collector and emitter are *p*-type while the base is *n*-type.

We mentioned before that transistors can be either signal-level devices or power devices. It turns out that, while these two types of transistors arise from the same semiconductor physics, they are fabricated differently. Figure 7.26 illustrates silicon cross-sections through a signal-level *npn* bipolar junction transistor and a power *npn* bipolar junction transistor. Plus signs on the *n* and *p* regions designate heavily doped areas (very conductive). A minus sign would denote lightly doped areas (slightly conductive).

In a signal-level bipolar device, all the terminals are patterned from the top side of the silicon wafer and the voltage between the base and emitter controls the flow of current from the collector to the emitter. For instance, in an *npn* device, when the base-emitter diode is forward biased, negative charge carriers "emitted" by the *n*-type emitter region travel toward the base but then are swept across into the collector region (before having a chance to get caught and recombine with any holes in the *p*-type base region) when a larger positive voltage is applied to the collector. Some small current must be supplied by the base to replenish any holes that did recombine with passing electrons, but this base current is much smaller than the collector current (which is why a bipolar transistor is a current amplifier). For signal-level devices, base, emitter, and collector all lie along the top surface of the silicon wafer and the backside is not

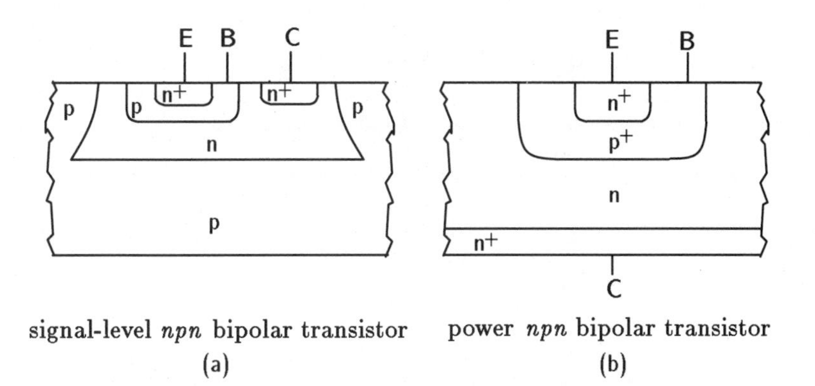

Figure 7.26. (a) In a signal-level device, all electrical terminals are on the top side of the silicon wafer and current flows along the surface, from collector to emitter. (b) In a power transistor, the backside is used for one of the electrical terminals and current flows vertically through the chip.

connected to anything. By having all terminals on the top side, it is easy to fabricate many different signal-level devices and interconnect them, allowing for very-large-scale integration (VLSI) for complex information-processing systems.

In a power device, on the other hand, the backside of the silicon wafer is used for one of the electrical terminals (the collector) and current flows vertically through the chip. Since power devices must handle more current and more heat, they are typically larger, often use backside connections and seldom integrate large numbers of different devices. More often, the tendency is to fabricate hundreds or thousands of vertical power transistors in parallel on one chip, creating in effect one very *big* transistor.

The cross-sections shown in Figure 7.26 are for *npn* bipolar transistors. The *pnp* bipolar transistors would have similar topologies but *p* regions would be replaced by *n* regions and vice versa. Since turning on a bipolar transistor requires forward biasing the base-emitter diode, turning on an *npn* version requires that the base be more positive than the emitter (at least 0.6 V more positive to be precise, as that is a diode's turn-on threshold). Conversely then, turning on a *pnp* version of a bipolar transistor requires that the base be 0.6 V more negative than the emitter.

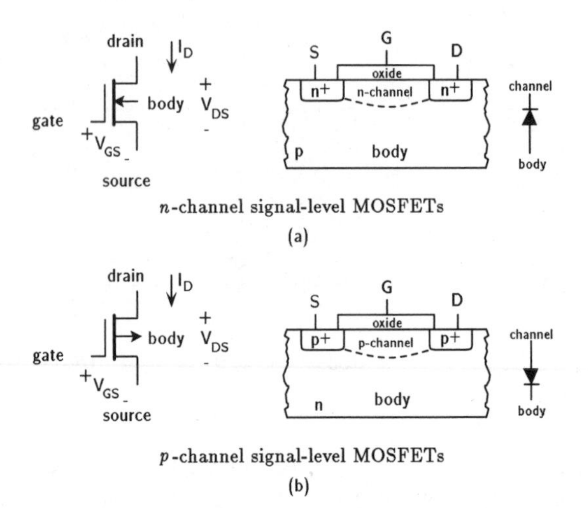

Figure 7.27. Signal-level MOSFETs also have all electrical terminals on the top side of the silicon wafer. (a) In an n-channel MOSFET, when the gate is positive with respect to the source, holes in the p-type body region move away from under the gate, leaving an n-type channel and allowing electron current to flow from drain to source. (b) In a p-channel MOSFET, when the gate is negative with respect to the source, electrons in the n-type body region move away from under the gate, leaving a p-type channel and allowing hole current to flow.

MOSFETs

Bipolar junction transistors rely on having two PN junctions in the main current path, which is why they are called *bipolar devices.* In contrast, a MOSFET has no PN junctions in the current path and is a *monopolar device.* Figure 7.27 illustrates symbols and cross-sections for n-channel and p-channel signal-level MOSFETs.

In the monopolar device, MOSFET junctions are fabricated to maintain separate regions of charge carriers when the device is off, but when an electric field is applied to the gate to turn on the device, the channel region separating two regions of like charge carriers is inverted making it the same "flavor" of charge carrier.

To be precise, we are speaking of enhancement-mode MOSFETs here (as opposed to depletion-mode MOSFETs) where, when the gate-source voltage is 0 V, the device is off. In this way, the entire current path is a region of the same type of majority carriers.

The two types of MOSFETs are then called *n-channel* and *p-channel* MOSFETs, and the three electrical terminals that correspond in many ways to the base, emitter, and collector of bipolar transistors are called the *gate*, *source*, and *drain*, respectively.

Notice that signal-level MOSFETs are similar to signal-level bipolar transistors in that the backside again is not used for any of the three electrical terminals. However, in the schematic symbol for a MOSFET, the body terminal is explicitly drawn in, whereas in the bipolar schematic, it is omitted.

One reason for this is that the body forms a PN junction with the channel when the MOSFET is on. The arrow on the body terminal connection is pointed in the direction that a diode's arrow points (from p to n), signifying the direction of the PN junction between the inverted channel and the body when the MOSFET is on.

Because of the formation of a diode when the channel is inverted, the body must be held at a voltage that will not allow it to conduct. The body can be tied to the source (as is done in a power MOSFET) or to a more negative voltage in the circuit for an n-channel MOSFET. (For a p-channel MOSFET, the body can be tied to a voltage more positive than the source.) Sometimes, schematics leave the body connection out and we must assume that it is tied to a voltage that will keep the body-channel diode from conducting. Note, however, that the arrow on the body terminal is the only way to distinguish whether the MOSFET is n-channel or p-channel.

The gate terminal in the schematic is drawn with a horizontal line extending from the source end of the gate. This is to clarify which end of the device is intended to act as the source and which end is intended to act as the drain. Actually, a signal-level MOSFET is symmetric and can be used reversibly (and often is used this way in analog multiplexors and pass transistors for memories). For this reason, some schematics use a symmetrical gate connection, where the gate terminal is midway between the drain and the source.

Another reason the body terminal is drawn explicitly is that, if a power device is fabricated instead of a signal-level device, the backside connection is used as the drain. The central body region is then connected to the source, and this connection creates another device, a *source-drain diode*. A power MOSFET then is not symmetric. Symbols and cross-sections for n-channel and p-channel power MOSFETs are illustrated in Figure 7.28.

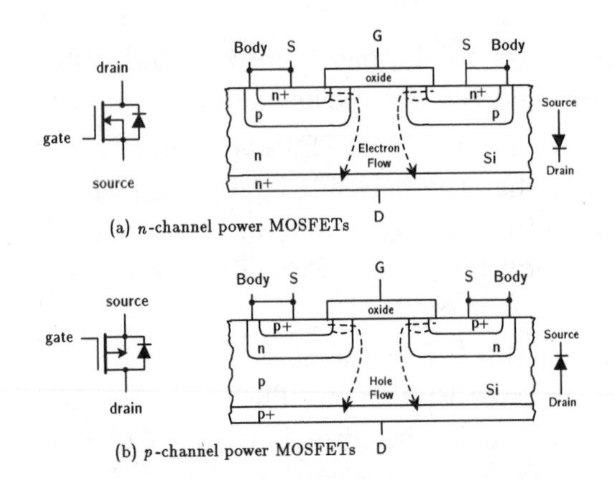

(a) n-channel power MOSFETs

(b) p-channel power MOSFETs

Figure 7.28. Power MOSFETs use the backside of the wafer as the drain. (a) When an n-channel power MOSFET is turned on, the p-type body region is inverted under the gate leaving a channel for *electron* current to flow vertically through the chip. (b) When a p-channel power MOSFET is turned on, *hole* current flows vertically through the chip.

Comparisons and Contrasts

Bipolar transistors and MOSFETs are similar in many respects, but a number of differences are worthy of note. First, though, let's take a moment to point out the general differences between n-type and p-type devices.

It turns out that the two types of charge carriers, holes and electrons, are not completely symmetric. Holes are not as mobile as electrons, and p-type devices, whether *pnp* bipolar transistors or p-channel MOSFETs, are never quite as good as n-type devices. In a bipolar transistor, a *pnp* device's high-frequency operation is poorer than a *npn* transistor's operation. In a MOSFET, a p-channel device does not exhibit as low on-resistance as an n-channel device. In fact, in the early days of MOSFETs, processes typically only gave the designer the option of having n-channel MOSFETs (often abbreviated as *NMOS* transistors). Later, when p-channel MOSFETs, or *PMOS* transistors, were introduced into the same process, the process became known as *CMOS* (complementary metal oxide semiconductor), since complementary n-type and p-type devices were then both available. Because p-type devices are poorer than n-type devices, this lack of performance has repercussions in the design of H-bridges for driving motors.

One of the main differences between a MOSFET and a bipolar transistor is that a MOSFET is essentially a voltage-controlled device while a bipolar transistor is a current-controlled device. In a MOSFET, the gate oxide creates a capacitor between the gate and the source, so the steady-state gate current is 0 (although some charging and discharging currents flow when turning-on and turning-off the device). Since very little gate current is required, MOSFETs are fairly easily driven from microprocessors or CMOS logic gates.

In contrast, bipolar transistors are current-controlled devices. Instead of having a capacitor between the gate and source, as in a MOSFET, the bipolar transistor has a diode between the base and emitter. Once the base-emitter diode is forward biased, the collector current is controlled by the base current. The ratio of collector current to base current is the current gain, β:

$$\beta = \frac{I_C}{I_B}$$

For signal-level bipolar transistors, the current gain might be 100 or 200, but for power bipolar transistors carrying large numbers of amps, current gains are typically much lower, possibly on the order of 20 or 50.

Data sheets for devices under consideration should be checked for more specific numbers, but even so, current gains can differ widely from piece to piece (for the same part number of transistor) due to process variations between manufacturers. In general, though, power bipolar transistors require significant amounts of base current. Since these magnitudes of base current cannot be delivered directly from microprocessors or logic gates, another level of interface circuitry is often needed to drive the H-bridges that are driving the motors. In addition to the added complexity involved in the bias network, the base current through the base resistor dissipates power (not to mention the power dissipated by the additional layer of interface circuitry).

In order to compare the efficiencies of bipolar power transistors and power MOSFETs for driving motors, return once again to the illustration of the H-bridge in Figure 7.21. For the ideal switches in that diagram, the voltage across the motor is always equal to the full magnitude of the supply voltage when opposite sets of switches

(S1 and S4, or S2 and S3) are closed. That is, there is no voltage drop across an ideal switch.

Real solid-state switches, however, do have finite voltage drops. The voltage drops associated with bipolar power transistors and power MOSFETs come about in different ways, however. In a power MOSFET, there are no PN junctions in the main current path from drain to source once the device has been turned on. Consequently, the only thing holding back charge carriers are factors such as their mobility, the width of the channel, and the like. These factors can be characterized as an effective resistance from drain to source. When the device is turned on as hard as possible, the channel becomes as wide as possible, giving the smallest on-resistance. This leads to the lowest voltage drop across the device, so this is the region where power MOSFETs should be run when switching motors. Figure 7.29(a) shows the I_D vs. V_{DS} characteristics for an n-channel power MOSFET.

The area to the left of the dotted line, where I_D increases with V_{DS}, is known as the *constant-resistance* or *linear region*. Typical MOSFETs have a threshold voltage on the order of 3.0 V-5.0 V, below which the MOSFET is cut off. To the right of the dotted line, for larger drain-source voltages and depending on V_{GS}, the channel becomes maximally opened and the current, I_D, reaches a saturation condition, where it remains constant even as V_{DS} is increased. If the gate voltage is high enough, usually around 10.0 V, the drain current stays in the constant-resistance region and the voltage drop from drain to source is minimal, as shown in the figure. This is the region in which a power MOSFET is run when switched to the "on" state, as the voltage drop, V_{DS}, across the switch is minimized.

The inverse of the slope of an I_D-V_{DS} curve in this linear region is the on-resistance ($\frac{1}{r_{DS}} = \frac{I_D}{V_{DS}}$) of a power MOSFET. The proper gate-to-source voltage should be chosen given the drain-source voltage and the desired current, so as to maintain the device biased in the constant-resistance region for the most efficient utilization of the power MOSFET.

The voltage drop across a turned-on bipolar power transistor comes about for a different reason. Whereas a turned-on power MOSFET has a continuous region of like charge carriers from drain to source, a bipolar transistor has two PN junctions in the current

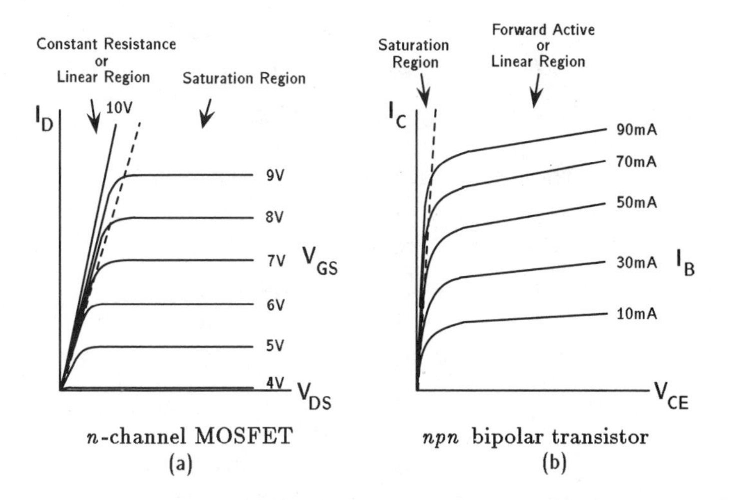

Figure 7.29. (a) An n-channel MOSFET shows these typical I_D-V_{DS} characteristics. When biased in the constant-resistance region, a MOSFET can be modeled as a resistor, where I_D varies linearly with V_{DS}. (b) An *npn* bipolar transistor is controlled by the value of the base current rather than the voltage as in case of a MOSFET.

path from collector to emitter, as was shown in Figure 7.25. In a turned-on bipolar transistor, the base-emitter diode is forward biased and the collector-emitter diode is reversed biased (at least, in the usual case of linear region operation). However, if the bipolar transistor is completely on, the collector potential should be close to that of the emitter potential (approaching the case of an ideal switch where there would be 0 V between collector and emitter). The closest a bipolar transistor can come is to have the collector-base diode no longer reverse biased but forward biased. With the base-emitter diode forward biased (transistor turned on) and the collector-base junction also forward biased, the bipolar transistor is in what is known as the *saturation region* of operation, where V_{CE} is almost constant and very small ($V_{CE} = V_{CE(SAT)}$) for any value of base current. Figure 7.29(b) illustrates the I_C versus V_{CE} characteristics for an *npn* bipolar transistor.

This saturation region is to the left of the dotted line in Figure 7.29(b) and is the region where a bipolar power transistor should be run when switched to the "on" state in order to provide the minimum voltage drop across the switch. The region to the right of the dotted line is known as the bipolar transistor's *forward active region*. The forward active or linear region is the region in which a bipolar transistor is used as a linear amplifier.

In comparing the graphs in Figure 7.29(a) and (b), note that the MOSFET is a voltage-controlled device, where I_D is determined by the value of V_{GS}, and the bipolar transistor is a current-controlled device, where I_C is determined by the value of I_B. Note, too, that the MOSFET's saturation region looks like what is called the linear region for a bipolar transistor, and the MOSFET's linear region looks like what is called the saturation region for the bipolar transistor. Again, this has to do with the MOSFET being a voltage-controlled device and the bipolar transistor being a current-controlled device and what parameter in each is actually being saturated.

Nevertheless, the point to be made is this: when transistors are used as switches, they should be biased in the regions to the left of the dotted lines in the figures so that they approach the function of ideal switches as closely as possible. That is, when an ideal switch is closed, it should have 0 V dropped across it. Solid-state switches cannot completely meet this goal, but when turned on hard enough, they can come as close as possible.

Any voltage drop appearing across a closed solid-state switch contributes to wasted power. For instance (referring again to Figure 7.21), if switches S1 and S4 are on and are implemented with bipolar transistors each having 0.3 V saturation voltage drop and if the supply voltage is 5.0 V, then only 4.4 V appears across the motor. Additionally, if the motors draws 500 milliamps (mA), then 2.2 W is delivered to the motor while 300 milliwatts (mW) is dissipated as heat in switches S1 and S4.

Deciding whether to choose power MOSFETs or power bipolar transistors when designing an H-bridge depends largely on which type of device will yield the most efficient solution. The answer depends on the power required by the motors and the choice of devices available. If MOSFET devices can be found that have low enough on-resistances and if, for the required current, they produce voltage drops less than saturation voltages of comparable bipolar devices, then power MOSFETs may be the right choice for designing an H-bridge.

When comparing and contrasting bipolar transistors and MOSFETs, another characteristic to take into consideration is how each type of transistor responds to temperature increases, as running large amounts of current through a transistor causes it to heat up.

Bipolar transistors are subject to a condition known as *thermal runaway*. When current flows through the device, it gets warmer and the temperature rise affects the bipolar transistor in such a way that more current flows. With additional flow of current, the device gets even warmer and the problem escalates. *Thermal runaway* means that bipolar transistors cannot share current when configured in parallel. If one bipolar transistor has slightly more current running through it, it will heat up, allowing more current to flow; it will eventually hog all the current, resulting in thermal runaway.

In contrast, MOSFETs do not suffer from thermal runaway and lend themselves nicely to parallel configurations. The on-resistance of a MOSFET increases with temperature, providing a negative feedback effect. As more current flows through a MOSFET, its resistance increases and the current through the device decreases until a stable operating point is reached. Consequently, MOSFETs do not suffer from current hogging.

This feature is often taken advantage of in motor drives for electric vehicles and solar cars. Instead of purchasing one very large power MOSFET to switch current from an electric vehicle's battery to its engine, designers often buy the most economical power MOSFETs and place them in parallel. Up to 150 discrete devices are often paralleled in this way.

For a small mobile robot, however, where space is a primary concern, power MOSFETs do have some disadvantages. Because typical power MOSFETs need 8 V to 10 V for full-on gate drive, it may be inconvenient to drive a power MOSFET from a battery-powered robot using a single battery pack. Alkaline batteries come in 1.5 V cells and nickel-cadmium batteries come in 1.2 V cells; many of the design decisions for a small mobile robot revolve around the issues of battery pack selection, motors, and motor drivers, as the weight of the robot is primarily composed of these elements. If four alkaline cells are used as a 6 V power supply for the electronics, either more batteries or a charge pump must be provided to create the 8 V gate drive.

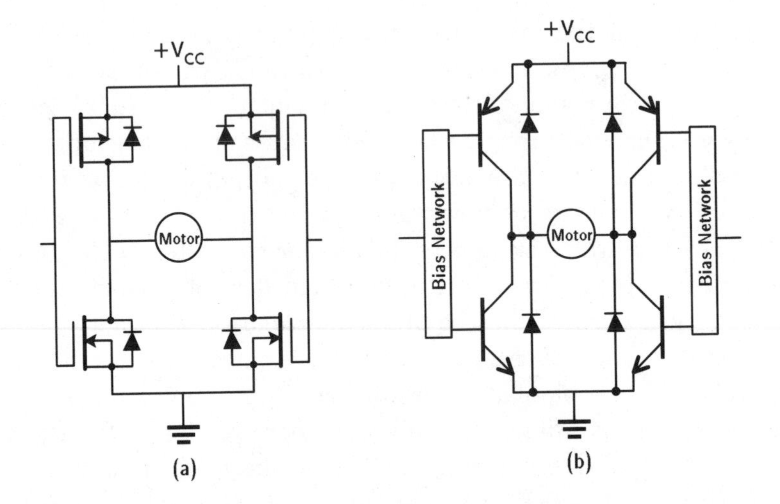

Figure 7.30. (a) A MOSFET implementation of an H-bridge, with *p*-channel devices on top and *n*-channel devices on bottom is shown here. (b) A bipolar transistor implementation of an H-bridge, with *pnp* devices on top and *npn* devices on bottom, requires more complex biasing circuitry to provide level shifting and base currents for the bipolar transistors.

One way around this problem is to use special low-threshold MOSFETs. These devices use very thin gate oxides to bring the turn-on voltages down to ranges from 1 V to 2 V. With threshold voltages that low, full-on gate drives can usually be achieved at 5 V. Such devices are called *logic-level* MOSFETs. Supertex makes a wide line of low-threshold MOSFETs. Motorola and International Rectifier also carry a variety of MOSFET devices.

H-Bridge Implementations

Whether MOSFETs or bipolar transistors are chosen to implement the H-bridge, the topologies are very similar. One convenient way to set up an H-bridge is to use *p*-type devices for the high-side switches and *n*-type devices for the low-side switches.

Figure 7.30 illustrates H-bridges in both bipolar and MOSFET technologies. If the gating signal on the left in each schematic is pulled low, the left-side bottom switches will be off and the left-side top switches will be on. If, at the same time, the right-side gating signals for each H-bridge are pulled high, the right-side bottom

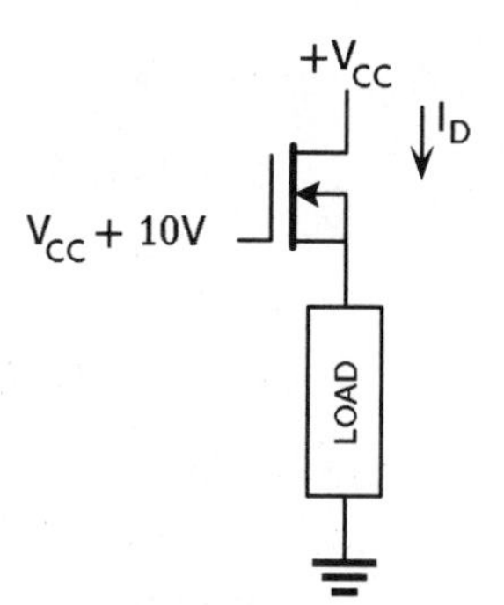

Figure 7.31. An n-channel MOSFET used as a high-side switch must have its gate voltage pulled higher than that of the positive supply in order to be on hard enough that the voltage drop between drain and source approaches 0 V.

switches will be on and the right-side top switches will be off. This configuration is exactly the scenario described in Figure 7.21 when switches S1 and S4 were on and switches S2 and S3 were off, allowing current to flow from left to right through the motor. Note that, in a MOSFET version of an H-bridge, flyback diodes do not have to be added discretely, as the built-in source-drain diodes provide the flyback function.

However, because p-type devices have higher on-resistances than n-type devices, it is possible to design more efficient H-bridges if n-type devices are also used for the high-side switches. The only problem with this design decision is that, if an n-type device is used for the high-side switch, the gating voltage to turn on the high-side switch must be pulled higher than that of the positive rail. For instance, in a MOSFET (see Figure 7.31), if an n-channel MOSFET is switching a load between the source and ground, the voltage at the source when the switch is on, should be very close to that of the positive supply rail. Since the gate turn-on voltage must be approximately 10 V higher than the source, the gate voltage must be $V_{CC} + 10\,\mathrm{V}$. Even if low-threshold devices are used, the gate voltage must be $V_{CC} + 5\,\mathrm{V}$, still requiring a separate power supply.

One solution to this problem is to add additional circuitry to the gate-drive network in the form of a charge pump. Charge pumps use switched capacitors to create voltages higher than the supply voltage. This type of design adds extra complexity to the input of the MOSFET implementation of an H-bridge, but fortunately, many manufacturers solve this problem by integrating all the required subsystems on a single motor-driver chip.

7.7.4 Motor-Driver-Power Integrated Circuits

Motor-driver-power integrated circuits (ICs) make it very convenient to interface motors to microprocessors. Typically motor-driver-power ICs also have circuitry that provides current-limiting and overvoltage protection. One single-chip solution is the MPC1710A motor driver from Motorola. This chip, whose block diagram is shown in Figure 7.32, uses an H-bridge composed of four n-channel MOSFETs. A level shifter and charge pump circuit are included on the chip to drive the high-side switches.

Three capacitors and an inverter are the only external components required to interface an MPC1710A to Rug Warrior's MC68HC11A0. We could use port D pin PD5 to set the forward or reverse direction of the motor and port A pin A5 to pulse-width modulate the enable input for speed control. The Motorola MPC1710A can deliver up to 1 A of current with a 0.4 Ω on-resistance when sourcing current and 0.2 Ω on-resistance when sinking current.

For the two motors on Rug Warrior, two MPC1710A chips would be needed, one for each motor. One motor could be controlled by pins PD5 and PA5 and the other motor by pins PD4 and PA6. The MPC1710A comes in a small 16-pin surface-mount package, which makes it very compact when used in a printed circuit board design but rather difficult to use when prototyping with Speedwire or Scotchflex wiring technologies. For this reason, on Rug Warrior, we chose to use a chip that would be more amenable for our readers, the SGS Thompson L293D.

The L293D was chosen because it comes in a normal 16-pin dual-inline package (DIP). This selection, which has two H-bridges on board, minimizes the parts count and delivers enough power for Rug Warrior's motors. The L293D, shown in Figure 7.33, uses a bipolar H-bridge instead of a MOSFET H-bridge. Again, all switches are made from n-type devices and a step-up circuit is incorporated on chip to drive the high-side switches. Flyback diodes are integrated

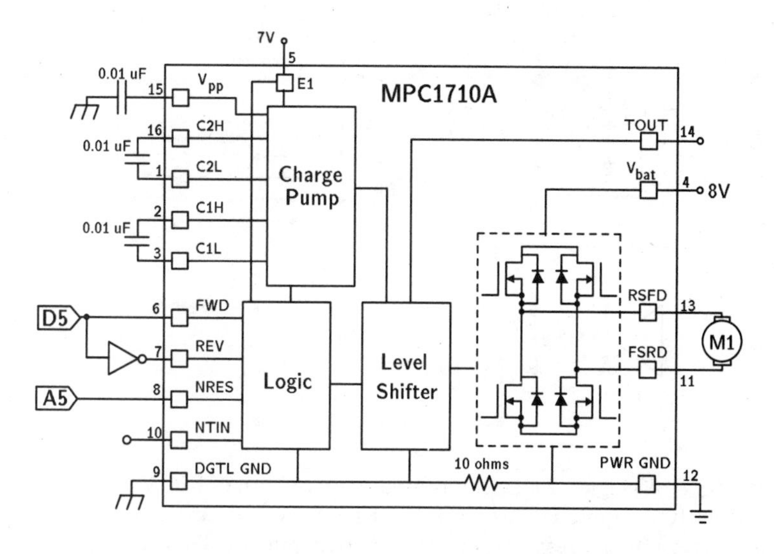

Figure 7.32. Two Motorola MPC1710As can be used to drive Rug Warrior's two motors. One MPC1710A is needed for the left motor and one MPC1710A is needed for the right. This surface-mount integrated circuit motor driver chip uses an H-bridge made from *n*-channel power MOSFETs.

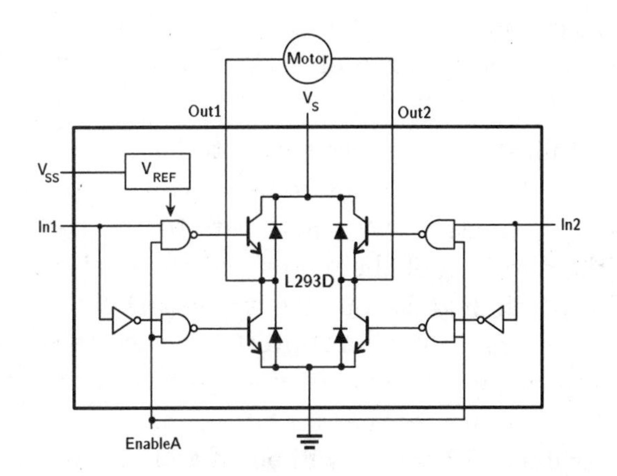

Figure 7.33. One motor-driver-power IC is an SGS Thompson L293D. This power IC incorporates a motor driver using an H-bridge made from bipolar transistors. While this illustration only shows one H-bridge, two full H-bridges are actually incorporated in an L293D.

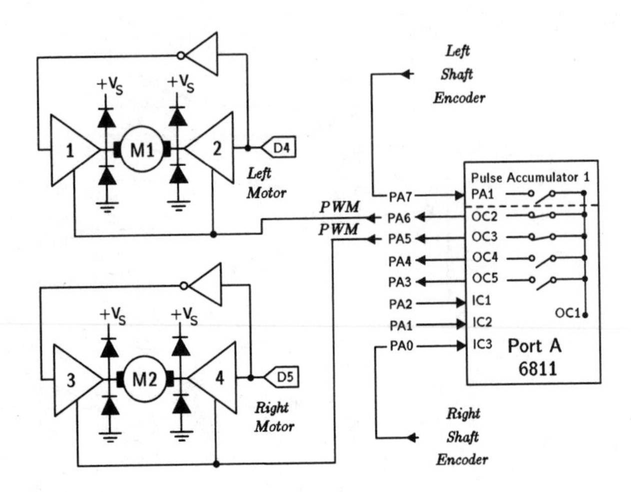

Figure 7.34. A single L293D chip is used to drive both of Rug Warrior's motors. The MC68HC11's port D pins PD4 and PD5 select forward and reverse for the left and right motors, respectively, while port A pins PA6 and PA5 pulse-width modulate the left and right enable pins. Note that OC1 here is used to also control OC2 and OC3.

on chip in this circuit. The L293D can deliver 600 mA to the motor, with a saturation voltage drop of 1.4 V when sourcing current and 1.2 V when sinking current.

Figure 7.34 illustrates how we have interfaced the L293D to Rug Warrior's 6811. The L293D has some on-chip logic that provides an Enable signal. In this way, the Inputs to the H-bridge can be used to set the direction of the motor and the Enable signal can be used for pulse-width modulation. We use port D pin PD5 to set the direction for the right motor. An inverter is used to set one side of the H-bridge to the opposite polarity voltage of the gating signal of the other side. This ensures that if switches S1 and S4, for instance, are on, then switches S2 and S3 will be off and vice versa. Note that this means that the motor is never actively braked. The H-bridge is pulse-width modulated by tying the right motor driver's Enable pin to port A pin PA5. The output compare function of PA5 is used to facilitate timing. One advantage of the L293D is that two full H-bridges are incorporated on chip, which means that only one L293D is needed to drive both of Rug Warrior's wheels. Port D pin

PD4 is used to set the direction of the left motor, and port A pin PA6 is tied to the Enable signal for pulse-width modulation.

Many other motor-driver-power integrated circuits are available. As mentioned in E, the place to begin searching is the *IC Master*. The *IC Master* lists integrated circuits both by part number and by function. Listings under "Motor Drivers" include a number of suppliers, such as Unitrode, Siemens, Motorola and International Rectifier, among others.

Another avenue to pursue is to purchase motor controllers for radio-controlled cars. These are often called *speed controllers*, which is a bit of a misnomer, since it is only the human who provides the speed control. However speed controllers do incorporate the power MOSFETs or power bipolar transistors in discrete H-bridges for driving larger motors. They are sold by Futaba, Tower Hobbies, and Sheldon's Hobbies, and are often advertised in radio-control hobbyist magazines.

7.8 Software for Driving Motors

The software for controlling Rug Warrior's motors must do two things. First, the software needs to control the speed of the robot in the manner desired by the programmer. For instance, a higher pulse-width ratio of voltage across the motor is needed to keep the robot moving up a ramp at one foot per second than would be required to make it move along a flat tile floor at one foot per second. To maintain a desired speed, regardless of terrain, means that the robot needs to count the number of pulses from one of the shaft encoders to see how fast the wheels are turning and then update the pulse width accordingly.

The second function that the software needs to perform is to make the two wheels actually revolve at the same speed so that the robot will move in a straight line. Recall that, in TuteBot, innate differences between the two motors caused TuteBot to move in an arc, even when the same voltage was applied to both motors. In that case, we simply added resistors in series with one motor until both motors went at the same speed. That analog solution was fine for TuteBot, but here, we implement a digital solution, since Rug

Warrior has a microprocessor right at hand. In this way, the solution is general, and if many Rug Warriors are manufactured, they do not all have to be individually tweaked with resistor trials. Again, the solution is to read the shaft encoders from each wheel and increase or decrease the speed of the right motor, say, to match its speed with that of the left motor.

7.8.1 Pulse-Width Modulation

The software we have configured for controlling Rug Warrior's motors takes advantage of timer-counter hardware associated with the MC68HC11A0's port A and succeeds in implementing a pulse-width modulation scheme without recourse to either polling or interrupts. Port A's eight pins have various output compare and input capture registers, as shown in Figure 7.34. Refer to the Motorola data books on the MC68HC11 for a more complete discussion than we will attempt here.

An output compare register can be set by the programmer so that, for instance, when the timer-counter's value matches the output compare register's value, a pin can be set high or an interrupt can be initiated. An input capture register has the opposite function. When a signal on a pin goes low for instance, the input capture register can store the value of the timer-counter register at the time that the event happened or initiate an interrupt.

Output Compare Registers

For pulse-width modulation, we will take advantage of the output compare registers associated with port A pins PA3-PA7, as shown in Figure 7.34. Pin PA7 happens to hold a dual role as either a pulse accumulator or as output compare register 1 (OC1). For Rug Warrior's right wheel, we have chosen to use PA5, which is associated with output compare register 3 (OC3) and for the left wheel, PA6, which is associated with output compare register 2 (OC2). We also take advantage of OC1 because it is a special output compare register in that it can control a given selection of the four other output compare registers. The closed connections between OC1 and OC2 and between OC1 and OC3 in Figure 7.34 illustrate how we intend to use the timer-counter capabilities to drive Rug Warrior's motors.

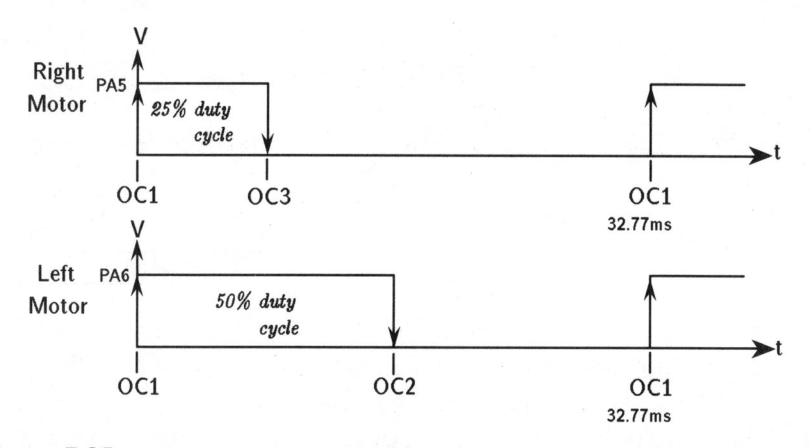

Figure 7.35. Pulse-width modulation can be conveniently accomplished using the MC68HC11A0's port A output compare registers. Here, we use three different output compare registers, where output compare register OC1 directs pins PA5 and PA6 to both go high at the beginning of each period. Output compare registers OC3 and OC2 each tell pins PA5 and PA6 when to go low, giving a programmable duty cycle for each motor.

Figure 7.35 illustrates the timing sequences for our algorithm that will be generated on PA5 and PA6 to implement pulse-width modulation.

The timer-counter itself is a 16-bit register, TCNT, where the high byte is at hex address $100E and the low byte is at $100F:

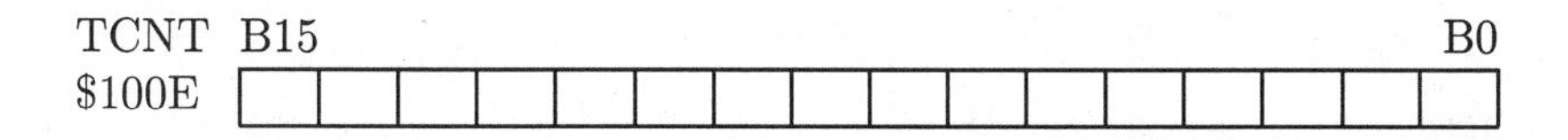

The timer-counter runs at a rate dependent upon Rug Warrior's crystal oscillator (and therefore the MC68HC11A0's E clock, which is on pin 5 of the MC68HC11A0 and can be checked with an oscilloscope). The E clock has a period of one-fourth that of the crystal oscillator frequency. TCNT is a free-running counter that starts at 0 when the MC68HC11A0 is reset and counts up to 2^{16}, which is 65,536 counts. The counter then overflows and starts again from 0. We use an 8 megahertz (MHz) crystal for Rug Warrior, giving the E clock a frequency of 2 MHz and a period of 0.5 microseconds (μs).

By default, the timer-counter counts at the same period as the E clock, but there is a way to prescale the timer-counter rate, which involves setting two bits in another register, TMSK2. The lowest two bits in the TMSK2 register, PR1 and PR0, are used to divide down the E clock for changing the rate at which the timer-counter runs.

TMSK2 $1024	Bit 7							Bit 0
	TOI	RTII	PA0VI	PAII	0	0	PR1	PR0

For our purposes, we will let the timer-counter run at its default setting and not bother with changing any values in TMSK2. This means that, after 2^{16} counts at $0.5\,\mu s$ per count, 32.77 milliseconds (ms) will have passed. We will use this standard overflow time as the period for pulse-width modulation, t_{period}, as was illustrated earlier in Figure 7.22.

Our plan is to create the pulse-width modulated signals for the left and right motors using waveforms generated by OC2 and OC3 associated with pins PA6 and PA5, as shown in Figure 7.35. In this case, we will use OC1 to set the bits high on both PA6 and PA5 when the timer-counter is at 0. We will use the OC2 and OC3 registers to clear the bits on both PA6 and PA5 when the value reached by the timer-counter matches the values stored in their 16-bit timer output compare registers, TOC2 and TOC3. So, for instance, if we want PA5 to have a 25% duty cycle, then we store 65,536 $\div$ 4 = 16,348 in TOC3. If we want PA6 to have a 50% duty cycle, we store 65,536 $\div$ 2 = 32,768 in TOC2:

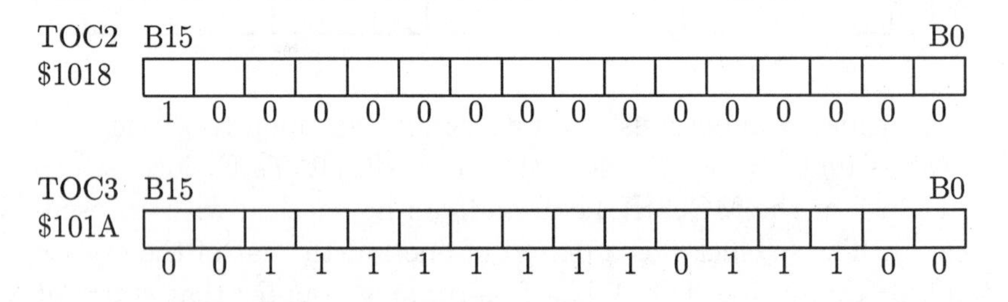

To specify what action should be taken on PA5 and PA6 when the corresponding output compare registers match the timer-counter, we must set some values in another register, TCTL1. The appropriate bit sequences are shown in Figure 7.36. For the way we designed

OMx	OLx	Configuration
0	0	OCx Does Not Affect Pin
0	1	Toggle OCx Pin
1	0	Clear OCx Pin
1	1	Set OCx Pin

Figure 7.36. The four actions possible by any output compare pin are to not change, to toggle, to go low, or to go high. Two bits in the TCTL1 register, the most significant bit (OMx) and the least significant bit (OLx), set the desired response for any successful output compare.

our algorithm in Figure 7.35, we want PA5 and PA6 to be set to 0 when OC3 and OC2 have successful output comparisons. To set this up, we store the two bits, %10 (which will make the pin go low), in TCTL1 in the locations associated with OC3 and OC2:

TCTL1	Bit 7							Bit 0
$1020	OM2	OL2	OM3	OL3	OM4	OL4	OM5	OL5
	1	0	1	0	x	x	x	x

The x's in any bit position represent **don't care**'s. With the falling edge of the pulse configured (the signal transitioning from high to low), now we just need to set up the OC1 rising-edge event (the transition of the signal from low to high). This is done by storing the value of time equal to 0 in TOC1, the 16-bit timer output compare 1 register:

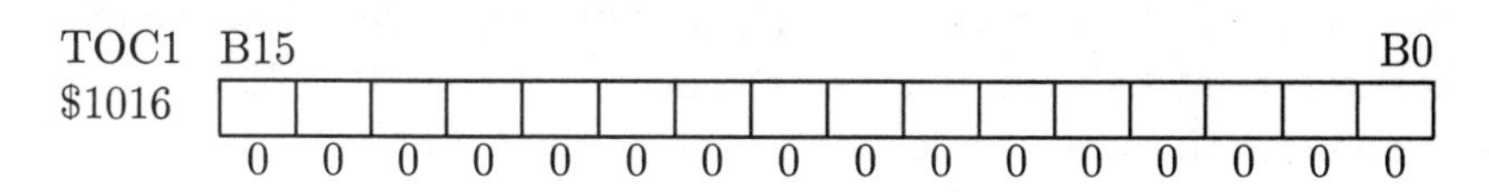

To configure the hardware so that OC1 will control PA5 and PA6, we set values in some auxiliary registers that control OC1. The output compare 1 mask (OC1M) register signifies which of the other four output compare registers OC1 will control. The high five bits in OC1M correspond bit for bit with a port A output pin. Therefore, we store the binary number %01100000 in OC1M to set up OC1 to control PA5 and PA6:

OC1M	Bit 7						Bit 0
$100C	OC1M7	OC1M6	OC1M5	OC1M4	OC1M3	-	-	-
	x	1	1	x	x	x	x	x

Once we have selected which pins will be active, we can program the action that we want to result when the timer-counter value matches the value of 0 in TOC1 by setting some values in the output compare 1 data (OC1D) register. By setting the bits corresponding to PA5 and PA6 to 1, whenever the timer-counter overflows and returns to the value 0, the PA5 and PA6 pins will be set to 1, which forms the rising edge of each pulse.

OC1D	Bit 7							Bit 0
$100D	OC1D7	OC1D6	OC1D5	OC1D4	OC1D3	-	-	-
	x	1	1	x	x	x	x	x

By using these built-in hardware features of the 6811, no interrupts or polling sequences are required to implement pulse-width modulation. We simply write to some registers in the timer-counter system and all actions on pins PA5 and PA6 take place in the background of other programs being run on the robot. The programmer merely writes new values to TOC2 and TOC3 when the speed has to be changed.

PWM Software Driver

The program below illustrates the **IC** code that implements this scenario of a 25% duty cycle signal asserted by pin PA5 and a 50% duty cycle signal asserted by pin PA6. This sequence will make one wheel rotate at half the speed of the other, causing Rug Warrior to move in an arc towards one side or the other.

```
int DDRD = 0x1009;   /* Port D data direction */
int OC1M = 0x100C;   /* Output Compare 1 Mask */
int OC1D = 0x100D;   /* Output Compare 1 Data */
int TOC1 = 0x1016;   /* Output Compare Timer 1, 16-bit reg */
int TOC2 = 0x1018;   /* Out Cmp Tmr 2, 16-bit reg (left motor) */
int TOC3 = 0x101A;   /* Out Cmp Tmr 3, 16-bit reg (right motor) */
int TCTL1 = 0x1020; /* Timer Control 1, 8-bit reg */
```

```
/* motor_index:  0 => Left motor, 1 => Right motor */
int TOCx[2] = {TOC2,TOC3};        /* Index for timer register */
int sign[2] = {1,1};              /* Sign of rotation of motor */
int dir_mask[2] = {0b010000, 0b100000};/* Port D direction bit */

/* Utility functions */
float abs(arg)                    /* Absolute value function */
{ if (arg < 0.0)
   return (- arg); else return arg; }

int get_sign(float val)           /* Find sign of argument */
{ if (val > 0.0)
   return 1; else return -1; }

/* Limit range of val */
float limit_range(float val, float low, float high)
{ if (val < low) return low;
   else if (val > high) return high;
   else return val; }

void init_pwm()                /* Initialize Pulse-Width Modulation */
{ poke(DDRD,0b110010);         /* D dir:  OUT 5,4,1; IN 3,2,0 */
  poke(OC1M,0b01100000);       /* Out Cmp 1 affects PA5 and PA6 */
  poke(OC1D,0b01100000);       /* Successful OC1 turns on PA5, PA6 */
  bit_set(TCTL1,0b10100000);   /* OC3 turns off PA5, OC2:  PA6 */
  pokeword(TOC1,0);            /* When TCNT = 0, OC1 fires */
  pokeword(TOC2,1);            /* Minimum on time for OC2 */
  pokeword(TOC3,1); }          /* Minimum on time for OC3 */


/* The sign is handled in a special way because */
/*  we have only a 1 channel encoder */
void pwm_motor(float vel, int motor_index)
{ if (sign[motor_index] > 0) /* Choose the direction of rotation */
     bit_set(port_d, dir_mask[motor_index]);
  else
     bit_clear(port_d, dir_mask[motor_index]);
  vel = limit_range(vel, 1.0, 99.0); /* 1 ≤ duty fctr ≤ 99 */
  pokeword(TOCx[motor_index], (int) (655.36 * vel)); }

/* Top-level open-loop PWM command */
void move(float l_vel, float r_vel) /* Vel range [-100.0, 100.0] */
{ sign[0] = get_sign(l_vel); /* Desired direction of rotation */
```

```
  sign[1] = get_sign(r_vel);
  pwm_motor(abs(l_vel), 0); /* Set pulse-width modulation cnst */
  pwm_motor(abs(r_vel), 1); }
```

Now let's walk through this program. First, all the necessary registers are assigned and three data structures are created. The arrays TOCx[ ], sign[ ], and dir_mask[ ] are all two-element arrays. TOCx[ ] is an array whose first element is the address TOC2, where left-motor velocities are stored, and whose second element is the address TOC3, where right-motor velocities are stored. The array sign[ ] is an array whose elements are bits representing which direction the left and right motors are commanded to go. The array dir_mask[ ] is an array whose first element holds the mask for Port D, required to select the left motor, and whose second element holds the mask for Port D, required to select the right motor.

The next three functions also just lay the groundwork for the main part of this program. The functions abs(), get_sign() and limit_range() are functions that C does not happen to supply: abs() simply returns the absolute value of its argument; get_sign() returns the sign of its argument; and limit_range() returns a maximum or minimum value for its argument if it is out of range.

The actual pulse-width modulation of the motors is accomplished by the functions init_pwm(), pwm_motor(), and move(). The timer-counter hardware is set up and started by init_pwm(). The three pokeword() commands set an initial (small) pulse width.

To change the duty cycle, the calling routine uses the pwm_motor() function, which takes two arguments: a velocity command and a motor index. pwm_motor() then pokes the new velocity into the address, either TOC2 or TOC3, as specified by motor_index.

The function move() is the interface the programmer has for directing the robot. move() takes two arguments, a velocity for the left motor and a velocity for the right motor. These velocities should be given in the form of percentages of full speed. That is, they should be in the range [-100.0, 100.0]. A move(25.0, 50.0) command would make the left motor move at 25% of full speed and make the right motor move at 50% of full speed, causing the robot to arc to the left.

Setting up the pulse-width modulation scheme for each motor then merely means writing some values to a few registers. Once this has been done, the hardware associated with the timer-counter system will run by itself - always setting pins PA5 and PA6 high when the timer-counter reaches zero, always setting PA5 low when the timer-counter reaches 16348, and always setting PA6 low when the timer-counter reaches 32768. The central processing unit of the MC68HC11A0 then is free to attend to other tasks, perhaps reading a sensor or calculating a new speed at which the robot should run. To change the speed, Rug Warrior's main program merely has to store new values in TOC2 and TOC3.

7.8.2 Feedback-Control Loops

The strategy we have just described for pulse-width modulating motors is known as an *open-loop control scheme.* In open-loop control, there is no feedback from the motors, telling the robot's program how fast the wheels are turning or how far the robot has gone. Rather, the motors are just given different commanded voltages. But depending on terrain, surface obstacles, slippage in wheel contacts, or load on the robot, the commanded voltages do not necessarily imply particular speeds.

To implement a true velocity- or position-control algorithm, the robot needs sensors on the wheels, such as the shaft encoders mentioned earlier. Such feedback enables what are known as *closed-loop control algorithms.* Figure 7.37 illustrates the simple control loop we will use on Rug Warrior, called a *P-I controller*, for *proportional-integral controller.*

The basic idea of a control loop is to take the desired velocity command (such as one created in the way just described for our pulse width modulation scheme), send that command to the motors, see how fast the motors actually spin, and then measure that speed and compare it to the commanded speed. The difference is called the error signal and it can be either positive or negative. There are three error signals (marked e_1, e_2, and e_3) in the P-I control loop of Figure 7.37.

What makes a control loop a proportional controller or an integral controller depends on what computation the loop performs on the error signal. For instance, if the loop multiplies the error by some constant to produce a new command, then the controller is a

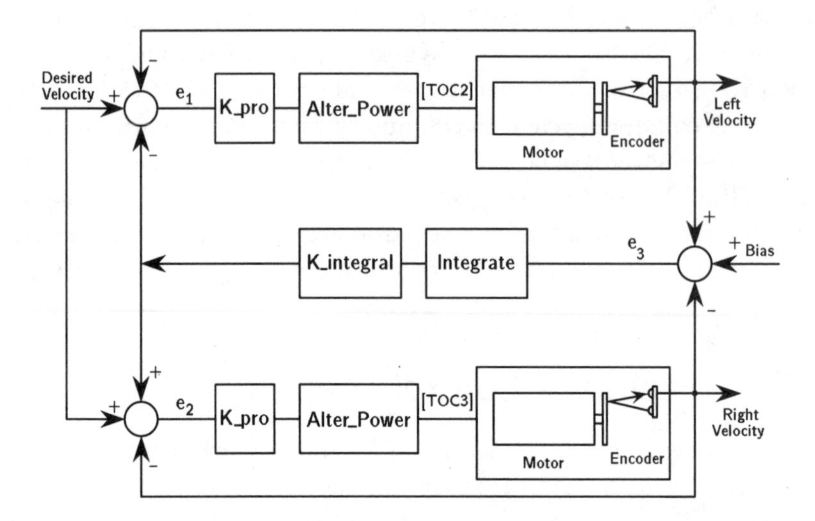

Figure 7.37. A simple proportional-integral control loop can be added in software to control the speed of the robot and to synchronize Rug Warrior's two wheels so that the robot will travel in a straight line.

proportional controller. In the controller shown in Figure 7.37, there are actually three separate feedback loops.

Imagine for a moment that the central feedback path is not there. The top loop, producing the error signal e_1 from the left motor, is identical in form to the bottom loop, producing the error signal e_2 from the right motor. Each of these loops is a proportional controller because the difference between the desired speed and the actual speed is multiplied by a constant, K_pro, and fed back to the motor to adjust the motor speed. If the actual speed is less than the desired speed, the difference is positive (as defined by the assignment of plus and minus signs on the feedback arrows), and if K_pro is also positive, a larger desired velocity is next sent to the motor. If the actual speed is greater than the desired speed, the error signal is negative and a smaller command is sent to the motor, slowing it down. This loop repeats until the error signal is small enough so that the motor is considered controlled at its desired speed.

We mentioned earlier that the software controlling Rug Warrior's motors should implement two things. The first was that Rug Warrior should be able to maintain a desired velocity (whether climbing a ramp or traversing a flat space, for instance). The two separate proportional controllers just described for each motor essentially fulfill that requirement. We said that the second responsibility of the software would be to oversee that the two wheels would be slaved to each other. That is, if the robot were commanded to go straight, the velocities of the two wheels would be synchronized so that the robot really *would* go straight. This feature is implemented via the central feedback path of Figure 7.37, the integral controller.

The integral controller looks at the actual speeds of both motors and compares them. The difference between the two actual speeds is the error e_3, as can be seen at the right in Figure 7.37 where e_3 = left velocity − right velocity + bias. The bias term is used for inputting the turn command. While the bias is 0, the error signal only changes over time if the robot is not going straight but swerving one way or the other. An integral controller integrates, or sums, the error signal over time, multiplies this sum by a constant, K_integral; and feeds that new command back into the proportional-control loops for each motor. In this way, one motor is sped up while the other is slowed down until they each reach speeds sufficiently close together.

In Figure 7.38, we focus on just the upper third of the P-I controller diagram, the proportional-control loop for the left motor. We can implement the computation that this illustration conveys with a few simple **IC** routines. First, the data structure we will rely on for the input desired velocity is the function move() described earlier, which we constructed for our open-loop PWM controller. If the desired velocity was commanded by calling move(25.0, 50.0), this piece of the control system would try to servo the left motor such that every time get_left_vel() was called, it would return 25 counts.

To assist in the computations necessary to control the motor, we will create a variable, K_pro, and the function alter_power(). The input to alter_power is computed by left_error() which is the product of the difference between the desired velocity and the actual velocity (both in units of clicks per interval) and some constant, K_pro. alter_power() just calls pwm_motor() with this new velocity command. The main program then waits for a time interval, calls get_left_vel() again, and repeats the adjustment continuously. In this way, if the robot is servoing along a flat floor at one speed but then approaches a ramp and begins to climb, more power will be supplied to the mo-

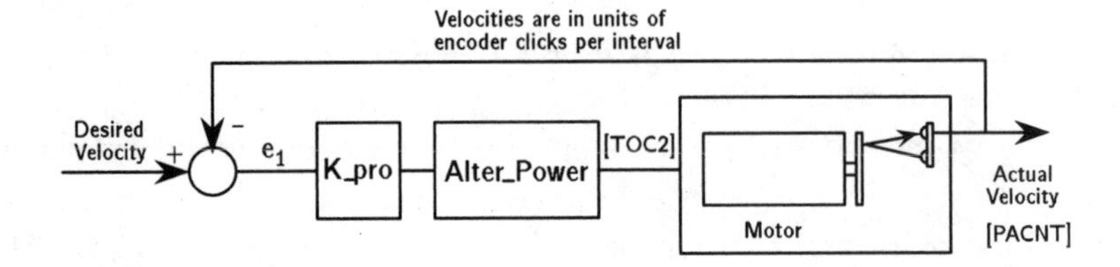

Figure 7.38. We focus on the top path of the P-I controller, which is the proportional-feedback control loop for the left wheel. The sequence of computations illustrated by this diagram are encoded in a few simple **IC** routines.

tor so as to keep the robot moving up the ramp at the same speed at which it was moving across the floor.

The **IC** program below illustrates both the proportional-control computation and the integral-control loop slaving the two wheels together:

```
float control_interval = 0.250; /* Run servo loop this often */
float des_vel_clicks = 0.0;      /* Desired vel, clicks/interval */
float des_bias_clicks = 0.0;     /* Desired bias, clicks/interval */
float power[2] = {0.0,0.0};      /* Power command to motor */
float integral = 0.0;            /* Integral of velocity difference */
float k_integral = 0.10;         /* Integral error gain */
float k_pro = 1.0;               /* Proportional gain */

/* Set and remember power level */
void alter_power(float error, int motor_index)
{ power[motor_index] = limit_range(power[motor_index]
     + error, 0.0, 100.0);
  pwm_motor(power[motor_index], motor_index); }

float integrate(float left_vel, float right_vel, float bias)
{ integral = integral + left_vel + bias - right_vel;
  return integral; }

void speed_control()
{ float left_vel, right_vel, integral_error,
    left_error, right_error;
  while (1)
    { left_vel = get_left_vel(); /* Get current vel */
    right_vel = get_right_vel();
```

```
    integral_error =
      k_integral *
      integrate(left_vel, right_vel, des_bias_clicks);
    left_error =
      k_pro * (des_vel_clicks - left_vel - integral_error);
    right_error =
      k_pro * (des_vel_clicks - right_vel + integral_error);
    alter_power(left_error, 0);
    alter_power(right_error, 1);
    sleep(control_interval); /* Run speed_control periodically */
    }}

void set_velocity(float vel, float bias) /* v,b:  [-100.0, 100.0] */
{ des_vel_clicks = k_clicks * vel;    /* Convert from vel as % */
  des_bias_clicks = k_clicks * bias; /*  to vel as clicks/interval */
  sign[0] = get_sign(vel - bias);    /* Sign of left vel */
  sign[1] = get_sign(vel + bias); }  /* Sign of right vel */


float k_clicks = 8.0 / 100.0;

void start_speed_control()
{ init_velocity();
  init_pwm();
  start_process(speed_control()); }
```

The integral controller works by representing the commanded robot velocity as two separate pieces of information, a common-mode desired velocity and a differential-bias velocity. That is, the desired velocity is the translational component and the bias velocity is the rotational component. Said another way, if the robot were commanded to go straight at 50% of full speed, its desired velocity would be [50.0, 50.0] and its bias velocity would be [0.0, 0.0]. This would coerce Rug Warrior to maintain a constant velocity of 50% of maximum speed, even as terrain or load on its wheels changed, as shown in Figure 7.39.

If the robot were commanded to spin in place about its right wheel at 35% of full speed, its desired velocity would be [0.0, 0.0] and its bias velocity would be [35.0, 0.0]. A command to arc forward and to the right would have both a desired velocity, say, [50.0, 50.0], and a bias velocity, say, [35.0, 0.0].

Figure 7.39. Rug Warrior is climbing up a ramp. Implementing a proportional-integral feedback controller keeps both wheels turning at the same speed and delivers more power to the motors as Rug Warrior begins to climb the ramp.

With this data structure for input, the integral-control loop adds the left velocity and the bias and subtracts the right velocity from that sum to calculate the error signal, e_3. The function integrate() accumulates this error over time, adding the new error to itself on each iteration. This running sum is multiplied by some constant, K_integral, and added into each motor's proportional controller. In this way, the new commanded velocity to each motor takes into account not only its own shaft-encoder's error signal but also the error signal between the two motors as it changes over time.

It becomes interesting now to play with the robot. Grab one wheel while the P-I controller is running, and try to keep it from spinning. The proportional control will try to raise the power level, and you will feel an increase in the torque output by the motor. If you hold the wheel tightly though, after a few moments, the other wheel will stop! This is because the program was not able to speed up the motor you were holding, and so the only way it could keep the two motors running at the same speed was to slow the other one down.

Try playing with the program in different ways. Change the values of the constants K_pro and K_integral. If these constants are made larger, the reaction time of the control loop will increase, but if you make them too large the system might become unstable and the motor will hunt, slowing down and speeding up but never converging on a steadily controlled speed.

Play with the time interval, too. The function speed_control is implemented as a process in **IC** that runs at a frequency specified by the variable control_interval. Changing the control interval modifies Rug Warrior's reaction time, also.

What we have implemented here on Rug Warrior, with a very minimal amount of hardware and an elegantly few lines of code, is a classical feedback-control system. These types of techniques have been well studied and are useful for a large number of problems. In Chapter 9, we will look at a different kind of control paradigm, a subsumption-style control system, which focuses on the problem of deciding which behaviors to select, given that many may be triggered from a large set of noisy and possible conflicting sensors.

7.9 References

A number of books which motor design and performance in great depth. *Electric Machinery*, by Fitzgerald, Kingsley and Umans (1990) gives a thorough treatment of the electromechanics of a wide variety of AC and DC motors. The three-volume set by Woodsen and Melcher (1985) *Electromechanical Dynamics*, delves into the physics behind the generation of electromechanical forces.

A comparative analysis of actuator technologies, spanning the range from electromagnetic motors to piezoelectrics and human muscle, can be found in the work of Hollerbach, Hunter, and Ballantyne (1991). They compare these alternatives from the point of view of applicability to robotics.

Our discussion of piezoelectric ultrasonic motors was rather brief. The piezoelectric ultrasonic motor of Figure 7.3 was made at the MIT Mobile Robot Lab by Anita Flynn. Further reading can be acquired in literature from a number of countries. Piezoelectric ultrasonic motors were invented by the Russians in the sixties (Ragulskis et al. 1988) and later commercialized by the Japanese (Sashida 1982). Recently, these motors have appeared in Japanese autofocus lens actuators (Hosoe 1989), paper-pushing actuators in copiers (Ohnishi et al. 1989) and as silent alarms in wristwatches (Kasuga et al. 1992).

Shape memory metals and artificial muscles are somewhat new to mobile robots. An informative booklet describing how to work with shape memory metals for small robots can be obtained from Mondo-tronics (1991). Artificial muscles also hold great promise for compact robotic actuators. Much of the pioneering work was done by Tanaka. Nice overviews can be found in Tanaka (1981) and Brock (1991).

For those interested in micromechanics, a review on silicon electrostatic microactuators can be found in the article by Howe, Muller, Gabriel, and Trimmer (1990). Progress in microfabricating ultrasonic motors and pumps can be found in papers by Moroney, White, and Howe (1989, 1990), Flynn et al. (1992) and Udayakumar et al. (1991).

Literature on power electronics, power MOSFETs, and motor-driver integrated circuits is available in application notes and data books of manufacturers such as Motorola, Supertex, Siliconix, and International Rectifier. The texts *Power MOSFETs*, by Grant and Gowar (1989) and *Power Electronics for the Microprocessor Age*, by Kenjo (1990) give excellent background on driving motors. For a practical guide to servo loops and interrupts, see Foster (1982).

Power

A mobile robot requires a power system that can meet several goals simultaneously. The power source must store energy sufficient to allow the robot to perform a useful amount of work. To ensure proper operation of the onboard electronic circuits, power must be provided at a constant voltage. Noise and power glitches produced by one circuit component must not be allowed to interfere with any other component.

8.1 Batteries

Batteries are by far the most common solution employed by mobile robots for the problem of energy storage. A battery converts chemical energy into electrical energy on demand. From the chemical nature of batteries stems a complex variety of properties. We begin with a synopsis of those properties and subsequently delve further into selected properties.

Rechargeability A battery that cannot be recharged is a *primary* cell. One that can be is a *secondary* or *storage* battery.

Energy density The maximum amount of energy per unit mass a particular battery technology is able to store is known as *energy density*. Energy density is usually measured in units of Watt-hours/kilogram (Wh/kg). Alternately, energy density can be measured in units of energy per unit volume.

Capacity Battery capacity is the energy stored in a cell. Capacity is usually reported in practical units of amp-hours or milliamp-hours. Capacity is the product of energy density and the mass of the battery.

Voltage The voltage produced by a single cell is characteristic of the particular chemical reaction occurring in the battery. Voltage also depends on the state of charge of the cell.

Internal resistance When short circuited, the current supplied by a battery is limited by its internal resistance. The internal resistance increases as the battery discharges.

Discharge rate This is the rate (in units of current) at which a battery is discharged. Maximum discharge rate is limited by the internal resistance of the battery.

Shelf life Batteries lose charge even when no external load is applied. Shelf life is a measure of how quickly this occurs.

Temperature dependence Most battery properties, in particular, available capacity and shelf life are affected by temperature.

An ideal battery would have very high energy density, maintain a constant voltage during discharge, have a low internal resistance, and therefore be capable of rapid discharge. It should also withstand temperature extremes, exhibit an unlimited shelf life, be rechargeable, and sell for a low unit cost. Unfortunately, no single battery technology exhibits all these characteristics. Thus, in practice, it is necessary to make trade-offs among these qualities, depending on the requirements of the task. The information in Figure 8.1 may serve as a guide when choosing the proper trade-off for your application.

Battery Chemistry	Recharge	Energy Density (Whr/kg)	Cell Voltage	Typical Capacity (mAh)		Internal Resistance (ohms)	Comments
Alkaline	No	130	1.5	AA	1400		Most common primary battery
				C	4500	0.1	
				D	10000		
Lead-Acid	Yes	40	2.0	1.2 - 120 Ah		C-size 0.006	Available in a wide variety of sizes
Lithium	No	300	3.0	A	1800		Execellent energy density, high unit cost
				C	5000	0.3	
				D	14000		
Mercury	No	120	1.35	Coin	190	10	
NiCd	Yes	38	1.2	AA	500		Low internal resistance, available from many sources
				C	1800	0.009	
				D	4000		
NiMH	Yes	57	1.3	AA	1100		Better energy density than NiCd, expensive
				4/3A	2300		
Silver	No	130	1.6	Coin	180	10	
Zinc-Air	No	310	1.4				High energy density but not widly available, limited range of sizes
Carbon-Zinc	No	75	1.5	D	6000		Inexpensive but obsolete

All numbers listed here are approximate. Precise values depend on the details of the particular battery.
Some values depend on the battery's state of charge, temperature, and discharge history.

Figure 8.1. Comparison of characteristics for selected batteries and sizes.

8.1.1 Chemistry

Choosing among the various battery chemistries may seem a daunting challenge. However, practical considerations dictate that most applications will use either alkaline cells, if primary batteries are required, or nickel-cadmium cells (NiCds) if rechargeables must be used.

Carbon-zinc batteries have been around for over 100 years, and although they have the lowest unit cost of all the batteries listed, they also have the lowest primary-cell energy density. Voltage changes by a large amount as these batteries discharge, internal resistance is high and performance at low temperatures is poor.

Alkaline manganese cells, commonly called *alkalines*, have higher energy density than carbon-zinc batteries. Internal resistance is also much lower. The cost of alkaline batteries is moderate, and they are widely available. They do, however, have a sloping discharge curve. (The discharge curve relates battery voltage to time as the battery discharges.)

The energy density of mercury and silver batteries is quite good, and they have other desirable properties as well. For example, they have very flat discharge curves. Their drawbacks are their generally higher prices and the fact that they are most readily available only in button or coin-sized cells.

Lithium batteries have by far the highest energy density of commonly available batteries—this alone makes them indispensable for certain applications. Lithiums also have a flat discharge curve and great shelf life—as much as 10 years. Lithiums power many autowinding cameras and have become easy to obtain; most photo stores carry them. These batteries do, however, have a higher internal resistance than alkalines and are much more expensive.

Sealed lead-acid cells are available in a variety of rectangular sizes. (Digi-Key has a good selection.) They are relatively inexpensive, have very low internal resistance, and can be recharged. Energy density is poor, however. Lead-acid cells have even less energy density than carbon-zinc cells.

Nickel-cadmium, or NiCd, cells are available in common AA, C, D and so-called 9 volt sizes. As such, they can be directly substituted for alkaline cells in most portable equipment. Cell voltage is less, however. A 9 V NiCd typically supplies only 7.2 V. NiCds have very low internal resistance, but energy density is comparable to that of lead-acid batteries. The operational constraints that must be observed when using NiCds are probably more severe than with other batteries. If a battery pack containing several NiCds is deeply discharged, the polarity of the weakest cells may reverse. NiCds also suffer from "memory" effect. If a NiCd is repeatedly discharged by, say, 50% of its rated capacity and then recharged, it will eventually begin to act as if it has only 50% of its original capacity. This condition can sometimes be fixed by discharging the battery completely (perhaps more than once) and then recharging it.

Driven in part by the demands of mobile computing, two new rechargeable technologies have recently become available. They are nickel-metal-hydride (NiMH) and Lithium-ion. NiMH batteries have many characteristics in common with NiCd cells. Although NiMH cells cannot supply surge currents quite as high as NiCd cells, the energy density of NiMH batteries is greater than that of NiCds. Unlike NiCd batteries, NiMH batteries show no memory effect. Also, NiMH cells, because they do not contain cadmium, pose less environmental risk when they are disposed than do NiCds. NiMH cells are more costly than NiCds but can withstand fewer recharge cycles than NiCds. Although voltage and charging characteristics are similar to those of NiCds, NiMH batteries will not necessarily work with battery chargers designed for NiCds. Lithium-ion cells have the highest energy density of rechargeable batteries and exhibit no memory effect. Lithium-ion batteries are also the newest and most expensive rechargeable batteries. Neither NiMH nor Lithium-ion batteries have yet seen widespread use in mobile robots.

8.1.2 Energy Density

The crucial parameter of any battery technology is *energy density*. Figure 8.2 graphs the energy density at room temperature of several commonly used batteries. To keep these figures in perspective, battery energy density is also compared with that of gasoline[1] and

[1]The comparison is marginally more favorable to batteries than it appears. Gasoline and nuclear fission both require a heat cycle to produce electricity—a process that is typically no more than 20% efficient.

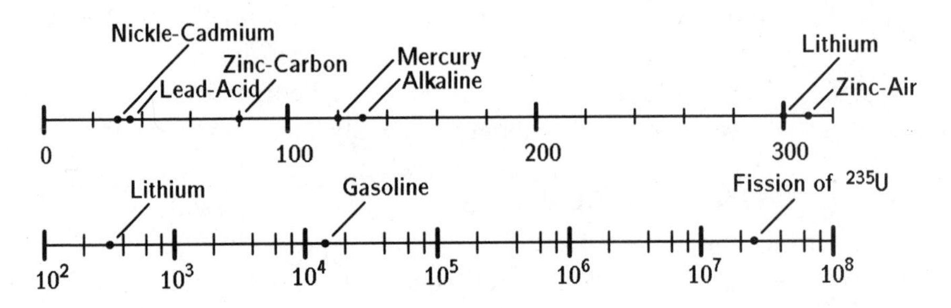

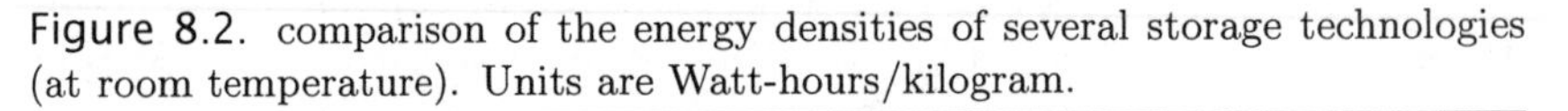

Figure 8.2. comparison of the energy densities of several storage technologies (at room temperature). Units are Watt-hours/kilogram.

the fission of ^{235}Uranium. This graph demonstrates the degree to which batteries are at a disadvantage compared to combustible fuels and nuclear energy storage techniques. It is also the case that the cost per kWh of energy delivered by batteries is much higher than that of other chemical storage mechanisms. Electrical energy supplied by a battery can easily cost 1,000 times as much as the same energy from your local electric utility. The redeeming feature of batteries is that they are highly mobile and provide the energy in the desired form: electricity. Other energy storage techniques require mechanical means (for example engines and generators) to produce electricity. Thus, batteries can be cost effective despite the high absolute costs of the energy they supply.

8.1.3 Voltage

Although it is desirable for a battery under load to maintain a constant voltage, typically, that voltage changes with the state of charge. How the voltage varies with charge is a property of the particular technology involved. Figure 8.3 presents a detailed, although still approximate comparison of the discharge characteristics of the four most common battery technologies. For example, as a one-cell, lead-acid battery discharges, its output goes from 2.1 V when freshly charged down to about 1.8 V when its capacity is effectively used up. A lithium battery, on the other hand, maintains a nearly constant voltage during discharge. The graph is normalized to the performance of the lithium battery. That is, if the lithium cell takes 1.0

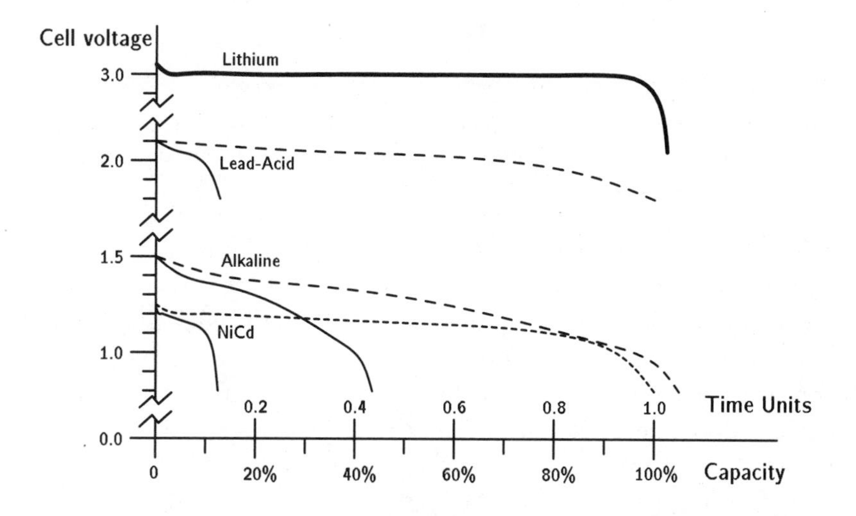

Figure 8.3. The discharge characteristics of the most common battery technologies are compared in two ways with those of a lithium battery. The figure assumes batteries of a similar size discharged at the same rate. The dashed lines show output voltage versus battery capacity consumed. The solid lines show voltage versus time. Time units are arbitrary, as we assume each battery discharges into the same load.

time units to exhaust its total capacity, the lead-acid battery is used up in about 0.1 time units. The dashed lines show more clearly how battery voltage changes as capacity is used up.

8.1.4 Capacity

Battery capacity, usually listed as some number of *ampere-hours* (informally, amp-hours) or milliamp-hours, can be misleading. Note that amp-hours is a practical term, not a proper unit of energy. (Amp-hours are equivalent to *coulombs*, the unit for charge.) To get energy, multiply the amp-hour rating by the voltage of the cell. This gives Watt-hours, which is a unit of energy:

$$(1\,\text{Watt} = 1\,\frac{\text{Joule}}{\text{sec}}).$$

Figure 8.4. A battery may be modeled as an ideal battery, one able to supply any current at constant voltage, in series with an internal resistance, R_i.

In general, the amount of energy that can be extracted from a battery depends on the rate at which the battery is discharged. At higher rates, the effective capacity will be reduced. The capacity published by the manufacturer assumes a favorable discharge rate—not necessarily a reasonable number for your project. Consult a battery data sheet for full information.

8.1.5 Internal Resistance

If the positive and negative terminals of a battery are shorted together, the current that flows is limited only by the internal resistance of the battery. A useful model of a real battery is a series circuit consisting of an ideal battery and a resistance, as shown in Figure 8.4. While the exact value of this resistance depends on a number of factors (such as battery age, charge, capacity, and temperature), different battery technologies have characteristic internal resistances. A small fresh alkaline cell, for example, may exhibit a resistance 10 times that of a similar-sized NiCd. Despite its lower energy density, this can make the NiCd more suitable for applications that require high surge currents. **It can also make the NiCd more hazardous to use. The current produced by a short-circuited NiCd may be enough to melt insulation and cause a fire.**

8.2 Recharging

Secondary cells are of particular importance to robots. A recharging circuit built into your robot can make it truly autonomous. All it needs to do when the power is low is to find an outlet and plug itself in. NiCd batteries offer advantages in this regard because of the simplicity of the circuits required to charge them. Figure 8.5

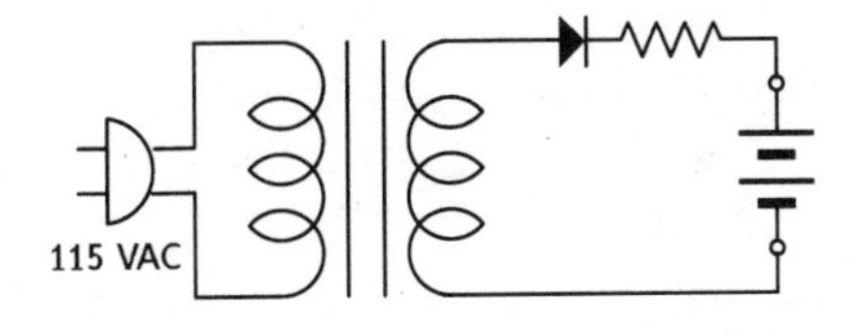

Figure 8.5. This figure shows a very simple charging circuit for a NiCd battery. Such a circuit is suitable for trickle charging.

shows such an example. This circuit may be safely used to charge a NiCd battery at a very slow rate; this is called *trickle charging.* If the capacity of the battery in amp-hours is C, then choose a resistor such that the current flowing into the battery is limited to $C/20$. Even if it is left connected to the charger indefinitely, the battery cannot become overcharged. Exactly this scheme is employed by many inexpensive rechargeable appliances.

A significant disadvantage of trickle charging is that it takes a very long time (several hours) for the battery to regain a full charge. More sophisticated battery chargers charge batteries at much higher rates. Battery manufacturers often include detailed instructions for designing charging circuits in their technical literature. A few companies manufacture integrated circuit chips that supervise the charging process; this makes battery-charger design very easy. Benchmarq Microelectronics and Integrated Circuit Systems produce such chips.

Stores and mail order companies that carry radio-controlled model cars and airplanes are a good source of both rechargeable battery packs and battery-charging equipment. Because of the mass market, prices are often lower for these items than for comparable generic batteries and chargers.

8.3 Power Regulation

As we saw in the previous section, the voltage supplied by a battery can change, sometimes by a large amount, as the battery discharges. One goal of any regulation scheme is to provide a constant output voltage, even if the input voltage varies over a wide range. Another goal of the regulator is to maintain a constant output voltage as the load changes. When motors and other actuators start up or reverse direction, they place a large transient demand for current on the power supply. The voltage supplied to the logic circuits must remain stable under these conditions. The requirements of the circuit may be such that several different voltages are necessary. It is generally desirable to supply all voltages from a single battery pack.

8.3.1 Avoiding Regulation

The regulation method employed by Rug Warrior is the simplest possible: it uses only circuit components that can operate satisfactorily over a wide range of voltages.[2] High-speed CMOS chips (the 74HC series of integrated circuits) are especially good in this regard. Such chips operate correctly when the positive voltage supply, V_{CC}, is between 2 V and 6 V. Many analog chips, such as Rug Warrior's LM386 microphone amplifier, can also accept a range of supply voltages. However, the operating characteristics of all chips (response times, for example) do vary with the voltage supply. Despite the choice made for Rug Warrior, it is good practice to include some form of voltage regulation in any circuit you design.

8.3.2 Linear Regulators

Because of its simplicity and low cost, the *linear regulator* is one of the most commonly used voltage regulators. A linear regulator, shown in Figure 8.6, is typically a three-terminal device: power-in, ground, and power-out. As long as the input voltage is higher than the required output voltage by a certain amount, called the dropout voltage, the output voltage will be constant as the supply voltage changes. For example, the LM7805 is a 5.0 V linear regulator capable of supplying 1.0 A of current. It has a dropout voltage of 2.0 V. As long as the input voltage is between a minimum of 7.0 V and a maximum of 35.0 V, the output will be a constant 5.0 V.

The relatively high dropout voltage of the LM7805 can cause a problem. Suppose we wish to power our robot using, say, five alkaline

[2] The new Rug Warrior Pro™ robot uses voltage regulation. Regulation is provided by a MAX603 low dropout regulator.

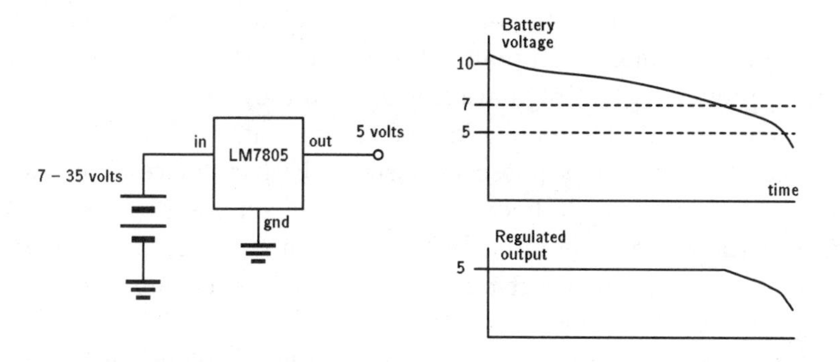

Figure 8.6. As long as the supply voltage is greater than the required output voltage by a characteristic amount, the linear regulator provides a simple solution to the problem of power regulation.

cells. The total voltage when the cells are fresh is 7.5 V. When exhausted, the voltage is about 5.0 V. However, when the LM7805's input falls below 7.0 V, it will no longer provide a regulated output. This occurs when only a small portion of battery capacity has been used up.

To solve this problem, we could simply use seven alkaline cells rather than five. Such an arrangement would give the LM7805 the input voltage it requires all the way to cell exhaustion. This, however, brings up another problem of the simple linear regulator: power loss.

If the supply voltage is V_{in}, the output voltage V_{out}, and the current output I, then the power, P, dissipated by the linear regulator itself is $P = I(V_{\text{out}} - V_{\text{in}})$.

In our example, the seven alkaline cells, when fresh, provide 10.5 V to the linear regulator. The regulator supplies 5.0 V to the robot's circuits, and 5.5 V is dropped by the regulator. If the current drawn by the robot's circuits is I, then the power consumed by these circuits is $5.0I$, and the power dissipated by the regulator is $5.5I$. This means that more than half of the power taken from the batteries is simply thrown away by the regulator.

Such waste poses no problem for a fixed installation, in which power comes from a cord plugged into a wall socket. But when batteries are used, we must take care to avoid wasting power whenever possible.

One improvement we could make would be to use a linear regulator with a smaller dropout. A standard LM2940CT-5.0, for example, provides good regulation when the input voltage is between 5.5 V and 26.0 V. The newer MAX603 offers an even lower dropout voltage and minuscule quiescent current. Either device allows us to minimize the difference between the voltage supplied by the batteries and the voltage required by the robot's circuits, thus reducing power waste. Low-dropout regulators are somewhat more expensive than the standard ones.

Power waste notwithstanding, a simple linear regulator is a good choice if the power requirement of the regulated circuit is only a small fraction of the total power the robot consumes. For example, if a 12.0 V battery supplies, say, 1 Ampere of unregulated current to the motors and maybe 50 mA at 5.0 volts to the microprocessor, the fact that $0.050 \times (12 - 5) = 0.35$ watts of power is wasted in the regulator will scarcely be noticed when compared to the power (12 watts) consumed by the motors.

8.3.3 DC-DC Converters

Linear regulators are capable only of supplying constant voltages that fall between 0.0 V and the battery supply voltage (less dropout). If a voltage higher than the battery voltage or a voltage with a polarity opposite that of the battery is required, another device must be used. Typically, this means a DC-DC converter.

Two distinct technologies are used to construct DC-DC converters. The flying capacitor, or charge pump-type converter, produces a voltage higher than or inverted with respect to the input voltage. It does this by charging capacitors in parallel and then discharging them in series to achieve a higher voltage or by connecting the charged capacitor with the polarity inverted to produce a negative voltage from a positive supply. Charge pump converters can thus produce an output voltage that is an integer multiple of the input voltage. The charge pump, however, only converts the voltage to a different value. Actual regulation may still require a linear regulator.

When current flowing through an inductor is interrupted, the collapsing magnetic field can induce a voltage much higher than that originally used to produce the steady-state current. With appro-

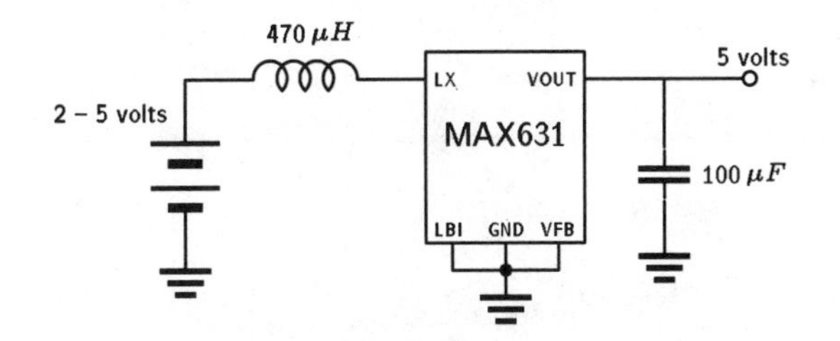

Figure 8.7. Maxim supplies a series of chips useful in building a simple DC-DC converter circuit.

priate switching and filtering, this transient voltage can be used to produce a constant output at a wide range of voltages either higher or lower than the battery voltage. This method is used in the other type of DC-DC converter, the switching regulator. By adjusting the switching parameters (how long current flows in the inductor versus how long the current is interrupted), the output voltage can be precisely regulated.

Switching regulators offer much better efficiency (often over 80%) than linear regulators, even when the input voltage differs greatly from the output voltage. The principle drawback of the switching regulator is cost. Although low-power devices (say, 20 to 100 mA) are comparable in price to low-dropout linear regulators, higher powered converters capable of delivering an Amp or more can easily cost over 10 times as much. Also, switching regulators can produce more electrical radio frequency noise than linear regulators.

There are two common ways to buy switching regulators: as discrete components or prepackaged units. Maxim produces a variety of integrated circuits that can simplify the construction of a switching regulator. The MAX631, for example, requires only two external components, an inductor and a capacitor, to produce a 40 mA, 5.0 V regulated output from an unregulated 3.0 V input, as illustrated in Figure 8.7.

The second approach to switching regulators is to purchase a unit with the switching circuit, inductor, and capacitor mounted in a convenient encased package. Such units are available from Digi-Key, Pico, and many other sources, as well. This format offers the same simplicity of design as is provided by a simple linear regulator

Figure 8.8. This photograph illustrates a number of types of regulators. On the left is an LM7805 5 V linear regulator. To its right is a Pico 5 V to 9 V DC-DC converter. Next is a Power Trends downconverting LMSX7805 DC-DC switching regulator. To the far right is a Pico IRE28D, a dual 5 V to 28 V DC-DC converter.

(although at a much higher cost). A number of different types of regulators and DC-DC converters are shown in Figure 8.8.

8.4 Isolation

Power supplies can often become noisy. For instance, when digital chips change state, they place a very brief demand for large amounts of current on the power supply. Similarly, each time the brush of a motor slides past a section of the commutator, a voltage spike is generated, which can find its way into the power supply circuitry. Often times, robots operate in an environment of electrical noise and changing magnetic fields (generated either internally or by external equipment).

All of these noise sources challenge the proper operation of the robot. The final goal of the power supply and distribution circuitry is, therefore, to isolate each component wired into the power supply from the interference produced by other components.

To combat the transient drain posed by state changes in digital chips, designers often connect small capacitors, say, 0.1μF, across the power and ground connections of each chip. It is generally more important to do this for high-speed memory chips. Figure 8.9 illustrates some different types of capacitors.

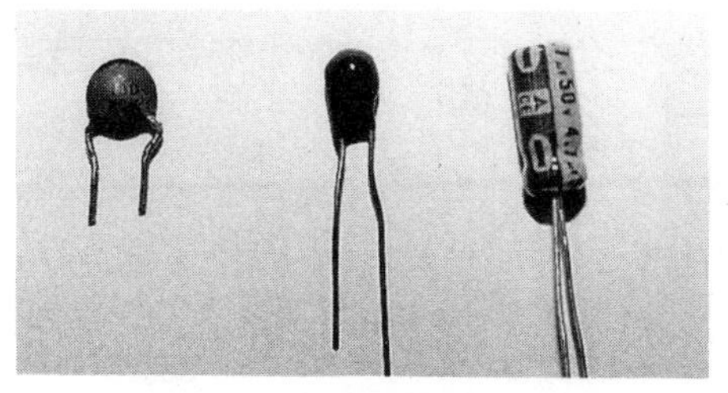

Figure 8.9. A $0.1\,\mu$F ceramic disc capacitor is at left. In the middle is a $10\,\mu$F tantalum capacitor, and at right is a $10\,\mu$F electrolytic capacitor. Tantalum and electrolytic capacitors are polarized and can be inserted into the circuit in only one direction.

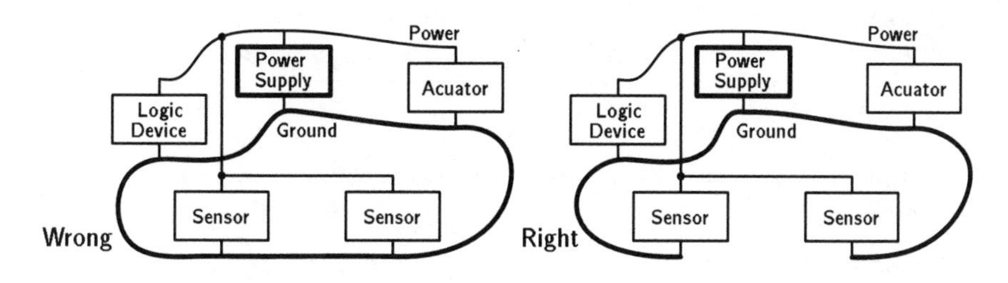

Figure 8.10. The layout of the power distribution circuitry should not contain ground loops.

The threat posed by stray magnetic fields can be countered by using what is called a *single-point ground*. Power distribution wires or printed circuit board traces must be laid out in such a way that no ground loops are formed, as shown in Figure 8.10. Changing magnetic fields induce a voltage in any wire loop they encounter. This can mean that components connected to different parts of a ground loop will not see a common reference voltage. That is, the "ground" of one component may actually have an instantaneous voltage higher or lower than that of some other component.

It is good practice to see that the power source separates the motor and logic components, as shown on the right side of Figure 8.10. This prevents any voltage drop in the distribution wires, caused by the high current demands of the motors, from affecting the logic components. Note also that, generally, it is not necessary to regulate the power going to the motors. To maintain a constant velocity from

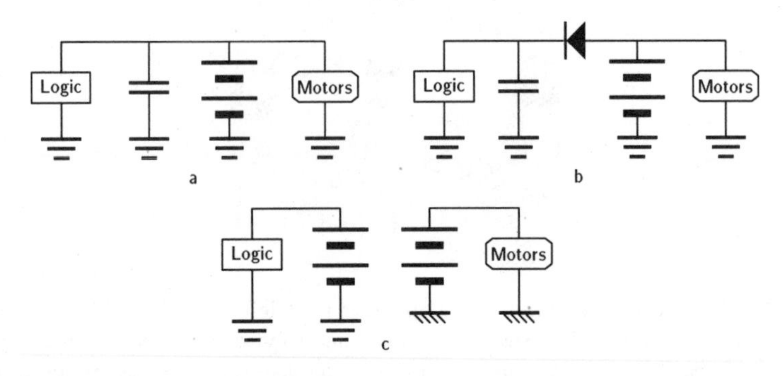

Figure 8.11. Some possible power supply configurations. (a) Logic power can be buffered from motor transients by a large capacitor. (b) The addition of a diode protects the logic against brief voltage dips when the motors demand high surge currents. (c) Having completely separate power supplies makes the robot bulkier but alleviates noise problems.

the motors when the input voltage falls, the pulse-width modulation circuit controlling the motors will simply remain on longer.

Much more difficult to solve than problems caused by switching digital chips and stray magnetic fields are the power glitches, voltage spikes, and voltage dips caused by motors. Motors act as virtual short circuits when they are first switched on. They try to feed power of the wrong polarity back into the circuit when their direction is reversed, and they can produce voltage spikes many times larger than the supply voltage each time a brush slides past a commutator section. Unless well isolated from the logic and sensor circuits, these effects will cause unreliable behavior. In general, the cheaper the motor, the more difficult the isolation problem.

Several resolutions are illustrated in Figure 8.11. In Figure 8.11(a) a capacitor protects the other circuits from motor spikes. This will work with high quality motors that produce little electrical noise. Figure 8.11(b) shows one possible way to guard against the voltage dips caused by a reversing motor. Even if the battery is unable to maintain a constant voltage under transient high-load conditions, the voltage seen by the logic remains high. The diode prevents the capacitor, in parallel with the logic circuits, from being discharged by the motors. This scheme may be helpful if the batteries in your robot have high internal resistance. If necessary, total isolation can

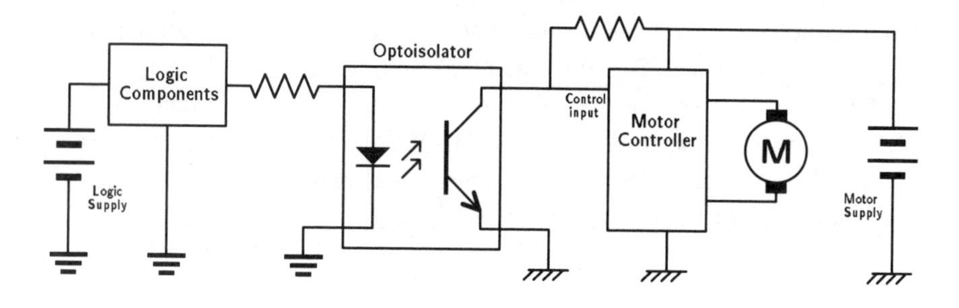

Figure 8.12. An optoisolator can provide complete electrical isolation among different parts of a circuit. Here, the output computed by the logic components is sent to a light-emitting diode (LED) embedded in an optoisolator package. The emitted light from the LED is detected by the optoisolator's phototransistor. The signal from the phototransistor is used to activate the motor-control electronics. There is no electrical connection between logic and motor power supplies.

be achieved by using separate power supplies for motors and logic, as shown in Figure 8.11(c).

For especially difficult isolation problems, the *optoisolator* offers an effective solution by making possible complete separation of the motor and logic power supplies. The optoisolator allows logic control circuits to be kept electrically isolated from the actuation circuits. Figure 8.12 illustrates the setup. The only connection between logic and power circuits is made by photons.

Clearly, in designing a power system for a battery-operated mobile robot, we must consider carefully the capabilities of the available battery technologies, the need for high-efficiency circuits and components, and the problems of isolating electrically noisy motors from sensitive logic components.

8.5 References

Information on the types of batteries available, along with details pertaining to recharging circuits, discharge rates, and the like, are usually found in manufacturers' catalogs. Check Appendix C for listings of suppliers.

Many of the power problems associated with mobile robots share the technology base with electric cars. Pratt (1992) gives an overview of issues and state-of-the-art electric vehicle design. Power supply and DC-DC converter design is covered extensively in the text by Kassakian, Schlect, and Verghese (1991). In a special issue of *Spectrum*, Riezenman (1992) discusses electrical vehicle efficiency and the shortcomings of battery technology.

9

Robot Programming

Advances on two fronts have brought mobile robots to the verge of what we believe will be a period of explosive growth. The first advance is in hardware. Progress in microcontroller and sensor technology has given us the hardware components we need to build useful and affordable robots. The second advance is in software. A new *behavior-based* approach to robot programming, called behavior control, is the subject of this chapter.

Behavior control allows us to tie together into a coherent whole all the elements of robot control we have discussed so far. An interesting side effect of this paradigm is that this method of integrating sensing and actuation can be accomplished using only modest computational resources.

The reason for this unforeseen benefit has to do with the way behavior control deals with sensors. The traditional approach to robot programming handles data in a manner known as *sensor fusion*, which is computationally intensive. A behavior control approach does not resort to sensor fusion but rather utilizes the notion of *behavior fusion*. In order to make these distinctions more clear, let us compare and contrast the two approaches. We begin with a historical perspective, the traditional approach to robot programming.

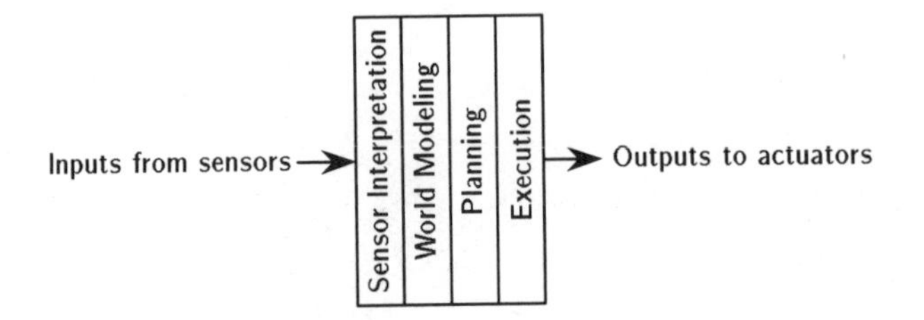

Figure 9.1. A robot program employing the modeling/planning paradigm is composed of a sequence of steps. These functional units transform a snapshot of sensory data into a series of actions intended to achieve a specified goal.

9.1 The Traditional Approach

A paradigm employed from the earliest days of robotics, and one that remains an active topic of research, is based on the ideas of world modeling and planning. This approach decomposes a robot program into an ordered sequence of functional components as illustrated in Figure 9.1.

First, data are collected from all sensors. Noise and conflicts in the data are resolved in such a manner that a consistent model of the world can be constructed. The world model must include the geometric details of all objects in the robot's world and their positions and orientations. Given a goal, usually provided by the programmer, the robot uses its model of the world to plan a series of actions that will achieve the goal. Finally, the plan formulated is executed by sending appropriate commands to the actuators. A sophisticated planner might even include sensory tests in the robot program it constructs. For example, "move gripper along the x-axis until a three oz. force is detected."

We will illustrate the world-model approach with a brief example using a modeling/planning system called HANDEY, which was dveloped at the MIT Artificial Intelligence Laboratory. HANDEY is a task-level planning system for manipulator-type robots that can solve the pick-and-place problem. The pick-and-place problem takes as inputs a model of the world, a part at some location and orientation, and the desired final location and orientation for that part. The pick-and-place problem then is solved if the program can compute

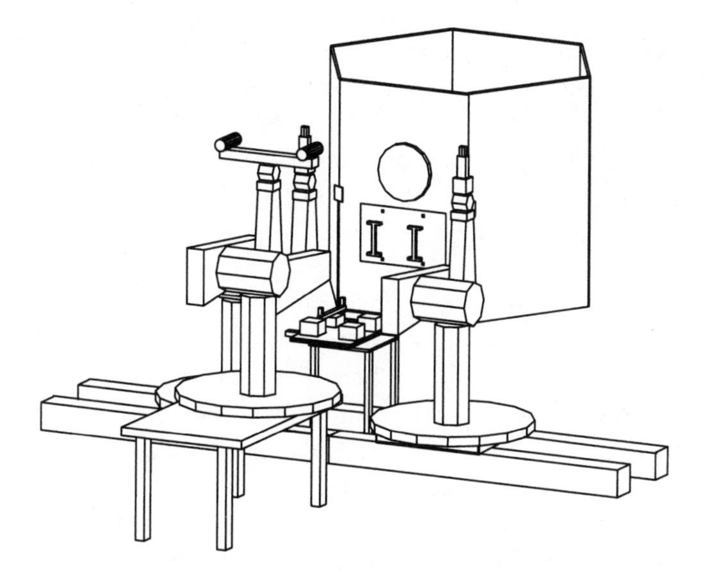

Figure 9.2. This is a world model used by the HANDEY task-level planning system. In this complex modeled environment (generated at the Jet Propulsion Laboratory), three manipulator-type robots cooperate to perform an operation on a satellite mock-up.

a detailed set of robot motions (and gripper openings and closings) that will move the part from its origin to its destination. Thus, the robot should be able to pick up the correct part and place it in its proper destination in some assembly.

Figure 9.2 illustrates a screen output from the HANDEY system as it directs two robot arms to geometrically reason about a pick-and-place problem. The real robot system first uses a laser scanner to identify the position and orientation of the part to be moved. It then incorporates this information into a geometric model of the world, provided by the programmer. The programmer also specifies the desired final position and orientation of the part. HANDEY then uses several sophisticated planners, first to plan how to grasp the part, and then to plan gross motions of the robot arm to move it into the vicinity close to the part.

It may be the case that no initial grasp of the part exists that is compatible with the geometrical and kinematic constraints at both the pickup point and the putdown point. For instance, it could be that the only available initial set of grasps would have the jaws of the two-fingered hand bump into another piece of the assembly when placing the part at its destination. If this condition is discovered, HANDEY plans a sequence of placements and regrasping operations. HANDEY directs the robot arm, first to put the part down on a clear part of the work table and then, to pick up the part in a more amenable position. The HANDEY system then generates another plan to direct the manipulator to finish the placement. HANDEY can also coordinate the motions of two robots using the same workspace.

Finally, after all plans have been formulated and all constraints satisfied, HANDEY executes the plan by sending a long series of commands to the robot. These commands specify precisely each small motion the robot must make and when to open and close the gripper.

This modeling/planning approach has strong appeal, principally because of the guarantees and optimizations it makes possible. There are planning strategies that, in a finite amount of time, will compute a sequence of motions *guaranteed* to accomplish the task or prove that the proposed task is impossible. In addition, a successful plan can be optimized before the robot makes any motions.

Such guarantees would also have appeal in the mobile robot domain. For instance, a mobile robot that used such global information about its world to formulate a plan would never fall into the trap of following a path to a dead end and then having to backtrack. Instead, it would always choose the most direct route from start to goal.

Unfortunately, the modeling/planning approach has some disadvantages. As the following sections will explain, these problems are accentuated for mobile robots that operate in natural or changing environments.

9.1.1 Computation

One drawback to world-model schemes is that they require large amounts of data storage and intense computation. This drawback is not necessarily a concern for a manipulator-type robot, but it can be for a mobile robot, which must carry its computational resources on its back. The HANDEY program is composed of over 100 high-level

Lisp files and requires a powerful computer with several megabytes of RAM to perform satisfactorily. Because the natural world is enormously rich in detail, schemes to represent it simply require a large number of bits. All world-model systems therefore simplify the world to make storage and manipulation of the model practical. HANDEY is restricted to dealing with polyhedra. Any curved surfaces in the world must be approximated by collections of flat surfaces.

9.1.2 Modeling

Many of the advantages of the modeling/planning approach come from its ability to use global information. A program that takes into account all relevant information can be expected to produce better results than one that makes all decisions based on local (i.e., only some) information.

It is the internal representation of the world that makes possible the use of global information—but problems occur in the construction of this model. For a plan to be reliable, the model on which it is based must be highly accurate. This requires high-precision sensors and careful calibration, both of which are expensive. Even the best available sensors suffer from several difficulties. Sensor data are unavoidably noisy. Sensors are subject to systematic errors, and different sensor technologies often produce conflicting results when measuring the same quantity. For example, sonar and infrared ranging systems may give different distance readings due to the surface properties of the objects at which they are aimed.

Typically, a modeling/planning algorithm must devote considerable resources to figuring out the most likely interpretation when presented with inconsistent data from a single sensor and conflicting data from multiple sensors. This general idea of combining data from multiple sensors into one data structure, the world model, is known as *sensor fusion*.

Some planning programs (HANDEY, for example) rely on the programmer rather than sensors for building most or all of the world model. Synthesizing a world model can reduce the burden of interpreting sensor data, but unfortunately doing so can also limit the robot's ability to respond autonomously to changes in its environment.

9.1.3 Time

The modeling/planning paradigm is by nature sequential. The approach first takes a snapshot of the world, then processes the acquired information, and then acts. If the world happens to change between snapshot and action, the plan may fail.

Trying to make the actions of such a program more intelligent may produce undesired results. The more time the program devotes to resolving conflicting sensor data, to refining its model of the world, and to optimizing its plan, the longer will be the delay between sensing and acting. This delay increases the chance that a significant change will occur in the world, thus invalidating the plan.

9.2 Behavior Control

As a result of work by Professor Rodney Brooks and the Mobile Robot Group at the MIT Artificial Intelligence Laboratory, a promising alternative to the modeling/planning paradigm has gained wide acceptance in recent years. Brooks' subsumption architecture provides a way of combining distributed real-time control with sensor-triggered behaviors. Subsumption architecture, instead of making explicit judgments about sensor validity, uses a strategy in which sensors are dealt with only implicitly in that they initiate behaviors. (Brooks labeled his approach subsumption architecture but this style of programming is now more commonly refered to as *behavior control* or *behavior programming*.)

Behaviors are simply layers of control systems that all run in parallel whenever appropriate sensors fire. The problem of conflicting sensor data then is handed off to the problem of conflicting behaviors. *Fusion* consequently is performed at the output of behaviors (*behavior fusion*) rather than the output of sensors. A prioritized arbitration scheme is used to resolve the dominant behavior for a given scenario.

Note that nowhere in this scheme is there a notion of one behavior calling another behavior as a subroutine. Instead, all behaviors actually run in parallel, but higher-level behaviors have the power to temporarily suppress lower-level behaviors. When the higher-level behaviors are no longer triggered by a given sensor condition, how-

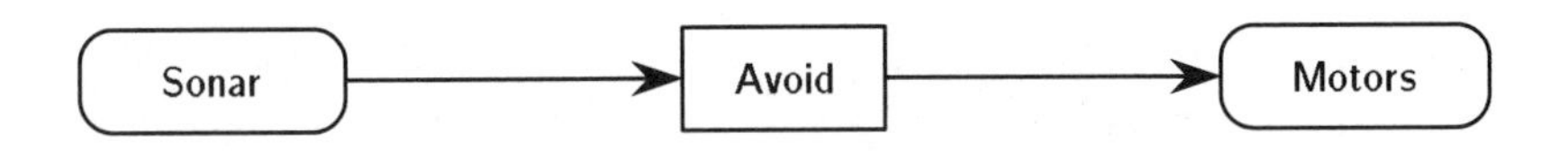

Figure 9.3. A block diagram of a simple behavior control program. Rounded boxes represent both the physical sensors or actuators and the software drivers that directly control them. The square-cornered box contains code that performs computations that transform sensor readings into actuator commands. Arrows can be thought of as wires that allow the components to communicate.

ever, they cease suppressing the lower-level behaviors and the lower level behaviors resume control. Thus, the architecture is inherently parallel and sensors interject themselves throughout all layers of behaviors. There is no unified data structure or geometric world model.

In order to understand these ideas more vividly, let us imagine some behaviors we could create on a mobile robot with a sensor and actuator suite similar to a suite we might implement on Rug Warrior.

9.2.1 Behavior Networks

Let's say our robot is equipped with a ring of sonar sensors, a top-mounted infrared detection system, and a low-powered microprocessor with a small amount of RAM. Let's also say that, at a minimum, we would like Rug Warrior to be able to avoid bumping into things. To achieve this goal we could create a behavior control program that consists of three parts as shown in Figure 9.3.

In the figure, Sonar is a software driver that operates the sonar sensors, continuously keeping track of the distance each measures. Motors is a software driver that sends the proper current to the motors in response to commands it receives. In between is a module called Avoid, which, based on the sonar data, constantly computes commands and sends them to Motors.

The Avoid module contains code that implements a simple reflexive behavior. If the reading from the front-pointing sonar is too short, Avoid stops the robot's forward motion. If a sonar other than the rear one measures the shortest distance, Avoid turns the robot until the rear sonar points in the direction of the shortest reading. When the rear sonar does measure the smallest distance, the motors are commanded to move forward. If all sonar readings are larger than some threshold, Avoid does nothing. It sends no commands to the motors.

These operations are simple enough that a processor no more powerful than an MC68HC11A0 can execute all the code in this structure many times each second. The behavior that emerges is one in which the robot tends to maintain a minimum distance between itself and all objects visible to its sonar sensors. With this tight coupling of sensing to actuation, Rug Warrior can respond quickly to changes in the world. If someone walks up to Rug Warrior, it will turn and move away.

Avoid is an example of a *task-achieving* behavior. Useful in and of itself, it provides the minimum level of competence we want Rug Warrior to exhibit. Next, we will illustrate how more sophisticated behaviors can be added on top without redesigning lower-level behaviors already in place.

Suppose we write a second behavior, called Dock, whose purpose is to drive the robot into its charging stand when the batteries are low. In this case, Dock takes input from the robot's sonar sensors, infrared detector, and battery-level indicator.

Let's assume that the charging stand is identified by a coded IR beacon placed on its top. When active, Dock computes motor commands that will steer the robot toward the charger, ultimately docking with it, while avoiding obstacles other than the charger.

We now have two task-achieving behaviors, Avoid and Dock, which, under some circumstances, will contend with each other for control of the motors. Figure 9.4 shows one way to resolve this conflict. We have broken the wire connecting Avoid with Motors and inserted a *suppressor node*, which is represented by the "S" in the circle.

A suppressor node allows messages from the original wire to pass through to the output, unless a message arrives at the same time from the new connection, the arrowhead. Figure 9.5 shows the series of messages that might be produced by Dock and Avoid as Rug Warrior first moves about its space and then approaches and docks with the charging stand. Note that messages sent to the inferior connection that are suppressed are not saved up and transmitted later. They are simply lost.

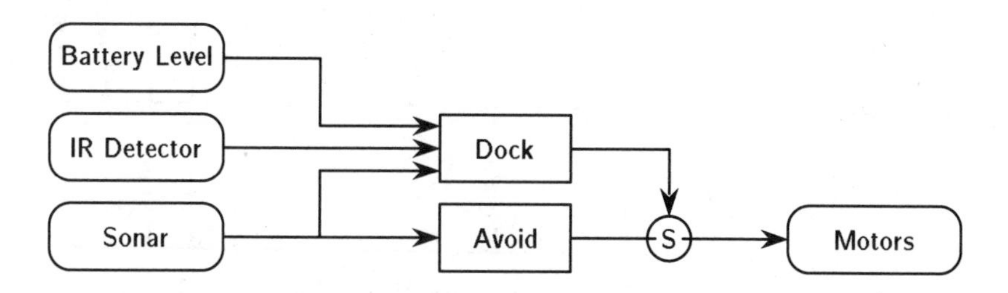

Figure 9.4. Dock is a program that looks for an IR beacon located atop a charging stand and drives the robot toward it. Additionally, Dock monitors the sonar sensors and suppresses the obstacle-avoidance behavior when close to the charger.

With Avoid and Dock connected, as shown, Rug Warrior behaves in the desired way. As long as the batteries are fully charged, the robot will avoid collisions with all obstacles as it moves about its environment. When the charge falls, Dock will become active. It will direct Rug Warrior toward its charging stand, responding to some sonar measurements but suppressing Avoid's attempts to turn away from the charger. Thus, Dock *subsumes* the function of Avoid in order to produce a higher-level of competence.

This style of robot programming, where the robot's control system is decomposed into a network of task-achieving behaviors, is the essence of behavior control.

Behavior control has a number of significant implications for programming robots. The tight coupling of sensing to actuation means that most behavior modules can be thought of as simple reflexes. This is important because such a system needs no world model. Because there is no world model, the robot needs very little memory. Most computations are uncomplicated and can be performed by simple microprocessors.

Another powerful feature of a behavior programming-style organization of a robot's intelligence system is that it can be improved incrementally. New layers of competence, in the form of additional behaviors, can be written and then simply wired into the existing structure. Basic capabilities are never lost as new ones are added.

Finally, the robot need not get slower as it gets smarter. Because the behavior control paradigm has all behaviors run in paral-

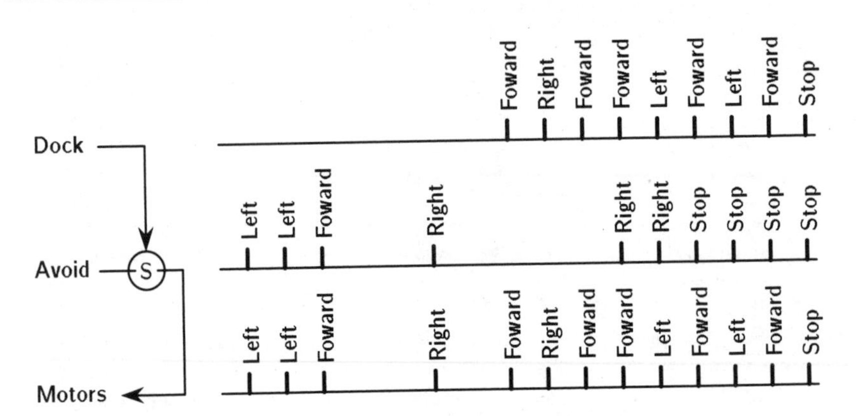

Figure 9.5. Behaviors Dock and Avoid produce a stream of messages that join at a suppressor node. Messages entering through the dominant connection (the arrowhead) suppress messages from the inferior connection. That is, only the dominant messages appear at the output. When no messages are present at the dominant connection, those from the inferior connection pass through.

lel, increased computational requirements of an improved behavior program can always be met by adding more processors to carry the load. The performance of the existing system need not be degraded. The robot designer is free to implement the behavior controller in several ways: as a number of behaviors in a single computer, as a single processor devoted to each behavior, or perhaps as a network of very large-scale integration (VLSI) gates.

9.3 Rug Warrior's Program

We will expand on the principles of behavior control further by using another example of how Rug Warrior might be endowed with a set of interesting behaviors. This time, let us assume that Rug Warrior has the set of sensors available on Rug Warrior Pro™: three bump sensors on a surrounding bump skirt, two near-infrared proximity sensors, two photocells, and a microphone. Figure 9.6 illustrates a number of behaviors we could program into Rug Warrior.

As implied by the diagram, the behaviors operate in parallel. Each module continuously examines its input and computes an output. The simplest module is Cruise, whose purpose is to make sure the robot always does something interesting. That is, Rug Warrior

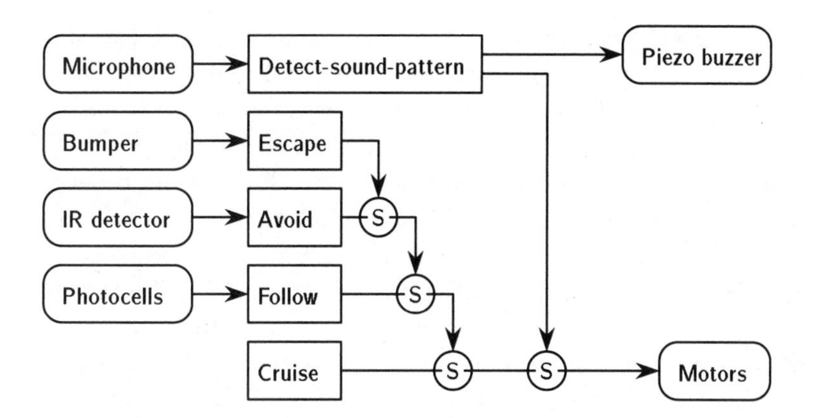

Figure 9.6. A possible behavior control program for Rug Warrior begins with a behavior called Cruise which merely causes Rug Warrior to move forward. Follow is triggered by photocells to move Rug Warrior towards light. Avoid suppresses Follow and Cruise when the near-infrared sensors detect an imminent collision and Escape also helps avoid obstacles if the near-infrared sensors were blind to an obstacle. The highest-level behavior, Detect-sound-pattern causes Rug Warrior to trigger on specific patterns of hand claps and then play a tune.

should always move. All that Cruise does is send messages to the motors, commanding them to go forward.

The Follow behavior is a higher-level behavior that monitors the output of a pair of forward-pointing photocells. When Follow detects a difference in intensity between the two photocells, it will send commands to the motors to turn Rug Warrior in the direction of the brighter side.

Commands from Follow fuse with those from Cruise via the dominant connection to a suppressor node. This means that, whenever Follow sends a command, it will take precedence over commands sent by Cruise. The behavior that will emerge from only this much network is that Rug Warrior will move forward until it senses a light and then will home in on the source of illumination.

The next behavior we add is Avoid which looks at the output of the near-infrared obstacle-detection system. When Avoid senses an obstruction to the left of the robot, it will command a turn to the right. When the obstacle is to the right, Avoid will turn Rug Warrior to the left. An obstacle straight ahead will cause Avoid to issue commands that will stop Rug Warrior, then turn it 90 degrees either to the right or the left. Again, commands from Avoid suppress commands from the behaviors below it. With only these three layers implemented, Rug Warrior will follow a light until it detects an obstacle in its path. Rug Warrior will then turn away.

If the infrared detectors fail, as they will for some objects, we add the Escape behavior. The Escape behavior will become activated when the bump skirt detects a collision. Escape reacts to collisions between obstacles and the robot's force-sensing skirt by commanding motions that will move Rug Warrior away from the obstacle. Messages from the Escape behavior are of the most immediate importance to Rug Warrior. The architecture of the behavior control system thus allows Escape to suppress commands from Avoid, Follow, and Cruise.

Lastly, we could implement a behavior that listens through the microphone. The Detect-sound-pattern behavior is programmed to detect specific sequences of hand claps and pauses. When Detect-sound-pattern notices such a sequence, it will send a command to the piezoelectric buzzer to play a particular tune. Detect-sound-pattern will also send a message to the motors, directing them to stop.

The overall effect of these five behaviors is that Rug Warrior will first speed forward, searching for the brightest source of illumination. As Rug Warrior heads toward the light, it will tend to avoid obstacles in its path. If Rug Warrior does collide with something, it will change direction and move away. When the designer claps his or her hands in a special sequence, Rug Warrior will stop, play a tune, and then resume wandering.

9.4 Implementing Behavior Control

How do you implement a network of many behaviors, all running in parallel on a small microprocessor that is inherently a sequential machine? The answer is to *multitask*, or run a loop that, when repeated, gives a small amount of time to each behavior. In this way, we can simulate the effect of all behaviors running simultaneously.

Before we jump into the details of explaining such a strategy, let us step back for a moment and understand more fully how we think of a behavior control network. In this section, we will describe a formalism for specifying a behavior control architecture. Then later, in Section 9.5, we will explain in more practical terms how you can apply the principles of behavior control while programming a robot in a conventional language, **IC**.

Before proceeding, though, we must first introduce three useful concepts: processes, schedulers, and finite-state machines.

9.4.1 Processes and Schedulers

First, we illustrate the concept of a process with an example of a robot flashing some light-emitting diodes (LEDs).

Suppose we have a software driver called `flash_leds` and a function called `sleep`. When activated, `flash_leds` turns on a set of LEDs briefly and then turns them off. The function `sleep` simply does nothing for some number of seconds. Using these tools, we could write the following **IC** function:

```
void multi_flash()
{ while (1) {              /* while (1) means loop forever */
    flash_leds();
    sleep(1.0); }}         /* Do nothing for 1.0 second */
```

The operation of `multi_flash` is easily understood. It will flash the LEDs, wait for one second, flash the LEDs again, and so on. There is a problem here, however. Once `multi_flash` begins to run, the microprocessor can do nothing else; `multi_flash` is the only code that can be executed. Since we would like Rug Warrior to do more than just flash its LEDs, we need to activate `multi_flash` in such a way that it does not consume all the resources of the microprocessor.

One way to do this is to make `multi_flash` into a *process*. A *process*, or *task*, is a piece of code that can be thought of as running simultaneously with other processes or programs. While the computer can only do one thing at a time, it is, nevertheless, possible to give the appearance that different pieces of code are running in parallel. This requires a supervisory program called a *scheduler*.

A *scheduler* is a master program that decides when all other programs are allowed to run. A scheduler gives exclusive control of the computer to one process for a brief period of time (typically, a small fraction of a second) and then gives control to the next process and so on. Each process is allowed to compute for a short time at regular intervals. This is known as *multitasking*.

A moderate level of sophistication is required to construct a scheduler capable of interrupting a task after a given time and then loading and executing another task (*preemptive multitasking*). A simpler strategy for switching between processes is called *cooperative multitasking*. In cooperative multitasking, it is *the process* that decides when to return control to the scheduler so that the next task may run.

In cooperative multitasking, the scheduler is simpler but the processes are more complicated. The reason that the processes must be more complicated is that each process must provide a way to resume computing at the place it left off when it last returned control to the scheduler. An effective approach for passing control between the processes and the scheduler (and the approach employed in Brooks' original subsumption architecture implementation) is to implement each process as a finite-state machine.

Next, we describe the concept of finite-state machines, as this mechanism will be useful for understanding how to construct your own cooperative multitasker. Later, we will use **IC**'s preemptive multitasker. (**IC** has its own scheduler for choosing when to run a process.) This makes it easy for us to describe behaviors, as each behavior then does not have to take care of the bookkeeping chores of releasing control back to the scheduler.

9.4.2 Finite-State Machines

In the absence of a sophisticated scheduler, it is possible to build a behavior control program by implementing the behaviors as finite-state machines. Even if such a scheduler is available, it may be helpful to think of behaviors in this way. A finite-state machine (FSM) is an abstract computational element which is composed of a collection of states. Given a particular input, a finite state machine may change to a different state or stay in the same state. The specification of an FSM includes rules that determine the relationship between inputs and state changes.

Figure 9.7 diagrams a possible finite-state machine representing the operation of a turnstile. The turnstile finite-state machine has

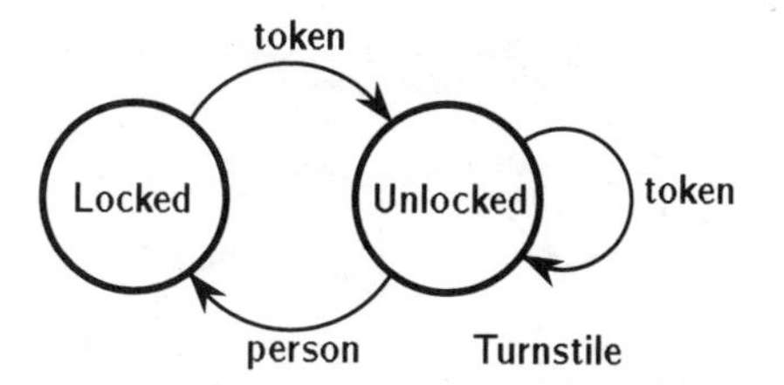

Figure 9.7. The finite-state machine diagram of a turnstile might consist of the states Locked and Unlocked, along with some conditions for transitions between the states. Here, circles in bold represent the states of the system, while labeled arrows indicate inputs and their corresponding effects on the system state.

two states, locked and unlocked. The turnstile finite-state machine accepts two forms of input, tokens and people. While in the locked state, inputting a token will change the FSM to the unlocked state. In the unlocked state, the turnstile finite-state machine will accept any number of additional tokens and remain in the unlocked state. From the unlocked state, the turnstile FSM will also accept the input of a person. This input will change the FSM back to the locked state and no further person inputs will be allowed.

9.4.3 A Behavior Control Formalism

The example of the turnstile gives us the general flavor of a finite state machine. To be more explicit, we can write a program that implements a finite-state machine. To do this, we begin by using *pseudocode*, which is not the syntax of any particular programming language. Pseudocode is used here to present a formal, explicit representation of how a finite-state machine should act.

Behavior as a Finite-State Machine

Suppose we wish to construct a behavior that causes Rug Warrior to respond appropriately when it strikes an object. Let's call this behavior Escape. Escape should monitor the bumper and cause Rug Warrior to back up and turn as required to move away from the object. We can implement this behavior as a finite-state machine. This behavior is illustrated in Figure 9.8 and described formally by the following structure:

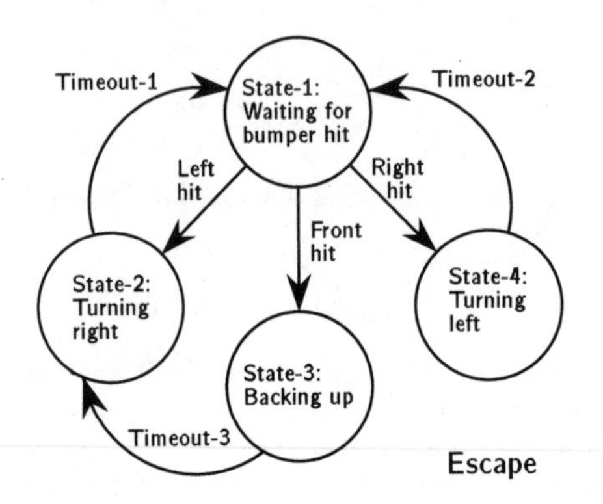

Figure 9.8. Rug Warrior's Escape behavior is diagrammed here as a finite-state machine (FSM). In State-1, the behavior waits for an input from the bumper. While in this state, the behavior issues no motor commands. When a bumper hit does occur, the FSM changes to State-2, 3, or 4, depending on which direction the bump came from. While in the backing up, turning left, or turning right states, motor commands are constantly issued. From any of these states, after a certain time has passed (Timeout-1, 2, or 3), the FSM switches states again.

```
Escape
  Outputs: (Motor-command)
  State-1: If Bumper-Hit = Nil
             Release
           else if Bumper-Hit = LEFT
             Switch to State-2
           else if Bumper-Hit = RIGHT
             Switch to State-4
           else Switch to State-3
  State-2: If time-in-this-state > timeout-1
             Switch to State-1
           else
             motor-command = turn-right
             Release
  State-3: If time-in-this-state > timeout-3
             Switch to State-2
           else
             motor-command = back up
             Release
  State-4: If time-in-this-state > timeout-2
             Switch to State-1
           else
             motor-command = turn-left
             Release
```

When initially called by the scheduler, the finite-state machine Escape will be in State-1. The code that implements this FSM checks a software driver called `Bumper-Hit` to determine what sort of collision, if any, has occurred. If a bump did not occur the Release statement will return control to the scheduler without changing the state of Escape. This means that, the next time the scheduler runs Escape, it will still be in State-1 and the same sequence of operations will occur again.

Eventually, `Bumper-Hit` will return a non-Nil value. When this happens, Escape will switch its state to one of the other states. For instance, if the left bumper hit, control would pass to State-2. The Escape FSM would then execute the body code associated with State-2. If State-2 were to keep control for an amount of time greater than Timeout-1, then control would switch back to State-1. Otherwise, the turn-right motor command would be issued, and control would be released back to the scheduler.

State-4 is similar to State-2 except that State-4 commands a left turn and remains active for a different amount of time, Timeout-2.

State-3 implements the back up phase of the Escape behavior. It commands the robot to backup for a period equal to Timeout-3 and then switches to State-2. State-2 then responds just as it would if activated from State-1. It makes the robot turn right until the period Timeout-1 expires.[1]

We can now represent the general format of a behavior module implemented by a finite-state machine as follows:

Behavior
    Inputs: $I_1, I_2 ... I_n$
    Outputs: $O_1, O_2 ... O_n$
    Local-variables: $L_1, L_2 ... L_n$
    State-1: {Body-code-1}
    State-2: {Body-code-2}
        ⋮
    State-N: {Body-code-N}

[1]Strictly speaking, these timed operations do not fit the definition of a finite-state machine. Rather, we must think of the structures described here as enhanced or *augmented* finite-state machines.

That is, the body code may compute any arbitrary function and may read local inputs and compute local outputs. The body code must compute the next state and explicitly release control back to the scheduler. This strategy places a burden on the code in each FSM. The code in each FSM must release control fairly quickly. Any FSM that hogs too much time will lock out all the other finite-state machines.

The Scheduler

Once finite-state machines have been defined, how does the scheduler manage to run them all in parallel? The scheduler of a cooperative multitasker is quite simple, as the following format illustrates:

```
Scheduler
   Call Behavior-1
   Call Behavior-2
         ⋮
   Call Behavior-N
   Call Arbitrate
```

The scheduler for a cooperative multitasker simply loops indefinitely, calling each behavior in turn. The active behavior computes for a certain time and then returns control to the scheduler. Once during each loop (at least), the scheduler calls an Arbitrate function to pass messages and resolve conflicts between competing behaviors.

Arbitration

The connections between behavior modules in a behavior control network are specified by a wiring diagram. To connect our Escape finite-state machine to our Motor finite-state machine, the behavior module that directly controls the motors, we might say:

```
Connect
     Output: Escape, motor-command
     Input: Motor, command-in
```

Escape and Motor both have inputs and outputs (called motor-command and command-in, respectively) which are stored locally.

The Arbitrate function sees to it that, whenever Escape computes a new value for motor-command, that value is transferred to the command-in variable of Motor.

The same output can be connected to any number of inputs, and we can implement suppression nodes by ordering the Connect statements. For example, if we order Connect statements as shown below, the second Connect statement will be given higher priority:

```
Connect
    Output: Behavior-1, B1-out
    Input: Behavior-2, B2-in


Connect
    Output: Behavior-3, B3-out
    Input: Behavior-2, B2-in
```

If, on the same scheduler loop, Behavior-1 computes a value for B1-out and Behavior-3 computes a value for B3-out, the arbitration code would make sure that only Behavior-3's value reached Behavior-2's B2-in input.

An implicit characteristic time has now entered the picture. If one passage through the scheduler loop is thought of as a single tick of the system, then one message may suppress a second message if the second message arrives within one tick of the first message. A careful implementation of behavior control will make this characteristic time explicit. Values other than one tick may be chosen.

There are other types of arbitration mechanisms than the suppression nodes described above. Brooks' original subsumption implementation also uses inhibit nodes. An *inhibit node* functions as a switch. Messages that enter through the dominant connection do not replace messages from the inferior connection; rather, they prevent the inhibit node from transmitting the message from the inferior connection.

We have now described the mechanism of a finite-state machine and the functioning of a cooperative multitasker. Next, we illustrate a preemptive multitasker and give an example implementation using **IC**.

9.5 Behavior Control in IC

In the previous section, we used pseudocode to present a formalism for thinking about behavior control programs and for instantiating robot behaviors as finite-state machines. Thinking of behaviors as finite state machines gives us a more or less simple way to program our own cooperative multitasking system on computers that lack true multitasking capability. With cooperative multitasking, the design of a task scheduler become effortless. The scheduler is simply a looping sequence of calls to the behaviors. The behaviors, although more difficult to program since they must cooperate with the scheduler, are still tractable.

In this section, the multitasking feature built into **IC** will act as our task scheduler. The behaviors will then be easier to follow, as they need contain no bookkeeping code to handle giving up and resuming program control. Let's now walk through how we might implement a behavior control program for Rug Warrior using **IC**.

Recall the earlier example of `multi_flash` on page 295. We wish to activate `multi_flash` as an **IC** process so that we can use the microprocessor to run more than just that one program. **IC** gives each process a unique identification number so that we have a "handle" for use later to terminate the process, if desired. To start `multi_flash` as a process in **IC**, we could say:

```
int id;
id = start_process(multi_flash());
```

Rug Warrior would then begin to flash its LEDs. While this process is running, we are free to type statements to the microprocessor to run a program or to start other processes. Through it all, the LEDs will continue to flash.

If at any time, we say:

```
kill_process(id);
```

the flashing will stop. This command could also be issued by another process.

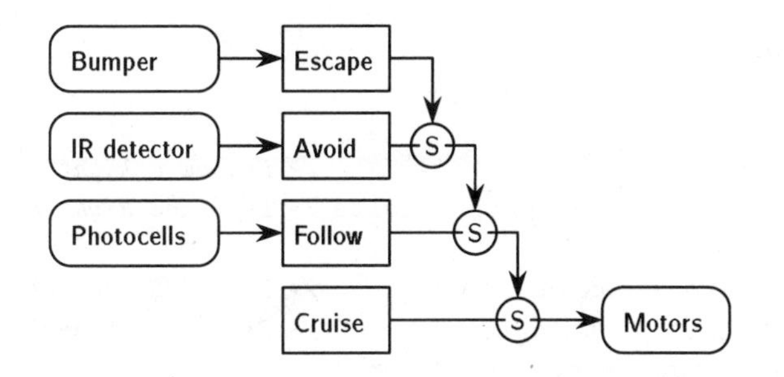

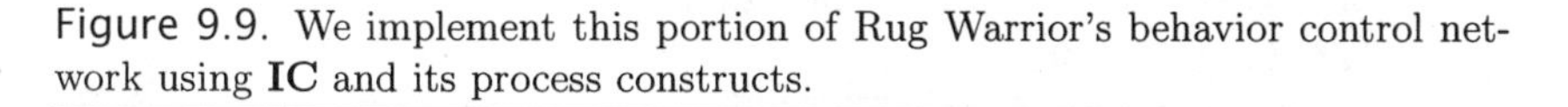
Figure 9.9. We implement this portion of Rug Warrior's behavior control network using **IC** and its process constructs.

Later, we will use the start_process function to turn on the Rug Warrior behaviors outlined in Figure 9.9. But first, we must describe the code that implements those behaviors.

The first behavior module we will implement is called Cruise. Like all the behaviors we will build, it declares two global variables for each quantity it wishes to output. Here, cruise_command holds the value of the command Cruise wishes to send to the motor controller and cruise_output_flag is a flag specifying whether Cruise is currently trying to send that value. Each time a module computes a new output, it will set the associated flag to 1. On any iteration, when a module chooses to produce no output, the flag will be set to 0. Observing this protocol is an essential part of our behavior control implementation:

```
int cruise_command;                    /* Command to motors */
int cruise_output_flag;
void cruise()
{ while(1) {
  cruise_output = FORWARD;             /* Rug Warrior goes forward */
  cruise_output_flag = 1; }}           /* Command is now active */
```

Cruise does nothing except output the FORWARD command and declare that this command is currently active as often as the scheduler will allow.

The Follow module implements light-source following and is somewhat more complicated:

```
int follow_command;
int follow_output_flag;

void follow()                            /* Follow a light */
{int left_photo, right_photo, delta;/* Left and Right Photocells */
  while (1) {
    left_photo = analog(1);              /* Read A/D channel 1 */
    right_photo = analog(0);             /* Read A/D channel 0 */
    delta = right_photo-left_photo;
    if (abs(delta) > photo_dead_zone)
      {if (delta > 0)
        follow_command = LEFT_TURN; /* Light on left, turn left */
      else
        follow_command = RIGHT_TURN;/* Otherwise turn right */
      follow_output_flag = 1;        /* Activate when detected */
      }
    else
      follow_output_flag = 0;        /* No difference, deactivate */
    }}
```

If Follow detects that the difference between what the left and right photocells measure is above the threshold, `photo_dead_zone`, it will turn the robot in the direction of the brighter side. Otherwise, Follow will compute no command.

Next, we implement the Avoid behavior, which gets sensor inputs from the near-infrared proximity sensors:

```
int avoid_command;
int avoid_output_flag;

void avoid()
{ int val;
  while (1) {
    val = ir_detect();
    if (val == 0b11) /* Both left and right see something */
      {avoid_output_flag = T;
      avoid_command = LEFT_ARC; }
    else if (val == 0b10)             /* Left IR sees something */
      {avoid_output_flag = T;
      avoid_command = RIGHT_ARC; }
    else if (val == 0b01)             /* Right IR sees something */
```

```
      {avoid_output_flag = T;
      avoid_command = LEFT_ARC; }
    else                                /* Neither sees anything */
      {avoid_output_flag = NIL; }
    }}
```

The Escape behavior is designed to allow the robot to escape from collisions with obstacles when Rug Warrior's bump sensors detect a collision:

```
int escape_command;
int escape_output_flag;

void escape()
{ while (1) {
  bump_check();                         /* Get state of bumper */
  if (bump_left && bump_right)          /* Bumped from the front */
    {escape_output_flag = 1;
    escape_command = BACKWARD;
    sleep(.2);                          /* Backup for a while */
    escape_command = LEFT_TURN;         /*  then turn LEFT */
    sleep(.4);}
  else if (bump_left)                   /* Bumped on the left */
    {escape_output_flag = 1;
    escape_command = RIGHT_TURN;
    sleep(.4);}                         /* Turn right for a while */
  else if (bump_right)                  /* Bumped on the right */
    {escape_output_flag = 1;
    escape_command = LEFT_TURN;
    sleep(.4);}                         /* Turn left a while */
  else if (bump_back)                   /* Bumped from behind */
    {escape_output_flag = 1;
    escape_command = LEFT_TURN;
    sleep(.2);}                         /* Confront attacker */
  else                                  /* No bumps so deactivate */
    escape_output_flag = 0;
}}
```

What we have so far is a collection of task-achieving behaviors. Each behavior may examine from none to several inputs and compute an output. The output of each of the behaviors above is what that behavior wants the robot to do.

Next, we must activate the behaviors as processes so that they will run simultaneously. We must also establish an arbitration structure that will decide which behavior gets control of the motors when a conflict arises. The behaviors are initiated by calling the following program. **IC** uses `main` as a special name. Once loaded into the battery-backed RAM, the `main` program begins to run whenever the robot is switched on:

```
void main()
{ start_process(motor_driver());
  start_process(cruise());
  start_process(follow());
  start_process(avoid());
  start_process(escape());
  start_process(arbitrate());
  }
```

The `motor_driver` function is a simple software driver that looks at a global variable `motor_input` and outputs the appropriate values to the ports connected to the motors.

The Arbitrate function implements message passing between the other processes. After the behaviors have been designed, a wiring diagram specifies how they are to be connected. Here, Arbitrate implements wiring instructions with an ordered list of statements. When multiple outputs are directed to the same input, those occurring later in the list of connections subsume (actually overwrite) earlier ones:

```
void arbitrate()
{ while (1) {
  if (cruise_output_flag == 1)
     { motor_input = cruise_output; }
  if (follow_output_flag == 1)
     { motor_input = follow_output; }
  if (avoid_output_flag == 1)
     { motor_input = avoid_output; }
  if (escape_output_flag == 1)
     { motor_input = escape_output; }
  sleep(tick);                     /* Message controls for one tick */
  }}
```

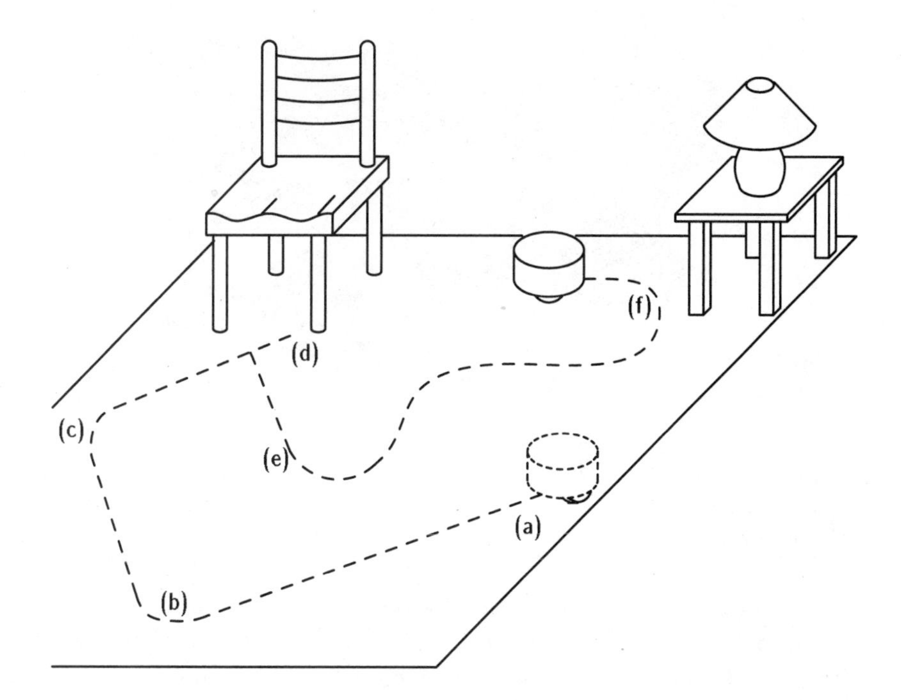

Figure 9.10. Rug Warrior executes the program described in the text. Initially, the room is dark. At point (a), Rug Warrior is switched on, waits a short while, plays its "alert" tune, and then begins to move. Since the room is dark, the signals from both photocells are the same and the light-following behavior issues no commands. Rug Warrior moves straight forward, as directed by Cruise. At points (b) and (c), the near-infrared sensors sense the wall and Avoid commands right turns. Because the leg of the chair is too narrow or perhaps the wrong color, the near-infrared sensors fail to detect it, and Rug Warrior collides at point (d). This collision activates the Escape behavior, which causes the robot to back up and turn. At point (e), the room light is switched on and the robot begins to execute its Follow behavior. When Rug Warrior gets too near at (f), it no longer sees the lamp directly and turns toward a brighter spot in its field of view.

Figure 9.11. When Rug Warrior's near-infrared detectors miss seeing an obstacle and Rug Warrior collides into a chair, the bump sensors trigger the Escape behavior, which causes Rug Warrior to back up or turn.

Emerging from these seemingly distinct sets of processes is an overall behavior for Rug Warrior, which is illustrated in Figure 9.10. Figure 9.11 shows a close up of one element of this behavior, namely a collision between Rug Warrior and a chair. (To see a somewhat different implementation of the program described above, refer to Section 10.1.1 and Section 10.1.1.)

It may seem pointless to go though the complicated exercise performed by arbitrate when the behaviors could have sent the messages to the motor driver themselves. When Behavior-1 computes an output for Behavior-2, why not just write it directly?

There are two reasons. First, we wish to maintain modularity. Suppose a robot control system has been written and debugged and that it contains no formal message-passing scheme. All the behaviors simply write their outputs to the correct inputs directly. If we now want to add a new layer of complexity, it is not possible to simply write new modules and wire them in by adding statements to the connection diagram. Instead, we must have a detailed knowledge of which behaviors pass what messages to which other behaviors and in what order. This is easy to do for a small system, but it becomes intractable for a large one.

Secondly, the relationship between the behavior control diagram and the code becomes difficult to understand. The connections, rather than being explicitly represented in the connection list, are now hidden in the code. If the order in which the behaviors are executed changes or new behaviors are added, the overall behavior of the robot may change in unexpected ways.

By observing certain protocols as we programmed our robot in **IC**, we were able to build a network of finite-state machines that could pass messages to each other, running in (what appeared to be) parallel operation.

9.6 What Did We Do?

What we have described in the previous section is a style of programming that implements behavior control principles. But what does organizing an intelligence system in this manner really buy us?

The approach of building networks of layered task-achieving behaviors that run concurrently has a number of advantages over the paradigm of sensing, world modeling, and planning. The first advantage is that behavior control grants real-time robustness to events in a changing environment. Taking the traditional approach of building a map of the world and updating it with fused sensor data leads to a computational bottleneck, which causes the robot to take a long time to plan a strategy about what to do. This common problem of artificial intelligence (AI) programs taking longer to run when more knowledge becomes involved was one of the original difficulties that subsumption architecture set out to solve. For instance, what if a robot were walking down the street and crossing a railroad track while contemplating "pawn-to-king-three" and a train began to come down the track? Should the robot finish searching the decision tree for possible moves or hustle off the track? Behavior control offers a way for pressing concerns to assume precedence.

Beyond the capability of real-time robustness, however, is the realization that, if the intelligence system is organized without a world model, then the hard problem of sensor fusion can be ignored. In fact, we can think of a programming technique using sensors as *sensor fission*, whereby different sensors interject themselves into the control system at various levels to trigger different behaviors. The problem of sensor fusion is then passed off to the problem of *behavior fusion*, which is much less computationally intensive. The problem of behavior fusion is arbitrated by the designer's prioritization scheme based upon his or her arrangement of suppressor nodes. Because no geometric world model is maintained, the robot requires less computational hardware. Rug Warrior does not have to haul a supercomputer around with it.

Even more interesting than the speed and space advantages of behavior control are the possibilities of what this paradigm might hint to us about models of intelligence. Because seemingly complex behaviors can be seen to emerge from what we know are very simple reflexive behaviors, perhaps complex mechanisms that we hypothesize exist in what we acknowledge as intelligence might actually just be combinations of much simpler mechanisms.

When we, as humans, look at the scene illustrated in Figure 9.10, we see Rug Warrior acting with interesting if not purposeful behavior. We also see walls, chairs, and table lamps. These images bring to mind the associations that people sit in chairs and that table lamps are useful for reading. Rug Warrior sees none of these things yet can operate effectively in this environment.

Building robots has helped us to stay on track and to keep focused on solving the problems that need to be addressed in creating machines that we would consider clever. Before actually building your own robot, it is easy to ascribe all sorts of complex structures to the putative thought processes of robots; you can hypothesize complicated networks, special architectures, and the need for lots of "computrons" to connect perception to action. One of the great advantages to building things is that you can see exactly how much machinery is required. Oftentimes, a priori intuition about what will be needed is completely wrong. After building a machine of your own, you can look at an already built system (such as Nature's) with greater insight and the hope of being able to discern the extraneous from the essential.

In our book, then, we have come full circle. From the myriad of details involved in learning about electronics, mechanics, motors, and software, we have seen that a robot can be much more than the sum of its parts. As a system, if organized in the proper way, intelligent behavior can seem to emerge from a collection of simple competencies.

9.7 References

There is a large body of work on robot intelligence systems in the artificial intelligence community. While we cannot go into all the threads here, we point to a few conferences and workshops whose proceedings encompass the broad field and then mention a few local pieces of work from which the robots of the Mobile Robot Lab, and later this book, evolved.

Over the past 15 or 20 years, different notions have taken form about how to go about organizing intelligent behavior in computer programs. Much of this work was chronicled in proceedings of conferences of the American Association of Artificial Intelligence (AAAI) and the IEEE Robotics and Automation Society (IRAS). Early work in intelligent autonomous robots led to the development of planning strategies (Nilsson 1984) and visual map making (Moravec 1981). These directions later found applicability, especially in fixed-base arm and manipulator-type robots for assembly (Lozano-Pérez, Jones, Mazer, and O'Donnell 1992).

During this time, new ideas were proposed, aimed at addressing issues that did not fit well with world-model paradigms. In an effort toward understanding common-sense reasoning, Minsky (1986) proposed *The Society of Mind* as the idea that the brain is composed of independent agents, collectively interacting to produce intelligent behavior. Simultaneously, Brooks, pondering why simple insects could perform feats that would be unimaginable even for the largest supercomputers, proposed subsumption architecture (Brooks 1986) for programming autonomous robots, where collections of simple behaviors and reflexive rules interact in such a way that seemingly more complex behaviors emerge. Connell (1990) extended this work, introducing a number of new ideas to the subsumption architecture approach. Brooks (1991a, 1991b) and Maes and Brooks (1990) discuss subsequent experiments in the Mobile Robot Lab with behavior-based robots that walk, climb, collect, wander, hide, and learn.

One fallout from the subsumption architecture/behavior control approach was that the resulting intelligence system did not have to deal with sensor fusion and world modeling and consequently compiled down to a very lean block of code. This breakthrough in software led to new opportunities in hardware. Brooks and Flynn (1989) outlined the possibilities.

Powerful ideas coinciding with a ripening of technologies has created a movement, *nouvelle AI*, where believers (there *are* skeptics) dabble in pursuing these distributed approaches to organizing intelligent systems. Maes (1991) is an edited collection of recent work along this avenue.

For a more rigorous explanation of behavior control and other methods of controlling robots combined with a broader and more historical perspective, consult (Arkin 98) and the forthcoming book (Brooks and Ferrell 98).

10

Robot Projects

Now that you have learned how to design a robot and (we hope) constructed one of your own, what can you *do* with your robot? How can you connect sensing to actuation in interesting and useful ways? In this chapter we sketch out a few ideas for projects suitable for Rug Warrior II class robots. For certain key concepts we also supply an analysis and sample code.

Robots have a knack for exposing the ignorance of their creators. Behaviors of elegance and sophistication, as seen in the mind's eye, often degenerate into mysterious bumps and gyrations when executed by the robot. Each time we seek to make a robot do something new, we as robot builders are obliged to learn something new. To be convinced of this, try a few of the projects in this chapter.

10.1 Projects for Individual Robots

The following projects are suitable for a single Rug Warrior robot. Even if your intent is to program several robots to operate in a team, you should first try some individual robot projects to build up your base of robot programming tools and skills. To give you a running

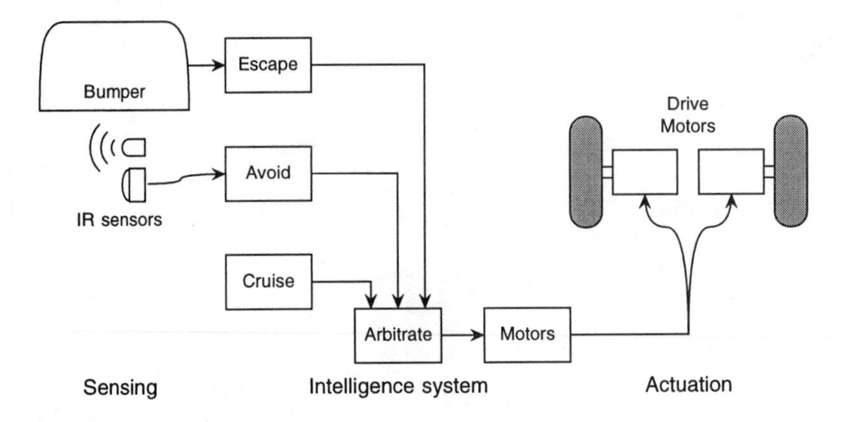

Figure 10.1. Behavior control diagram for Lewis and Clark. In conflicts over which behavior gets to control the robot's motors, Escape has the highest priority, Avoid the next highest, and Cruise the lowest.

start, we will work through the first project below from behavior diagram to Interactive **C** code.

10.1.1 Lewis and Clark

The first imperative for a great many robot projects can be stated as "move about and don't get stuck." The Lewis and Clark program embodies that notion in its essence—the robot's only purpose is to keep moving, exploring new territory.

One way to program Lewis and Clark is to use the collection of five behaviors diagrammed in Figure 10.1. These behaviors running together satisfy the problem statement. Cruise provides the "move about" part while Avoid and Escape give us "don't get stuck."

Cruise is a single-minded behavior. Ignoring all sensing, Cruise's only purpose is to make the robot go forward.

Avoid tries to steer the robot away from obstructions. Using Rug Warrior's infrared (IR) detector, Avoid determines if any obstacles are in the robot's path. If so, Avoid issues motion commands that will turn the robot so as to avoid hitting the obstacle. When deciding which behavior gets control of the motors, Avoid must have a higher priority than Cruise.

Sometimes things go wrong however, and collisions happen despite Avoid's best efforts. To recover from this, an Escape behavior is needed. After the bumper detects a collision, Escape takes control and commands the robot to backup and turn a bit, so that the robot will go off in a new direction, away from whatever object it just struck.

These three simple behaviors, plus an arbitration behavior to enforce priorities, and a motor-driver behavior to make the motors move, are all that is needed to keep Rug Warrior exploring its surroundings indefinitely.

Now consider the **IC** code that actually implements these behaviors. A complete listing of the code follws the analysis.

Each behavior in the behavior control diagram corresponds to a similarly named **C** function in the code. To control robot motion, a behavior function must have three associated global variables, *X*`_trans`, *X*`_rot`, and *X*`_act` (where *X* is a mnemonic for the behavior's name). To make the robot move, behavior *X* computes translation and rotation values, stores these numbers in *X*`_trans` and *X*`_rot`, and sets *X*`_act` to 1. When behavior *X* is ready to give up control, it sets *X*`_act` to 0.

Behaviors run constantly in parallel, thus behavior functions all contain a `while(1)` expression. The code in the body of the `while` expression runs as often as **IC**'s scheduler will allow.

The Cruise behavior repeatedly sets `cru_act` to 1 and commands a motion whose translation component is `cru_def_vel` and whose rotational component is 0. The Arbitrate behavior ultimately decides if these commands ever reach the motors.

Code implementing the Avoid and Escape behaviors also sets active flags, `av_act` and `es_act` respectively, and assigns translational and rotational velocities. Unlike Cruise, Avoid and Escape compute these values based on sensor readings.

In the Lewis and Clark program, the Arbitrate behavior implements a strict priority scheme. If Escape wants control (if `es_act` = 1), Arbitrate passes Escape's motion commands on to the Motors behavior.

If Escape does not need control, Arbitrate checks to see if Avoid is active. If so, Avoid's motion commands are passed on; if not, the commands issued by Cruise are allowed to control the robot.

The Motors behavior is aware only of the values output by Arbitrate, `mot_trans` and `mot_rot`. Motors constantly calls the function `drive` with these values, making the robot move. (It would be a violation of the behavior control paradigm if any other behavior were allowed to call `drive`).

Avoid functions by checking to see if the IR detectors have sensed a reflection from an obstacle (`ir_detect` is a library function that returns the state of Rug Warrior's obstacle detectors). If the detectors have seen nothing (if `ir_hit` = 0) then Avoid is finished; it sets `av_act` to 0. If there was a detection on both sides of the robot (`ir_hit` = 3), Avoid arbitrarily decides to turn left (the rotation variable is set to a positive value). The remaining possibility is that there was an IR detection on either side of the robot. In this case the utility function, `lr_rot`, uses the value of `ir_hit` to compute a positive or negative number, as appropriate, to make the robot turn left or right.

Avoid computes a motion command only during the time the IR sensors detect a reflection. No reflection means Avoid gives up control immediately. Escape functions differently; it continue to issue motion commands for a brief time after the event that triggered Escape has ended.

The Escape behavior monitors the bumper using the library function `bumper`. If there is a collision on the left or right, Escape commands a negative translational velocity, then waits for a time, `es_bf`, allowing the robot to backup. Next, based on which side the collision occurred, Escape commands a rotation to the left or right, and waits again. Finally, Escape commands a forward motion. The final forward motion ensures that when Escape terminates, the robot is in a different position and does not immediately bump into the obstacle again. If bumped from the back, Escape spins a bit, then releases control.

To run in parallel, each behavior must be instantiated as a process. This is what the `main` function does. Another purpose of the special function `main` is to automatically start the Lewis and Clark program after a reset or power up.

A Report behavior is included for debugging purposes. Report displays on the liquid crystal display (LCD) screen which behaviors are active and shows what velocity is actually being commanded. Knowing which bit of code causes the robot to act in a particular way is essential when your program does not behave as you expect.

From these primitive behaviors, a global behavior emerges. The robot moves forward until it nears an obstacle. Rug Warrior then steers so as to avoid the obstacle. If it collides with an undetected obstacle, Rug Warrior backs, turns, and proceeds off in a new direction.

Lewis and Clark Program

```
/* Lewis and Clark
 * Purpose: Move and don't get stuck */


/* UTILITY FUNCTION */

/* Decide to spin left or right based on IR or bumper hit
   1 => detection on right, 2 => detection on left.
   Return +1 if robot should spin to left, -1 if it should
   spin right
*/


int lr_rot(int detect)
{ return ((detect & 1) << 1) - 1;
}


/* BEHAVIORS */

/* Cruise Behavior:
   Move forward always */

int cru_trans = 0;        /* Cruise translational velocity command */
int cru_rot = 0;          /* Cruise rotational velocity command */
int cru_act = 0;          /* Cruise active flag */
int cru_def_vel = 100;    /* Cruise default velocity */

void cruise()
{ while (1)
    { cru_act = 1;        /* Cruise always wants control */
      cru_trans = cru_def_vel;  /* Move at the standard velocity */
      cru_rot = 0;        /* Don't rotate */
    }
}
```

```
/* Avoid Behavior
   Detect obstacles with the IR sensor and arc away */

int av_trans = 0;        /* Avoid translational velocity command */
int av_rot = 0;          /* Avoid rotational velocity command */
int av_act = 0;          /* Avoid active flag */
int av_def_trans = 100; /* Avoid default translational velocity */
int av_def_rot = 50;     /* Avoid default rotational velocity */

void avoid()
{ int ir_hit = 0; /* Local variable for obstacle detection */
  while(1)
    { ir_hit = ir_detect();
      if (ir_hit == 0)          /* No IR detection */
        av_act = 0;          /* Avoid behavior has nothing to say */
      else if (ir_hit == 3)     /* Obstacles on both sides */
        {
          av_act = 1;           /* Avoid wants control */
          av_rot = av_def_rot;  /* Arbitrary rotate left */
          av_trans = 0; }       /* Don't go forward */
      else                   /* Obstacle on one side or the other */
        { av_act = 1;   /* Avoid wants control */
          av_trans = 0;         /* Don't move forward */
          av_rot = av_def_rot * lr_rot(ir_hit)
                                  /* Rotate left or right */
      }
  }
}



/* Escape Behavior
   Move away from a collision */

int es_trans = 0;        /* Escape translational velocity command */
int es_rot = 0;          /* Escape rotational velocity command */
int es_act = 0;          /* Escape active flag */
int es_def_trans = 100; /* Escape default translational velocity */
int es_def_rot = 50;     /* Escape default rotational velocity */
float es_bf = 0.25;      /* Backward/Forward time */
float es_spin = 0.25;    /* Spin in place time */

void escape()
{ int es_hit = 0;              /* Variable to save collision dir */
  while (1)
    { es_hit = bumper();     /* Get and remember the bumper */
```

```
      if (es_hit & 3)        /* If left or right collision... */
        { es_act = 1;        /* Escape wants control */
          es_trans = (- es_def_trans);  /* Backup */
          es_rot = 0;        /* Don't turn while backing up */
          sleep(es_bf);      /* Backup for a while */

          es_trans = 0;      /* Rotate left or right */
          es_rot = lr_rot(es_hit) * es_def_rot; /* Rotate L or R */
          sleep(es_spin);    /* Spin for a while */

          es_trans = es_def_trans;  /* Forward past obstruction */
          es_rot = 0;        /* Don't rotate */
          sleep(es_bf);      /* Wait for motion */
          es_trans = 0;      /* Stop */
          es_act = 0;        /* Relinquish control */
      }
      else if (es_hit & 4)  /* Bumped from the back */
        { es_act = 1;        /* Bump wants control */
          es_trans = 0;      /* Stop forward motion */
          es_rot = es_def_rot;  /* Spin in place */
          sleep(es_spin);    /* Spin a while */
          es_rot = 0;        /* Stop spinning */
          es_trans = 0;      /* Stop */
          es_act = 0;        /* Relinquish control */
      }
  }
}


/* Motor Driver Behavior
   Constantly tell the motors how to move */

int mot_trans = 0;
int mot_rot = 0;

void motors()
{ while(1)
    drive(mot_trans, mot_rot);
}


/* Drive Motor Arbitration
   Prioritize the outputs of the behaviors */

void arbitrate()
{ while(1)
```

```
    {
      if (es_act)                    /* If Escape wants control... */
        { mot_trans = es_trans; /*     pass along Escape's commands */
          mot_rot = es_rot; }
      else if (av_act)               /* If Avoid wants control... */
        { mot_trans = av_trans; /*     pass along Escape's commands */
          mot_rot = av_rot; }
      else if (cru_act)              /* If Cruise wants control... */
        { mot_trans = cru_trans; /*     pass along Cruise's commands */
          mot_rot = cru_rot; }
      else                           /* If no one is active, just stop */
        { mot_trans = 0;
          mot_rot = 0;}
    }
}



/* Report Behavior
   Reporting what the robot is doing to the LCD screen aids
   debugging the program */

void report()
{ while (1)
    { printf("L&C  E:%d A:%d C:%d", es_act, av_act, cru_act);
      printf("  Tr:%d Rt:%d\n", mot_trans, mot_rot);
      sleep(0.2);       /* Don't overtax the LCD */
    }
}



/* Automatic startup behavior */

void main()
{ printf("Lewis and Clark\n");  /* Announce the program */
  sleep(1.0);                   /* Give user time to read LCD */
  start_process(escape());      /* Start each of the processes */
  start_process(avoid());
  start_process(cruise());
  start_process(motors());
  start_process(arbitrate());
  start_process(report());
}
```

10.1.2 Moth

A Moth program makes the robot seek out the brightest light (or dimmest light if you like). By pointing a flashlight at a robot that is executing a Moth program, you can make the robot come to you from across the room.

One way to write the heart of the Moth program is given in the following code:

```
int mth_trans = 0;
int mth_rot = 0;

void moth_point(int light_dark)     /* Aim at light (or dark) */
{ int diff;
  while(1)                          /* Left/Right difference */
    { diff = analog(photo_right) - analog(photo_left);
      mth_rot = light_dark * diff; /* Light diff => Rot rate */
  }
}
```

This very simple function computes rotational velocities that point the robot toward (or away from) a light source. The argument to `moth_point` (`light_dark`), controls whether the robot seeks light or darkness. A positive number makes the robot turn toward the light; a negative number makes the robot point toward the dark. The magnitude of the argument determines how strongly the robot responds to light differences. The code is an elementary proportional control system. If the magnitude of `light_dark` is made too large, the robot's motion will be unstable.

Using the structure of Lewis and Clark you can easily add the Moth behavior to make a new program. With this done, Rug Warrior follows the light but avoids obstacles and escapes from collisions along the way. (This will give you an alternate implementation of the program described in Section 9.5.)

10.1.3 Baryshnikov

Rug Warrior can play tunes and it can move in interesting ways. The Baryshnikov program puts these two talents together. You are the choreographer—write a program that enables your robot to dance to its own accompaniment.

A much more difficult variation on this theme is to have two or more robots coordinate their actions, perhaps playing tunes in harmony.

10.1.4 Mouse

An interesting and useful exercise is to program your robot to follow walls. Such a behavior is often a vital component of other more complex programs.

One simple approach to a Mouse program is to use the IR detector to track the wall. Suppose you wish your robot to move parallel to a wall on its left. Have the robot arc to the left when no wall is present on the left and arc to the right when a wall is detected. (This is one of the behaviors TuteBot exhibited in Chapter 2.)

The behavior that emerges from this simple programming scheme is that the robot wiggles back and forth as it moves forward along the wall. If the parameters that control the arcing are chosen correctly, the robot will even go through doorways and find its way out of dead ends.

In Mouse, as with most simple behaviors described in this chapter, it is helpful to use the bumper as a fail safe. Regardless of what else it is doing, when the robot detects an object against its bumper in the front, the robot should stop. Without this feature, your robot will frequently find itself stuck against an obstacle with motors at full power—needlessly wasting battery charge and stressing components.

10.1.5 Magellan

Decide on some closed path through your home or classroom and then write a program that will make Rug Warrior navigate the course so that it returns to the starting point. Depending on the course you choose, there are a great many possible ways to accomplish this task.

A popular approach is to have your robot repeatedly drive forward for some time or for some number of encoder clicks, turn for some time or number of clicks, drive forward again, and so on until the robot arrives back at the starting point. This method is called dead reckoning. With this approach, your job is to discover the pa-

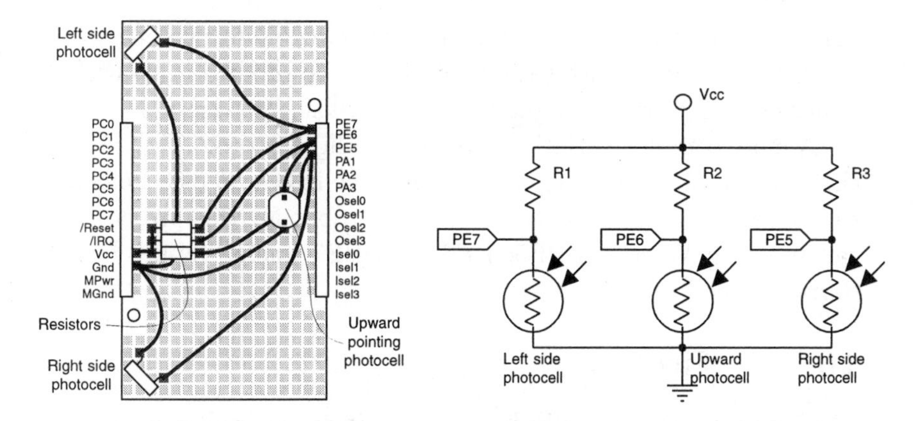

Figure 10.2. Here three additional photocells are connected to Rug Warrior via a RugEx^{TM} board. Photocell light sensors are implemented as voltage dividers. The proper value for resistors R1, R2, and R3 is about 10 kΩs assuming the same photocells are used for this circuit that are used for the Rug Warrior brains.

rameters of the path. How far should you command the robot to go, how long should you have it turn at each segment of the path?

Unless the path you have selected is very short, however, small positional errors will build up as the robot moves. Quickly the robot will become lost and no amount of tuning will make the robot follow the path reliably.

A more robust approach is to use some feature of the environment that the robot can sense to help guide the robot. Perhaps there is a wall the robot can follow for most of the way, or maybe the robot can home in on a strong light. Consider carefully which features along your proposed path the robot can exploit to find its way.

10.1.6 Apollo 13

If you place a single bright light source in a dark room, it is easy to have the robot home in on the light. The robot can do this by monitoring the difference in intensity between left and right photocells. (See Section 10.1.2.) The robot moves forward while turning left or right so as to minimize the difference.

An additional interesting behavior is possible if you also have the robot pay attention to the absolute intensity of the light. When far away, the robot can move toward the light. But when close to the light, the robot can turn away. By balancing these two effects you can get the robot to orbit the light. Have the robot do this without either crashing into the light or becoming lost in space.

A more complex variation is to use two light sources. Try to make the robot perform a figure eight around the lights. It may be helpful to connect additional photocells to your robot for this project. You can use Rug Warrior's PE5, PE6, and PE7 analog inputs for this purpose. See Figure 10.2.

10.1.7 Fire!

The Fire Fighting Home Robot Contest (www.trincoll.edu/~robot/) is a popular event held annually at Trinity College in Hartford, Connecticut. Robots compete in a maze of four rooms connected by hallways. A lit candle is placed in one of the rooms. The robot must navigate the maze, find the candle, and extinguish it. The winning robot is the one that does this in the least amount of time.[1]

Simpler variations of this contest are possible without adding fire extinguishing equipment or ultraviolet (UV) flame-detecting sensors to Rug Warrior. You could, for example, replace the candle with a small lamp. Consider the fire extinguished when the robot bumps the lamp.

10.2 Multi-Robot Projects

Employing more than one robot in a project tremendously expands the possibilities. For two or more robots, new avenues open for both competition and cooperation between robots and between builders.

10.2.1 You're It!

Robot tag is an exciting game that a group of two or more Rug Warriors can be programmed to play.[2] One simple implementation requires adding a small lamp to each robot. You can use Rug Warrior's batteries to power the light.

---

[1] A contestant using a Rug Warrior brain is a past winner of this contest.

[2] We thank Professor Susan Hruska's graduate seminar class at Florida State University for developing this robot project and presenting it to the authors.

Start the game with the room darkened and with a lamp in place on each robot. One robot is designated "it" and the rest "not-it." The "it" robot executes a "seek light" behavior. Thus it tends to chase any "not-it" robots that it sees. The "not-it" robots execute an "avoid light" behavior—they tend to move away from the "it" robot (and each other).

When a collision does occur, an ambiguity arises which the robots must resolve. Did an "it" collide with a "not-it," did a "not-it" collide with another "not-it," or did a robot just strike the wall? To solve this problem, each robot monitors its bump sensor. When a collision happens the robot stops. A "not-it" robot plays a tone on its piezo buzzer, then listens. An "it" robot listens first, then plays a tone. If a robot hears a tone, then it assumes it collided with a robot of the opposite type and it switches state. That is, a "not-it" becomes "it" and an "it" becomes "not-it." If the robot does not hear a tone, it assumes it collided with a wall or a robot in the same state as itself and it does not change state.

10.2.2 Ready, Set, Go!

Interest has grown in robot racing. (For more on this topic, see for example, www.teleport.com/~raybutts/.) The traditional way to implement robot racing is line following; a black line on the floor defines a path for the robot to follow. Special sensors on the robot identify the robot's position relative to the line and the robot moves forward as fast as it can without losing the line.

You can implement your own system to perform line following with Rug Warrior. Downward pointing photocells, photodiodes, photoreflectors, etc. can be used for this purpose.

Other clever courses are also possible. You may define a race course consisting of a wall that must be followed. Or, you can arrange a series of lights that the robots must approach sequentially (see Figure 10.3.). Suppose each light is placed on a low table, somewhat taller than a robot. A robot on the race course will approach the nearest light until it comes into the shadow of the table. The robot will then lose sight of the first light and will begin to travel toward the next nearest light in the series. In this way the race course can be extended indefinitely or formed into a closed loop.

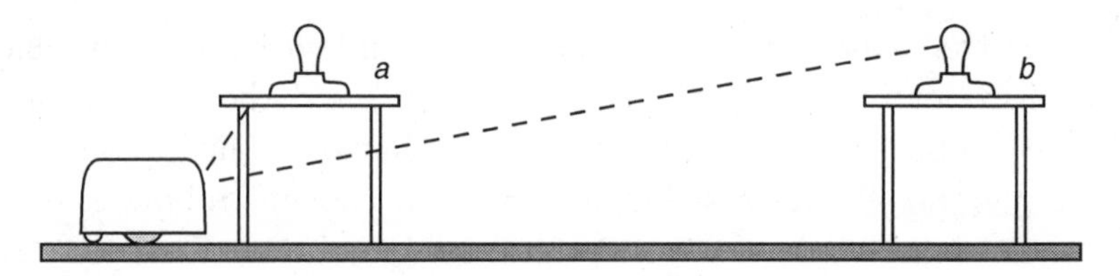

Figure 10.3. Lights and shadows can be arranged to a robot race course. The robot approaches lamp a until the robot drives into the shadow of the table. The robot then begins to travel towards the next nearest lamp, lamp b.

10.2.3 Ready or Not, Here I Come

With a bit of cooperation, robots can play hide-and-seek.

The maximum range of Rug Warrior's IR obstacle detection system is only about 18 inches to 2 feet. However, when the emitter of one robot is directed toward the detector of another, detection across a room of moderate size is possible. This feature can be exploited to allow robots to play a game of hide-and-seek.

At the start of the game, the seeker robot and one or more hider robots are together at the center of the room. The seeker robot begins to count, beeping 20 times with its piezo buzzer. When the hider robots hear the count start, they run away, perhaps stopping when they bump into something. When they stop, the hiders turn to face the direction they came from and they turn on their IR emitters.

The seeker finishes its count and begins to search for the hiders. The seeker spins slowly in place, monitoring its IR detector. If the seeker senses an IR source, it approaches that source. When the seeker bumps a hider robot, the hider acknowledges by sounding its piezo buzzer. The seeker then begins looking for the next hider and proceeds in this way until all the hiders have been found.

Hardware for Beacon Following

Because the notion of *beacon following* is an important one, we will describe in depth one method by which a robot can follow a beacon. Rug Warrior nominally has only one IR detector. This is adequate for avoiding obstacles as the left and right emitters allow us to determine on which side an IR-reflecting obstacle is located. The single

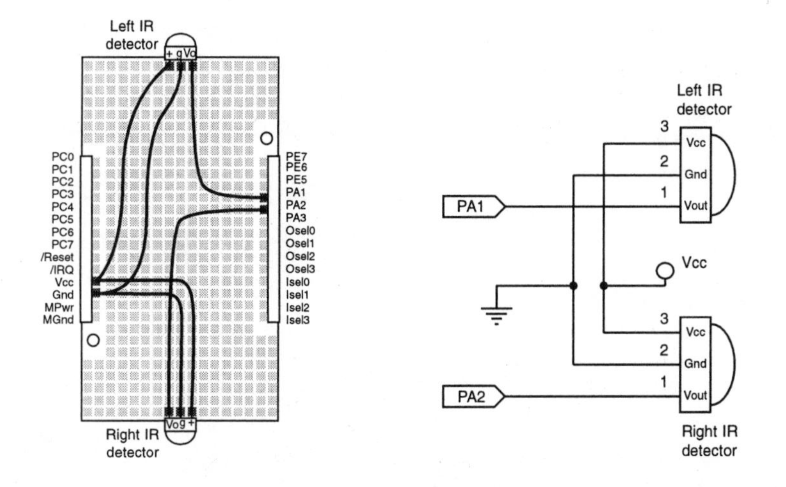

Figure 10.4. The schematic diagram on the left and the pictorial layout on the right, show one simple way to add extra IR detectors to Rug Warrior. The detectors used are Sharp IS1U60s.

detector poses a problem, however, when trying to follow a beacon. Thus, we need to make a small addition to Rug Warrior's complement of sensors.

Adding extra sensing is a simple matter for Rug Warrior (see Figure 10.4). In this example we have made use of *RugEx*™, the Rug Warrior Experimenter's board.[3] Signals from the IR detectors go to unassigned digital I/O pins PA1 and PA2.

Building a Beacon

Creating a beacon requires no new hardware; Rug Warrior's standard IR emitters can be used. All we need is a snippet of code to turn the emitters on and off at detectable intervals:

```
void main(long msec)          /* Process for the Hiding robot */
{ printf("Hiding...\n");      /* Report which program is running */
  poke(0x1009,0b111100);      /* Set port D for output */
```

[3]RugEx™ can be plugged directly into the new Rug Warrior Pro™ robot. A connection to the original Rug Warrior can be made through *RugUp*™, the Rug Warrior upgrade board. Alternatively, the IR detectors can be wired to a connector plugged into the expansion socket. RugEx™ and RugUp™ are available from A K Peters, Ltd.

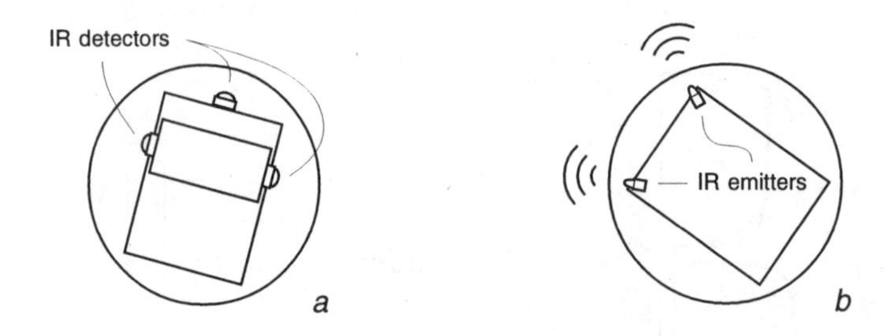

Figure 10.5. Top view of two robots playing a game of hide-and-seek. The hiding robot b cooperates by emitting an IR signal. The seeking robot a uses three IR detectors to determine the direction of the signal.

```
  while(1)                    /* Broadcast forever */
    { poke(0x1008,0b1100); /* Turn on PD2 and PD3 */
      msleep(1L);             /* Wait for 1 millisecond */
      poke(0x1008,0);         /* Turn PD2 and PD3 off */
      msleep(1L); }           /* Wait for 1 millisecond */
}
```

This simple program is all that is required for the "hiding" robot. The program informs the user that the robot is hiding (the `printf` statement does this), sets the bits of port D to be outputs (the IR emitters are controlled by PD2 and PD3), then at one millisecond intervals turns the modulated IR emitters on and off.

Code for Beacon Following

The seeking robot's job is somewhat more difficult. The basic idea is that we will constantly monitor Rug Warrior's three IR detectors, counting how often each turns on and off. We make use of the fact that the more directly a detector is pointed at an emitter, the more reliably it will detect the transmission. For example, if the hiding robot is to the right of the seeking robot, the seeker's right IR detector will count a large number of detections in some interval, the forward detector will count a few detections, and the left detector will count no detections. See Figure 10.5.

First we must have a function to tell the robot which IR detectors have sensed an on-off signal from the beacon. That is accomplished by `ir_rec3`.

```
int ir_rec3(long msec)
{ int val1, val2;
  val1 = (peek(0x100A) & 0b00010000)|(peek(0x1000) & 0b00000110);
  msleep(msec);
  val2 = (peek(0x100A) & 0b00010000)|(peek(0x1000) & 0b00000110);
  return (val1 & ~val2) | (~val1 & val2);
}
```

The current value of Rug Warrior's standard IR detector (connected to PE4) is bitwise ORed with the output of the two new detectors connected to PA1 and PA2. This value is stored in `val1`. Then `ir_rec3` waits for a time (typically 1 millisecond) to let the beacon change state. Next, the states of the IR detectors are collected again into `val2`. Finally, `ir_rec3` returns a bitwise accounting of what happened. A detection occurred if the signal from a detector was high and then, one millisecond later, low. Or a detection is counted if the signal was first low, and then one millisecond later, high. For example, if the binary values of `val1` and `val2` are 0b10010 and 0b00110 respectively, then the result will be 0b10100. The implication is that the middle and right detectors have sensed the beacon, and the left detector has not.

Now that we can tell when a single detection of each detector has occurred, we need a way to keep a constantly updated count of how many detections by each detector have occurred within some interval. This is accomplished by the `ir_measure3` function. `ir_measure3` calls `ir_rec3` a number of times equal to `ir_measure3`'s argument `count`. Following each call, `ir_measure3` increments the local counter associated with each detector. After `count` iterations, these internal values are copied to global counters associated with each detector and the process repeats.

```
int r_rate = 0;              /* Right detector detection count */
int m_rate = 0;              /* Middle detector detection count */
int l_rate = 0;              /* Left detector detection count */

void ir_measure3(int count)
{ int i, rec;
  int li_rate, mi_rate, ri_rate; /* One local ctr per detector*/
  while(1)
    { li_rate = mi_rate = ri_rate = 0; /* Init local ctrs */
```

```
      for (i = 0; i < count; i++)       /* Check COUNT times */
        { rec = ir_rec3(1L);            /* Get reading */
          if (rec & 0b00000100)         /* If a detector has */
            ri_rate++;                  /*  detected something */
          if (rec & 0b00000010)         /*  then increment the */
            li_rate++;                  /*  associated counter */
          if (rec & 0b00010000)
            mi_rate++;
      }
      r_rate = ri_rate;  /* Copy local variables to global */
      l_rate = li_rate;  /*  detector counters */
      m_rate = mi_rate;
   }
}
```

At this point we know which detector recorded the most detections within a standard interval. We can now define the function seek to use this information to guide the robot toward the beacon.

```
int bias_abs = 25;          /* Default homing rotation rate */
int spin_speed = 40;        /* Default find-beacon rot rate */
int seek_trans = 35;        /* Default translation speed */
int lr_dead = 5;            /* Left/Right difference dead band */


void seek()                 /* Make a robot follow an IR beacon */
  int bias = 0;
  int trans = 0;
  while (1)
    { if (bumper() != 0)    /* If the robot bumps into anything */
        { trans = 0;        /*  make it stop */
          bias = 0;
      }
      else if ( (r_rate + m_rate + l_rate) < 5)  /* If the */
        { trans = 0;           /*  detectors see nothing, spin */
          bias = spin_speed; /*  in place looking for beacon */
      }
      else if ((r_rate > l_rate) && (r_rate > m_rate))
        { trans = seek_trans;  /* If most hits are on right */
          bias = (- bias_abs); /*  then arc to the right. */
      }
      else if ((l_rate > r_rate) && (l_rate > m_rate))
        { trans = seek_trans;  /* If most hits are to left, */
          bias =  bias_abs;    /*  then arc to the left */
      }
```

```
      else                         /* Beacon is straight ahead */
        { trans = seek_trans;
          bias = 0;
      }
      drive(trans,bias);           /* Send vels to motors */
      printf("Seek  tr%d rt%d", trans, bias);  /* Report  */
      printf("   L%d M%d R%d\n", l_rate, m_rate, r_rate);
      sleep(0.25);              /* Drive a bit before switching */
    }
}
```

The function **seek** finds the detector with the most hits (above a minimum number, 5) and computes a rotational velocity **bias** that will turn the robot in that direction. Here, for simplicity, we allow **seek** to act as both a task achieving behavior and as the arbitration behavior. This is possible because there are no other behaviors that wish to control the robot. If you adopt this code into a larger program, you should break arbitration out into a separate behavior.

Finally, we define a **main** function for our seeker robot. In this case **main** starts up the detection counter process and the seek process. Seeker begins the search for hider.

```
void main()
{ poke(0x1008,0);                     /* Seeker's IRs must be off */
  start_process(ir_measure3(50)); /* Start beacon hit counter */
  start_process(seek());              /* Then go find the beacon */
}
```

10.2.4 Couch Potato

The IR emitter and detector systems built into Rug Warrior can be used to enable inter-robot communications. To do this, you must define some number of distinct codes—combinations of IR pulses of specific duration. Connect some sensory input of the transmitting robot to a specific code. For example, when you push the rear bump switch, the `110` code is transmitted.

The receiving robot constantly samples its IR detector, classifying patterns of IR detected and not detected into one (or none) of the predefined codes. When the receiving robot identifies a valid

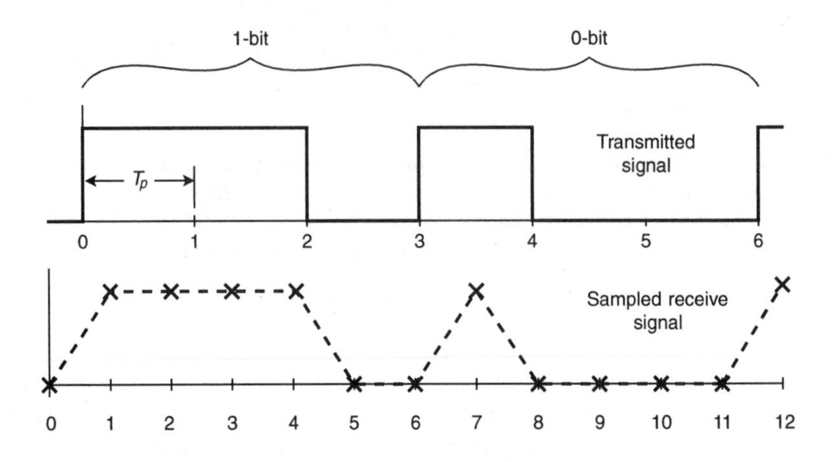

Figure 10.6. To signify a 1-bit, the transmitter turns the emitter on for two transmit periods, T_p, then off for one transmit period. A 0-bit is indicated by leaving the emitter on for a time equal to T_p and off for $2T_p$. To decode the data, the receiving robot must sample the signal from the transmitter at a rate of at least twice the highest frequency of the transmitter. That is, the sampling period, R_p must satisfy the relation, $R_p \leq T_p/2$.

code, it performs the action associated with that code. Perhaps code 110 means "spin 90 degrees to the left."

With a scheme such as this, one robot can perform as a sort of remote control for another robot.

The initial challenge in a Couch Potato program is to identify when a signal has been received from the transmitting robot. You must encode the bits of the transmitting signal into periods of varying length indicating when the IR emitters are on or off. For one example of how this can be done, see Figure 10.6. Here the transmitter has a fundamental period of T_p. A 1-bit is encoded by leaving the emitter on for a time $2T_p$ and off for T_p. A 0-bit keeps the emitter on for T_p and off for $2T_p$.

The receiving robot samples the incoming signal with a period of R_p. According to the Nyquist theorem we must have $R_p \leq T_p/2$. Note that since there is no connection between the clocks of the receiving and transmitting robots, sampling by the receiver will not ordinarily be exactly in phase with transmissions from the sender. Thus, if $R_p = T_p/2$ we can expect the sampled 1-bit to consist of three or four high samples followed by one or two low samples, while a 0-bit will be received as one or two high samples followed by three or four low samples. In general, other sequences of samples must be discarded as noise.

Sampling in this way yields a high communication speed; other simpler ways are possible if slower communication can be tolerated. In Section B.7 we control the robot by counting the number of loud sounds during an interval of predetermined duration. The same scheme is possible with an IR transmission. You can use a modified form of the IR detection code in Section 10.2.3 combined with the sound counting code in Section B.7 to decode IR transmissions.

10.2.5 Out of Africa

There is a species of termite found in parts of Africa that builds large, meter-high nesting mounds. The mounds are elaborate constructions, internally incorporating true arches. Individual termites, arguably, have no more neural computing power than Rug Warrior and yet, without explicit communication, termites are able to cooperate in the construction of their homes.

Partly because of observations such as this, robot cooperation has become an active area of robotics research. Box pushing is an elementary cooperation task you can program your robots to do. Boxes can be made heavy enough that one robot acting alone cannot successfully move the box, but two robots acting together can. To decide when and how to cooperate, robots must monitor their bump sensors and shaft encoders. Robots move until they contact a box (perhaps the boxes are identified with lights on top).

The bump skirt tells the robot that it is in contact with an object. The robot applies power to its motors and monitors its shaft encoders. If the shaft encoders indicate that the robot is moving, it continues to push the box. If the robot is pushing but not moving, then it can either give up and look for another box to push or the robot can try to attract other robots to help it push the box. The robot might try to attract others by, for example, turning on its IR emitters.

Although Rug Warrior cannot easily climb to build a tower, you may find ways that groups of robots can cooperate to arrange boxes or other objects in some useful way. Or you may simply have the robot collect scattered objects, pushing them into a central area.

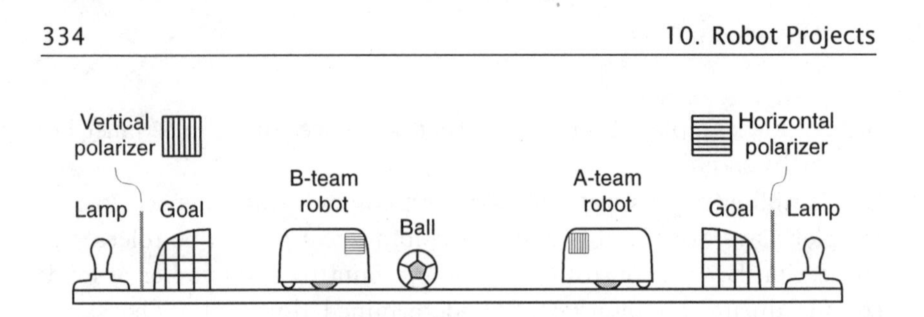

Figure 10.7. Robots competing in a robot soccer game identify the ball using their IR sensors. Robots locate the proper goal using polarized light. The robot with vertical polarizers covering its photocells, for example, sees only the vertically polarized light of its opponent's goal.

10.2.6 World Cup

In recent years, soccer tournaments for robots have gained in popularity worldwide. The participating robots in these contests are typically custom-designed, centrally-controlled, special-purpose machines that incorporate some sort of vision system. However, a team of Rug Warriors can be programmed to play a credible game of soccer; all it takes is a bit of ingenuity and possibly a specially engineered "ball."

To play soccer, robot players must be able to locate the ball, their opponent's goal, and maybe each other. It is possible to use a beacon transmitting a particular code to identify each goal. (You can use a spare robot or a Rug Warrior brains as the beacon.) The identifying code can be quite simple since there are only two goals.

Robots can detect the ball using their photocells. You can make a special-purpose photocell-detectable ball from a small lamp and battery. The ball need not be able to roll (the game may work better if it doesn't) but it should slide easily when pushed. The ball should not tip over and it should be rather rugged.

An alternative that allows the use of an unlit rolling ball is to arrange your playing field such that the ball is the only IR-reflective object. Robots detect the ball with their IR sensors, while goals are identified with polarized light. Each goal has a lamp with a polarizer between the lamp and the playing field. (See Figure 10.7). One goal is identified by horizontally-polarized light, the other by vertically-polarized light. Each robot wears small sunglasses made of either horizontally- or vertically-polarized material over its photocells. In this way, each robot can see only the light coming from the goal toward which the robot should dribble the ball.

Begin the game by blowing a whistle. Robots race to the ball and try to push the ball toward their opponent's goal. The simplest game needs only two players; more players increase the opportunity for subtle strategy.

Robot soccer is a complex undertaking and should probably not be your first robot project. Hone your programming and design skills by completing other projects. When you then tackle robosoccer, your arsenal of robot behavior programs will match your imagination.

10.2.7 The Undiscovered Country

Do not be constrained by the small number of projects described here. We strongly encourage you to invent robots and projects that appeal to you.

The most pervasive use of any technology is almost always one that the original practitioners did not foresee. No one dreamed, for example, in the early days of computers that the automotive industry would become the world's largest consumer of computers. Now there is a microprocessor in every car. It is up to you to take mobile robotics where we cannot, to create the applications we have not imagined.

10.3 References

The best reference for robot projects is the World Wide Web. Most robot builders are eager to share their experiences with whomever is interested, and many builders have designed web pages that reveal their creations in great detail. Use a search engine to look for phrases such as "my robot" or "robot project," and you will get a large number of hits.

Cooperating robots and robots working in swarms have become widespread research topics in recent years. For information on some groups involved in such research, see the following sites:

- avalon.epm.ornl.gov/~parkerle/coop_robotics.html,
- www.cs.brandeis.edu/~agents/, and
- www.cs.ualberta.ca/~kube/crip.cgi.

Information about a prominent robot soccer competition held in Korea can be found at the site: mirosot.org/MIROSOT98/index.html. Information on these and other robot contests can be found in Appendix F.

11

Robot Applications

Robots are tools. We use robots, as we use other tools, to reduce the amount of human effort required to satisfy our needs and desires. It is often stated that robots are especially appropriate for jobs that are dirty, dull, or dangerous. This chapter presents examples of robots, both commercial and pre-commercial, in each category.

The DC-3 of mobile robots has yet to appear.[1] However, many compelling demonstration projects and pioneering products hint that commercial success is tantalizingly close. The past few years have seen mobile robots descend into a volcano, explore the surface of Mars, and clean dirty floors. In this chapter, we briefly review a few of these robotic firsts. Our list is by no means complete. We seek only to illustrate some key elements from the broad and growing area of robotic applications.

The road to the everyday robots of the near future is not without bumps and detours. The successes and failures described in this chapter can provide useful caveats for robot designers.

[1]The Douglas DC-3 was the first aircraft to become an unequivocal commercial success. First introduced in 1935, the DC-3 accounted for 95% of US civil air commerce by 1938.

Figure 11.1. RoboScrub, a floor scrubbing robot, was developed by Denning Mobile Robotics in cooperation with Windsor Industries. (photograph courtesy of Jim Maddox)

11.1 Down and Dirty

When, in casual conversation, one admits to being a roboticist, the almost universal first question one is asked is "Can you make a robot that will clean my floors?" The appeal and market potential of floor-cleaning robots have not been lost on the commercial world. Numerous attempts to build vacuuming and scrubbing robots have been made, and efforts continue.

11.1.1 Denning RoboScrub

In 1991 Denning Mobile Robotics, a company dedicated to commercializing mobile robots, and Windsor Industries, a manufacturer of cleaning equipment, joined forces to build RoboScrub, a floor scrubbing robot. RoboScrub was billed as a "large area cleaner."

The approach taken by Denning seemed sensible; begin with an existing manual floor scrubber and add robotic elements to eliminate the need for an operator. Obstacle-detecting sonar transducers, a forward-looking infrared (FLIR) cliff detector, a tape-switch bumper, and a high-precision laser navigation system were added to the existing Windsor floor scrubber.

Figure 11.2. This navigation system recently marketed by Intelligent Solutions, Inc., allowed RoboScrub to be accurately positioned within its cleaning area. A rotating laser reflecting from bar coded targets determined the robot's position by triangulation. (photograph courtesy of Jim Maddox)

RoboScrub was programmed to follow a path through the space to be cleaned, testing as it went for the presence of obstructions. RoboScrub responded in a sophisticated way to such unexpected encounters. Obstacles at some distance caused the robot to slow down but otherwise maintain its path; closer obstacles triggered an immediate avoidance behavior, and obstacles detected only at the closest range made the robot stop and wait. In the later case, RoboScrub would not move until the obstacle departed. RoboScrub's programming scheme might best be described as *ad hoc*; the robot did not adhere strictly to either classical modeling/planning or to behavior control.

The outcome of the RoboScrub development was unsatisfactory. The robot performed as it was designed to perform, but RoboScrub never became a commercial success. The reason for this likely has to do with the mismatch between the needs of the marketplace and RoboScrub's technology.

In the performance of their duties, robot cleaners add value slowly. Usually the rate is no more than a few dollars per hour—approximately the hourly rate an employer would have to pay to have a human worker do the same job. If a cleaning robot should bump into something, knock something over, or tumble down the stairs, it might damage merchandise, the facilities, or the robot itself. In an instant, the robot cleaner could easily undo all the economic good it had ever done. That is, the cost of the damage caused by the robot could exceed the money saved by using a robot in the first place. This concern makes designers of cleaning robots very cautious.

RoboScrub's designers understood this; they also understood the limits of their technology. In particular, sonar, as the primary obstacle avoidance sensor, has shortcomings. Large objects some distance away are sensed adequately. This is especially true if the obstacle presents a surface perpendicular to the direction of the sonar beam. Unfortunately, small obstacles (say, the slim metal pegs on which merchandise hangs) or large smooth objects approached obliquely (a glass case containing watches for example) are not reliably sensed by sonar.

Because of these and other blind areas, RoboScrub's designers designated the robot for use in large open areas only. Unfortunately for RoboScrub, customers do not normally segment their cleanable space into open areas versus areas near obstacles. A robot that could clean the middle of an aisle, but not the sides of the aisle, or one that could scrub the entrance foyer but none of the cluttered corridors connected to the foyer, held little interest to most potential customers. Still other cleaning possibilities were off limits because RoboScrub's navigation system required attaching large bar code targets (about 6 $\times$ 12 inches) at eye level. RoboScrub never found its market niche.

11.1.2 RoboKent

In 1988 the Kent Corporation, a manufacturer of cleaning equipment, began development of a product called RoboKent. RoboKent has the distinction of being perhaps the most successful cleaning robot to date.

The basic technology of RoboKent is similar to that of the Denning RoboScrub. RoboKent uses sonar as its primary obstacle sensor, and the robot has a method of detecting cliffs and sensing collisions. RoboKent, like RoboScrub, includes a mode that allows direct

Figure 11.3. The Kent company manufactures and sells both a robotic floor scrubber, pictured here, and a robot vacuum cleaner of similar construction. (photograph courtesy of Kent)

operator control—it's robotic brain can be disconnected to allow it to operate as a piece of manual equipment.

RoboKent is designed to work in either hallways or large open areas. To initiate cleaning, an operator guides the robot to the area to be cleaned and sets it in motion along a long straight wall. The robot follows the perimeter of the space until it returns to its starting location (as determined by dead reckoning). The robot then cleans inward from this point until it has covered the entire area.

There is no need to install beacons, coded targets, or other artificial markings as RoboKent does not rely on these things. Neither does RoboKent need to be programmed with a map of the area it is to clean. Instead, RoboKent restricts the areas it can clean to places with simple, rectangular geometries, and it requires an operator to move the robot from one area to another.

RoboKent has had a position in Kent's product line for a number of years, and has enjoyed modest success. Arguably, RoboKent's technology is no better than that of RoboScrub—the limitations of both robots makes them attractive to only a specialized segment of the cleaning market. No floor cleaning robot to date has demonstrated the advances in technology needed to breach the boundaries of this small niche.

Figure 11.4. The Denning Sentry, a security robot, patrolled a fixed area defined by active beacons. (photograph courtesy of Phil Veatch)

11.2 Making the Rounds

11.2.1 Sentry

The first product Denning Mobile Robotics attempted was a robot called the Denning Sentry. Sentry was envisioned as an all-purpose security robot. Bristling with intruder-detecting sensors, Sentry would tirelessly patrol a warehouse or other facility. When power was low, Sentry would automatically return to its charging station and recharge its batteries without operator assistance.

Sentry incorporated a ring of obstacle-detecting sonar sensors, infrared motion detectors, and a microwave motion detector. It also used a TV camera, microphone, and transmitter to transmit information back to the security station. To find its way around, Sentry required that some number of active IR beacons be installed in the space to be patrolled. Sentry followed one beacon until it reached an intersection where another beacon was detectable. Depending on the route stored in its program, Sentry could then follow the intersecting beacon, or note its position and continue following the first beacon. Given proper electrical modifications to the facility, Sentry could even call the elevator by radio and ride to a different floor.

Like RoboScrub, Sentry could claim both technological success and commercial failure. Without studying the security industry, it might be imagined that security personnel spend their time pursuing intruders and actively preempting theft. But in practice such events are very rare. Typical duties of a security guard include such things as testing the doors to make sure they are locked, checking that the coffee maker is turned off, and adjusting temperature and lighting to their after-hours settings. The Denning Sentry could do none of these things. Few customers felt that rolling about the hallways, transmitting back picture and sound justified the approximate $75,000 cost of the Denning Sentry.

11.2.2 HelpMate

The effectiveness of nurses and other hospital workers are reduced when they spend their time on incidental tasks. Ferrying paper work, medications, and specimens from station to station is not high on the priority list of any caregiver. The HelpMate robot from HelpMate Robotics, Inc. was designed to provide relief from these tedious tasks.

HelpMate navigates the corridors of a hospital using a combination of dead reckoning and sonar location. By pressing a button, a user can dispatch HelpMate to follow a stored map and navigate from one programmed station to another.

For obstacle avoidance, HelpMate relies on its sonar sensors, an interesting strobed-light triangulation system (Everett 95), and a sensitive bumper mounted near the floor. Like the Denning Sentry, HelpMate can be wired into elevators and automatic doors so that it can travel anywhere within a building.

HelpMate seems to be a robot that has found a profitable niche; it has been installed in nearly 100 hospitals worldwide. HelpMate Robotics offers the robot for sale or lease. Currently lease rates are a few dollars per hour.

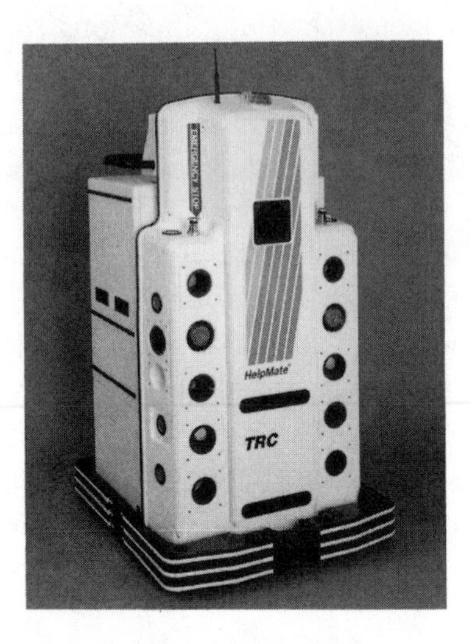

Figure 11.5. HelpMate Robotic's HelpMate robot transports medications, samples, and paperwork from one hospital station to another. (photograph courtesy of HelpMaster Robotics)

11.3 In Harm's Way

There are frequent reports in the media of robots being used in hazardous situations. Teleoperated robots were used in the cleanup following the incidents at Three Mile Island and the Chernobyl nuclear plants. Police routinely employ remotely-controlled devices to pick up, probe, or destroy bombs and suspicious packages.

So far, no danger-seeking autonomous robots have attained product status. However, one-of-a-kind autonomous robots have been used in high-risk situations, and other robots that may become products are under development.

11.3.1 Sojourner

The Pathfinder mission to Mars in the summer of 1997 stands as one of autonomous mobile robotics most compelling success stories. The robot Sojourner captured the imaginations of millions on Earth as it explored an alien world (see Figure 6.1).

Sojourner faithfully executed commands from its controllers at the Jet Propulsion Laboratory in California, but Sojourner was capable of more. For the first time, a robot sent to another planet had the power to take actions on its own. As Sojourner rolled about what proved to be the site of an ancient flood, the robot was alert for obstacles and cliffs. Sojourner could countermand instructions from Earth if it determined that it was in danger. Several times in the course of the mission, the robot did exactly that. In the event that communication from the Earth to the robot was lost, Sojourner even had the ability to conduct the exploration mission by itself.

The reduced scale and cost of Sojourner's mission represented a significant change of course for NASA where formerly bigger was better. But the choice of a small, autonomous robot was deliberate and carefully reasoned (Gat et al. 94). The cost of placing a payload on the surface of another planet is proportional to the weight of the payload, but the constant of proportionality is very large. The only way to limit the cost of a space mission is to severely limit the weight of the payload; the robot must be very small.

Mostly because of power considerations, however, a very small robot can be allowed only a very small computer. The roundtrip communication time between Earth and Mars is as much as 40 minutes. This delay prevents the possibility of direct teleoperation. Thus, a robot used in space with a modest onboard processor must nevertheless be capable of autonomous operation.

NASA recognized that behavior control is the logical software architecture for an autonomous robot of this type. The robot Sojourner was developed through a series of earlier prototypes and was programmed according to behavior control principles. Behavior control allowed Sojourner to get by with a processor some have described as "computationally challenged."

Because qualifying hardware for space operation can take years, Sojourner used a proven, but ancient, processor. At the heart of Sojourner's control system was an Intel 80C85, 8-bit processor able to execute 100,000 instructions per second. A total of 160K of memory was available to Sojourner, but the core of the program fit into only 16K of high-reliability memory. Remarkably, the robot that amazed the world had a processor much slower than that of Rug Warrior and only half as much main memory!

Figure 11.6. IS Robotics' Fetch robot was designed for explosive ordnance remediation. Fetch is a tracked, differentially driven robot. The mast holds the antenna for the onboard positioning system, the compass, and inclinometer. An electro-magnet on a one degree-of-freedom arm enables Fetch to lift and transport unexploded munitions made of ferrous material. A force sensing bumper protects the robot's front, infrared obstacle detecting sensors look out on all sides, and a munition sensing metal detector is mounted underneath the robot. (photograph courtesy of IS Robotics)

11.3.2 Fetch

The Fetch robot from IS Robotics[2] is designed for a serious purpose—the removal of small unexploded cluster bombs from a battlefield. (See Figure 11.6.)

Anti-tank cluster munitions are dropped from the air by the thousands during a battle. When the battle is over, munitions that didn't explode (up to 25% of the total) create a difficult and dangerous problem. Removal of these small but deadly munitions is an obvious task for a robot. Fetch was designed to prove the concept that a robot could perform this task.

In the operational scenario, the battlefield is returned to productive use in the following way: An explosive ordnance disposal

[2]Members of the original Fetch design team included Joseph Jones, Art Shectman, and Rosario Robert from IS Robotics and Richard Myers of ISX Corp.

technician arrives at the contaminated site in an all-terrain vehicle. The technician unloads and powers up 10 to 15 robots, each about the size of a carry-on suitcase. The robots comb the hazardous area, searching for unexploded munitions. The robots pick up any munitions they discover and transport them to a disposal point. After the entire area has been searched and all munitions collected, a robot carries an explosive charge to the disposal point. The charge is detonated remotely and all unexploded munitions are destroyed at once.

To implement these tasks, Fetch requires a number of functional subsystems and behaviors.

Fetch needs a means to detect munitions. Fortunately, the munitions are made of ferrous metal and are thus easily located using commonly available metal detectors. By bringing an inert munition and other metal objects into a local Radio Shack store and performing experiments in the aisle, a member of the Fetch team was able to identify and purchase a suitable metal detector for the robot.

To retrieve munitions, Fetch uses a one degree-of-freedom arm. The arm can lower the attachment mechanism to the ground and lift up the munition for transportation. The attachment mechanism, a simple electromagnet, exploits the fact that the munitions are made of ferrous metal. A break beam sensor shines along the bottom of the electromagnet. With the arm lifted, a broken beam indicates that a munition is attached.

Fetch's munition search strategy needs to efficiently cover the area being searched while respecting the fact that obstacles will be present in the search area. It was discovered that a reflecting spiral pattern performs well in both regards (see Figure 11.7).

Fetch begins its search at some selected point (indicated by the checkered circle in Figure 11.7). From this point, Fetch moves outward in an arithmetic spiral. The distance between successive rotations of the robot is chosen such that the new area swept out somewhat overlaps the area of the previous rotation. This ensures that Fetch does not leave any accessible area unsearched. Upon encountering an obstacle that it cannot surmount, Fetch stops, spins in place 180 degrees, and continues the spiral in the opposite direction. This is a simple, yet robust, strategy for dealing with most obstacles in the search area.

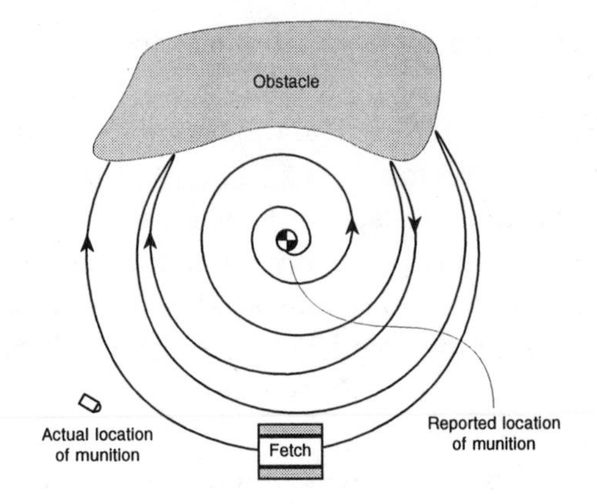

Figure 11.7. Fetch searches for munitions using a reflecting spiral strategy. Fetch begins by spiraling outward from its starting point. When an obstacle is encountered, Fetch rotates in place 180 degrees then continues the spiral in the opposite direction.

Systematic search would not be possible without a positioning system to tell Fetch where it is. Fetch uses a carrier phase-differential global positioning system (GPS) device for this purpose. This product delivers 2 cm positioning accuracy. Other search strategies that do not require a high precision navigation system to ensure search area coverage were investigated. However, such strategies, most of which involve random search, were rejected as inappropriate for the munition removal application.

A carrier phase-differential GPS device operates by, (incredibly) counting the number of wavelengths between the receiving mobile antenna and four or more GPS satellites. All this, while the satellites whiz by overhead at greater than 17,000 miles per hour. Unfortunately, such a system cannot compute its position instantly, but only converges on a highly-accurate estimate after integrating for many minutes. If the link between receiver and satellite is broken, the solution is lost, and the integration process must begin again. For this reason, anytime it becomes necessary to approach Fetch during testing, members of the Fetch team must crawl along, belly to the ground. This is the only way to avoid coming between the antenna on the robot and passing satellites!

Unlike most robots, Fetch cannot simply detect objects at a large distance and avoid them. Were it to do this, Fetch would leave an unsearched area, possibly harboring unexploded munitions, around every obstacle. Rather, while searching, Fetch must slow down when it comes near an obstacle, then proceed until it bumps the object and can go no further. For this reason Fetch has no need of long-range obstacle sensors but rather makes do with short-range reflective, near-infrared obstacle sensors. These sensors are mounted on the front, back, and sides of the robot. Fetch also uses a collision sensor that covers its entire front. An inclinometer enables Fetch to avoid tumbling over on slopes that are too steep.

The Fetch system incorporates an operator control unit (OCU) that allows the operator to intervene in the otherwise autonomous operation of the robots. The OCU makes it possible for an operator to direct the robots at any level of detail, reassigning robots to new tasks on the fly or extricating robots from situations their onboard programming cannot handle. An onboard video camera and transmitter relay the view from each robot back to the operator.

Behaviors

Fetch is controlled by a large number of behaviors. A simplified version of Fetch's behavior control diagram is shown in Figure 11.8.

At the top level is the Exec behavior. The Control Radio behavior handles the interaction with the radio modem. Control Radio feeds commands from the operator to Exec. Exec, like many of the other behaviors, can issue robot motion commands. Exec also sets the mode of the Drive Motor Arbitration behavior. Drive Motor Arbitration transfers motion commands from other behaviors to Drive Control. When the arbitration behavior receives conflicting commands, it selects for transfer the command whose behavior has the highest priority. Priority depends on the mode selected by the Exec behavior. Drive Control is the behavior that actually makes the motors move. The dark line in the Figure 11.8 is meant to indicate Exec's mode-setting capability.

The Escape behavior monitors the front force-sensing bumper and the rear-pointing IRs. Anytime the bumper senses force above a certain threshold, Escape commands motions to free the robot. These motions consist of backing up until either the robot has moved a certain distance, a timer has expired, or the rear IRs indicate an obstacle. Then Escape commands the robot to spin in place away

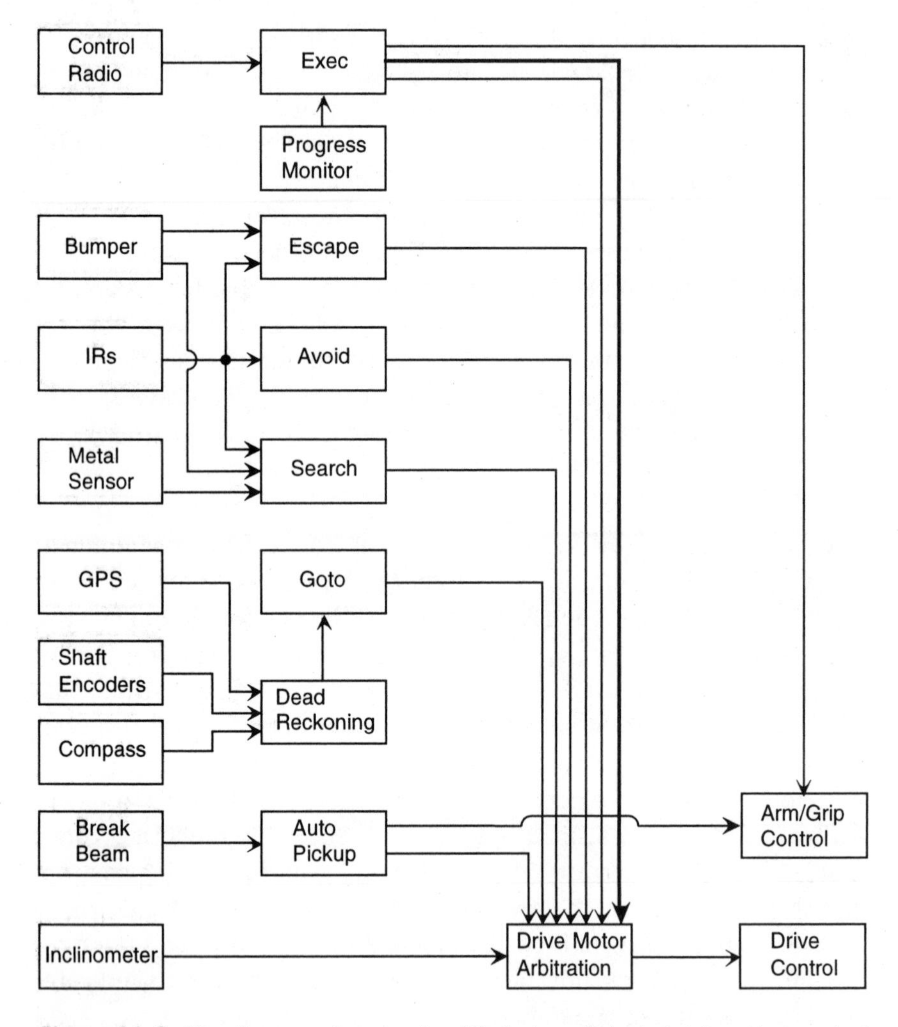

Figure 11.8. The diagram shows a simplified view of the behaviors that control Fetch.

from the side where the collision occurred. Finally, Escape moves the robot forward a distance equal to the backup motion.

While navigating to the starting point of a search operation or to the munition drop-off point, Fetch must avoid obstacles. But it is not acceptable for Fetch to simply bounce away from obstacles it encounters—this would make reaching the commanded point more difficult. Fetch instead must hug the perimeter of any object it finds, turning toward the commanded point when the obstacle has been cleared. The perimeter-hugging feature is embedded in the Avoid behavior.

Avoid and Escape are allowed to gain control of the robot's actuators only when Fetch is navigating to a point, either to begin a search or to put down a munition. While conducting a spiral search for a munition, the Search behavior subsumes Escape and Avoid. As stated earlier, during a search it would be an error for Fetch to turn away from an obstacle detected remotely. To discover as many munitions as possible, Fetch must search up to every obstacle until it can physically go no farther. The Search behavior is complete when either the Metal Sensor behavior indicates that a munition is present or when the robot has reached the boundary of its local search region.

The GPS, Shaft Encoders, and Compass behaviors all output values to the Dead Reckoning behavior. Dead Reckoning integrates this data to compute the robot's current position. This positional information is sent to the Goto behavior. Goto's purpose is to compute commands that drive the robot toward a given point. The connection is not shown in the drawing, but this goal point can be provided by the operator and sent to Goto via the Exec behavior.

When Search terminates successfully, that is, when the metal detector indicates the presence of a munition, the Auto Pickup behavior becomes active. The metal detector and electromagnet gripper are separate from each other; one is under the robot, the other is at the end of the arm. Thus, there is some uncertainty as to exactly where the robot should put down its gripper when the metal detector indicates a munition.

Auto Pickup commands the arm to go down, the magnet to turn on, and the arm to lift up. If the break-beam sensor that shines across the electromagnet is unbroken, then the munition has not been successfully acquired. In this case, Auto Pickup commands a localized search procedure. The arm moves up and down as the robot moves back in an effort to acquire the munition. If the beam of the break-beam sensor remains broken with the arm lifted, the munition is assumed to be attached. Goto is then allowed to drive the robot toward the disposal point.

The Arm/Grip Control behavior, as the name implies, has control over the arm and the electromagnetic gripper. Arm/Grip Control also has a bit of embedded arbitration, giving commands from the Exec behavior priority over commands from Auto Pickup.

Drive Motor Arbitration uses the readings from the inclinometer to limit the maximum velocity the robot is willing to go. When moving over rough terrain, the bouncing of the robot generates large transient readings from the inclinometer. This causes the robot to slow down. Also, when the robot is on a steep slope or has upended itself, the inclination reading causes the robot's velocity to go to zero. If the robot finds itself not moving, it alerts the operator. The operator can override this behavior and extricate the robot from its predicament.

Finally, the Progress Monitor behavior seeks to avoid behavior loops. Unfavorable arrangements of obstacles can cause the robot to become stuck while executing the Avoid behavior. Occasionally, avoiding one wall of an inside corner causes the robot to turn to face the other wall, which causes the robot to turn back to face the first wall, and so on. Progress Monitor measures the time the robot has been near the same location. If the robot dwells too long near one spot, Progress Monitor, forces a 180 degree turn in an attempt to break the cycle.

Results

In evaluation-testing, Fetch achieved its design goal, demonstrating that a robot can dispose of explosive ordnance. Fetch succeeded in autonomously navigating to a point, conducting a spiral search, locating a munition, picking the munition up, transporting the munition to a disposal point and placing the munition on the ground. However, more development, ruggedization, and cost reduction must be accomplished before Fetch is ready for operation in the field.

11.4 Summary

Many individuals, research groups, and companies have built impressive robots. The Denning Sentry, RoboScrub, and several other robots we have not mentioned were technological triumphs but failed in the marketplace. The RoboKent and HelpMate robots as well as a few others have carved out a small niche for themselves. But the mobile robotics industry still awaits the appearance of the equivalent of desktop computing's "killer app."

It should be clear that technology alone cannot ensure a robot's success. As builders, we may appreciate and value robots for their elegance and technological sophistication, but a customer purchases a robot because the robot accomplishes a needed task and does so at a cost lower than competing methods. Robot builders may be unused to thinking in these terms, but if our creations are to have any impact they must go beyond elegance and beauty and actually fill an economic need.

11.5 References

The best source of information about new robot applications is the World Wide Web. Thousands of pages on the web are devoted to robotics. More pages are added daily. An excellent starting point for most queries about robotics is the Robotics FAQ (Frequently Asked Questions) page at www.frc.ri.cmu.edu/robotics-faq.

NASA was famously successful with its web site devoted to Sojourner during the 1997 Mars exploration. Archived information about the Sojourner robot can be found by following the links at mpfwww.jpl.nasa.gov. Check the site for future developments as follow-up robotic missions to Mars are planned. (Gat et al. 94) has an illuminating account of the reasons behavior control architecture was chosen for the Mars robot. The paper also recounts the series of Earth-bound experimental robots leading up to Sojourner.

Before Sojourner, the robot *Dante* took on a dangerous exploration job on Earth. On eight metal legs, Dante walked into an active volcano. See img.arc.nasa.gov/new/projects/dante/dante.html for the details.

Information about RoboKent can be found at www.kentco.com. For a look at another interesting floor cleaning attempt and to see how the Paleozoic era has influenced robotics, try the website www.electrolux.fi/robot.

More information on the HelpMate robot can be found by visiting the HelpMate Robotics site, www.helpmaterobotics.com.

The Honda Motor Company has produced an impressive if somewhat eerily humanoid robot. The Honda Humanoid can be found by following the links starting at www.honda.co.jp/eng.

Cog, a robot that attempts to act like a human, if not look like one, is the current project of Professor Rodney Brooks at MIT. Check out www.ai.mit.edu/projects/cog.

Fetch and a plethora of other interesting robots are on display at the IS Robotics' website, www.isr.com.

Finally, because the web is so dynamic, you can expect any references given here to become obsolete at some point. In any case, to get the latest information from the web you should be prepared to search.

12

Robot Design Principles

You have by this point learned a great many technical details of robot creation. Take a moment now to step back and consider how all these lessons can be applied. What principles should guide you as you design and construct value producing, task achieving, real world robots? In this chapter we offer a few ideas.

12.1 Complexity Kills

Every system on a robot and every component of every system must be designed, built, and tested by the roboticist. Individually, each element of a robot can fail or misbehave in one or more ways. Additionally, every element has the potential to interact with every other element causing failures in combination that could not be produced by any element alone.

Every system and component you add increases your robot's complexity. If the added complexity is not justified by a corresponding increase in essential functionality, your project will suffer. Unnecessary or unjustified elements increase the time needed to design and build your robot. Once in place, unnecessarily complex systems add new possibilities of failure and increase the likelihood your robot will exhibit unexpected (and undesirable) behaviors. Such problems increase the time needed for debugging and repair.

Early in your project, as you decide exactly what you will build, attempt to strip away embellishments and non-essential features. Continue in this manner until only the "atomic" robot is left. An atomic design is one that cannot be further reduced; removal of any of the remaining elements would leave the robot unable to perform its task. Once the atomic robot functions robustly, you can confidently add other systems that complement and improve the functionality of the basic unit.

Complexity is the scourge of robust performance and the nemesis of project deadlines. A well-worn engineering adage is especially *apropos* for robots: "KISS–Keep it simple stupid."

12.2 Holistic Robotics

Every robot is built to accomplish a particular task. To succeed, the robot's various systems and components must all work together in support of this purpose. A common pitfall among robot builders (especially novice builders) is to become focused on a single aspect of the robot to the detriment of the system as a whole. Resist this temptation. If you want to build a robot that performs proficiently, you must think of the whole system not just your favorite part.

Strive for a balanced design. Overall performance is often limited by the robot's weakest subsystem. Concentrate your efforts on improving the most poorly performing subsystems. This will yield a much bigger payoff than making improvements to a subsystem that already performs adequately.

12.3 Code versus Reality

The experience that carefully planned code often yields unplanned responses from a robot, is a common one. This dissonance between the expected behavior of a robot and its actual behavior often stems from the programmer's failure to appreciate certain complexities. Situated in the real world, neither sensors nor mobility systems behave as our idealized models would have them do. Commanded motions may not be executed faithfully; sensor readings may be interpreted in more than one way. The programs you write must respect these ambiguities. The less sensitive your code is to inaccuracies in sensing or errors in motion execution, the more robustly your robot will perform. Experience will be your best guide.

12.4 Computer Program $\neq$ Robot Program

If you are used to writing computer programs targeted to run on a desktop computer, you will appreciate the advantage of faster processors and algorithms as they improve the performance of your program. The program that you write must compute a value (or perhaps a large set of values) and return the correct result to the user.

In contrast, a robot control program must compute an adequate response to incoming sensor data. The program must do this sufficiently quickly to assure a timely response to avoid irreversible errors or damage. This response is limited by the physical characteristics of the robot, such as mass, motor power, wheel traction, etc. A program that can make a decision in 1/1000 of a second is no better than a program that reacts within 1/100 of a second if the robot requires 1/10 of a second to actually respond. While a faster processor on a desktop computer can compensate for a host of programmer sins, this is not true for robot programs. If the robot programmer has failed to solve a problem correctly, using a faster processor to arrive at the solution more quickly is of no help.

12.5 Magician's Bag

Mobile robots face a harsh reality. Their motions are imprecise, their sensors are often in error, and the real world is too complex to be adequately modeled. What's a roboticist to do? The answer is redundancy.

When no one element can be relied on to work all the time, you must employ many elements that each work some of the time. To achieve robust performance in the real world a robot must, in effect, draw from a bag of tricks. When, for example, a specular reflection prevents the robot's sonar from seeing a wall at an oblique angle, the infrared sensor will discover the wall when it gets close enough. If the wall is a shiny black and the IR misses the wall too, the collision sensor will inform the robot.

12.6 Avoiding "Usually"

When analyzing the problem your robot will solve, beware of depending on the average case. Suppose you design your robot in a certain way based on the (correct) observation that: "Usually X is true." If X is usually true, it follows that sometimes X is false. You can be sure that your robot will find all those situations where X is false. What will it do then?

Never dismiss a potential problem by appealing to "usually." Designing a robust robot requires giving more consideration to what will happen when your assumptions are violated than to how the robot will behave when everything goes as planned.

12.7 Design Steps

We recommend asking the following questions in the order given whenever you design a new robot.

1. *What is the robot supposed to do?* It is vital that you develop a clear and detailed understanding of exactly the task your robot is to accomplish. The statement of this task is the source from which all other decisions about how to design your robot should flow.

2. *What is the simplest way to accomplish the task?* How you answer this question can largely determine whether your design job is easy or impossible, so be scrupulously honest. It may seem heresy but sometimes the simplest solution is to use a mechanism other than a robot.

3. *What mechanical platform is needed?* Knowing the problem and having chosen an approach to solving it, what power supply, body, and arrangement of actuators is required? The practical realities introduced at this step will often force you to rethink the previous steps.

4. *What information does the robot need?* Before considering the sensors that will supply data to the robot, decide in the abstract what the robot needs to know. This will drive your choice of sensors.

5. *What sensors can supply this information most effectively?* The robot uses data from its sensors to answer certain questions about its environment. The robot chooses between possible actions based on the answers to these questions. Thus, in general, a sensor that can answer a relevant question directly is preferred over a sensor whose output requires more interpretation.

6. *How can the problem be decomposed into behaviors?* Once the problem is understood, the simplest approach to a solution has been devised, and appropriate actuators and sensors have been chosen the problem of writing the robot control program should be greatly simplified.

An effective robot can almost never be built in a single iteration. You should expect multiple cycles of design, prototype, test, and redesign. Your project will benefit if you plan for these cycles in advance.[1]

Finally, following the heuristics given here can help you avoid some common errors. A text book, however, cannot build a robot, only you can do that. We wish you good fortune.

[1]Some will argue that adherence to good systems engineering practices will allow robot construction in a single iteration. We believe this is incorrect. In a mature technology, bridge building, for example, systems engineering has great benefit. However, for a technology in its infancy, such as robotics, many of the mathematical models and rules-of-thumb on which good systems engineering practices are founded, do not yet exist.

13

Unsolved Problems

Rug Warrior has been an exercise in both engineering and artificial intelligence (AI). We have seen that building a robot involves many issues. We have had to deal with bias in our circuits, bugs in our code, slip in our wheels, noise in our sensors, and transients in our power supplies. The process has forced us to take lessons from electrical engineering, mechanical engineering, computer science, and artificial intelligence.

The fortunate part is that we have been able to demonstrate this system on a minimal budget with eight chips, a few connectors, and some inexpensive sensors and actuators. Using only this simple system, we have also been able to teach modern theories of AI, which preach combining simple behaviors in programs that are embodied in the real world with real sensor data for input and real actuators for output. Rug Warrior then is an input/output (I/O) device for those wishing to study the issues involving the interplay between intelligence and embodiment.

We have seen some examples of the range of behaviors that Rug Warrior was able to achieve. Of course, if you expected that Rug Warrior would be as talented as R2D2 or C3P0, then perhaps you were disappointed. The gap between expectations and experiences for the beginning roboticist can be daunting.

The crux of the problem is that humans are just very good. We take many things for granted in our own biological selves: the acuity of our eyesight, the fine dexterity of our fingertips, the amazing power-to-weight ratio of our muscles, and the efficiency of our energy conversion system, to name a few. Instilling human-level equivalence in a robot is quite a challenge!

In fact, the disparity between expectations and experiences grows even wider if we think about the tiniest insects. Even their perceptual-motor skills are amazing. Common houseflies can land upside down on ceilings, spiders can assemble the most intricate homes, and ants can carry loads many times their weight. Robots have a lot of catching up to do.

13.1 Navigation

There are a host of unsolved problems in mobile robotics. One open question has to do with what is involved in endowing a robot with the ability to navigate its environment. Salmon can locate their spawning grounds from thousands of miles away, pigeons can find their destinations on either sunny or cloudy days, and bumblebees can make a beeline to a food source in quick response to another bee's dance. By contrast, few robots can make it down the hallway without recourse to humans modifying the environment with beacons and bar codes.

The underlying issue here is one of *representation*. What sorts of computational structures are required to grant competence in navigation? How far can reactive systems be stretched? Are world models and sensor fusion required at some point?

13.2 Recognition

Another problem that goes along with navigation is *recognition*. Landmark recognition in a generally unstructured environment is a very hard problem. Whether using cameras, pyroelectric sensors, force sensors, or microphones, recognizing patterns in the environment is not trivial. Recognition problems can be computationally intensive and subject to complexities due to lighting, occlusion, and noisy data.

13.3 Learning

As our mobile robots become more and more complex and as we attempt to make them more sophisticated, perhaps by incorporating more sensors for greater perceptual acuity or by adding more actuators for finer dexterity, the software can become severely strained in trying to deal with so many inputs and outputs. One area of research is to investigate how, when, and where learning algorithms can be incorporated into a robot's intelligence system to alleviate the programmer's burden. What are the right types of things to learn? Can a robot learn to calibrate its sensors? Can a mobile robot learn new and better behaviors?

13.4 Gnat Robots

The versions of Rug Warrior described in this book were rather small, simple machines. However, from another viewpoint, they were actually awfully *big*. Our Rug Warriors did not have any manipulators, did not do any assembly or heavy lift operations. They simply wandered around, looking, listening, and reacting.

There was really no good reason that they had to be as big as they were. The sensors and the silicon on board actually took up a fairly small amount of space. The motors and batteries took up most of the heft and bulk. The motors used were picked because they offered the cheapest configuration available at a rating suitable to carry the weight of the chassis and batteries.

With recent advances in hardware technology, we should be able to do much better in the future. The idea of gnat robots is to scale all the components of a robot down to a single piece of silicon, where motors, sensors, computers, and power supplies can all be printed in a single process very cheaply in a batch fabrication manner, much like integrated circuits.

It may be quite a while before a chip gets up and walks, but the technologies of microsensors, micromotors, and microbatteries are moving in that direction.

13.5 Cooperation

In the drive to make ever cheaper mobile platforms, both for useful robot applications and for wider availability of embodied machines for AI research, *single-chip* gnat robots have provided one image for a future goal. Rug Warrior, a machine approaching the concept of *single-board* robot, is a step toward this goal, but closer to today's technology.

The process of contemplating the notion of a single-chip robot makes building single-board robots look easy. Indeed, we have shown this to be the case with Rug Warrior and have walked all our readers through it. Rug Warrior is not exactly a single-board robot. While we have managed to put all the computational hardware and almost all the sensors on a single board, we still required a few hand-assembled connectors to attach the two motors, two encoders, and two batteries to Rug Warrior's circuit board. It is not a single-board robot, but it is close, and it is fairly simple and inexpensive.

Imagine, then, swarms of robots. What kind of intelligence systems could we build now? Again, intriguing questions arise from biology. How is it that colonies of small termites can work together to devour something as large as a house? How can a society of bees survive as an organization without (we presume) explicit detailed communication? What are the constraints that communication (or the lack thereof) imposes on the communal intelligence of a swarm of independent agents? How might collections of simple robots aid us in our human endeavors?

13.6 Thoughts

Twenty-five years ago, nobody would have believed a computer would be used as a fuel-injection controller in every car. Back then, when computers filled entire rooms, the people who built the first microprocessor were laughed at. They took a little bit of CPU, a little bit of memory, and a little bit of control logic and got a little bit of nothing, as far as most people were concerned—except that lots and lots of those tiny little computers eventually changed the way the world works.

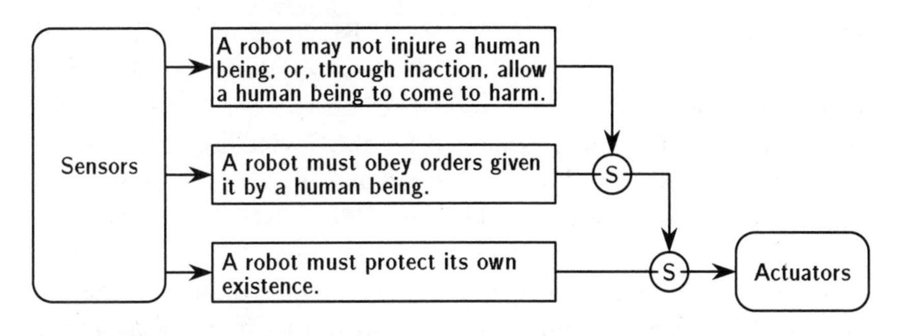

Figure 13.1. Asimov's familiar laws of robotics fit naturally into behavior programming.

Single-board robots are not very different—a little bit of brain, a little bit of brawn, a little bit of battery, and a little bit of sensing creates a little-bitty robot. But lots and lots of little-bitty robots may just change the way we think about solving problems.

The knowledge that surprising and novel unforeseen applications lie down the road makes the dream of building these mobile robots even more compelling. From inspiration to implementation, we hope our readers take their dreams to their limits.

13.7 Exercise

Many years ago, Issac Asimov listed three laws of robotics that declared how robots should behave. These behaviors fit well into the style of behavior programming (see Figure 13.1.) A robot's most basic behavior should be to protect its own existence, but given orders by a human, it should obey. Of highest priority is a behavior that prohibits a robot from harming a person or, through inaction, allows a human being to come to harm.

We leave the details of the implementation as an exercise for our readers. A mere matter of programming!

13.8 References

The problems of navigation and recognition are widely worked on, and many papers can be found in the IEEE journals *Robotics and Automation*, *Computer Vision* and *Pattern Analysis and Machine Intelligence*. Horn (1986) is a good text on the essentials of computer vision. Navigation is explained for the novice in (Miller, Winkless, Bosworth 98). The book includes a disk of simulation software. Another book (Borenstein, Everett, Feng 96) delves into robot navigation on a more rigourous level.

Machine learning is also a popular research interest at the present time. McClelland and Rumelhart (1986) introduce many of the ideas involved in neural networks and parallel processing.

Microfabrication technologies have expanded in the last few years from integrated circuits to microsensors and more recently to microactuators and microbatteries. A series of IEEE workshops on microelectromechanical systems [MEMS] was begun in 1987, and the proceedings from the ensuing years give broad coverage of this emerging field. Demonstrations of subsumption architectures during a similar timeframe illustrated that control systems for mobile robots could compile straightforwardly to a small number of silicon gates. This realization led Flynn to the idea of gnat robots (1987), which was expanded on in Flynn, Brooks, and Tavrow (1989). A journal, begun in 1992, the *IEEE/ASME Journal of Microelectromechanical Systems*, provides a forum for research results in this area of micromechanics and integrated systems.

With possibilities for smaller, cheaper robots, many people have begun to contemplate the possibilities of swarm intelligence. New conferences, such as the *International Conference on Simulation of Adaptive Behavior* (SAB), have emerged to draw these people together.

The most up-to-date information on a given topic in robotics can often be found on the web. To pursue an interest in any of the unsolved problems, fire up your favorite search engine and begin your quest in cyberspace.

Schematics

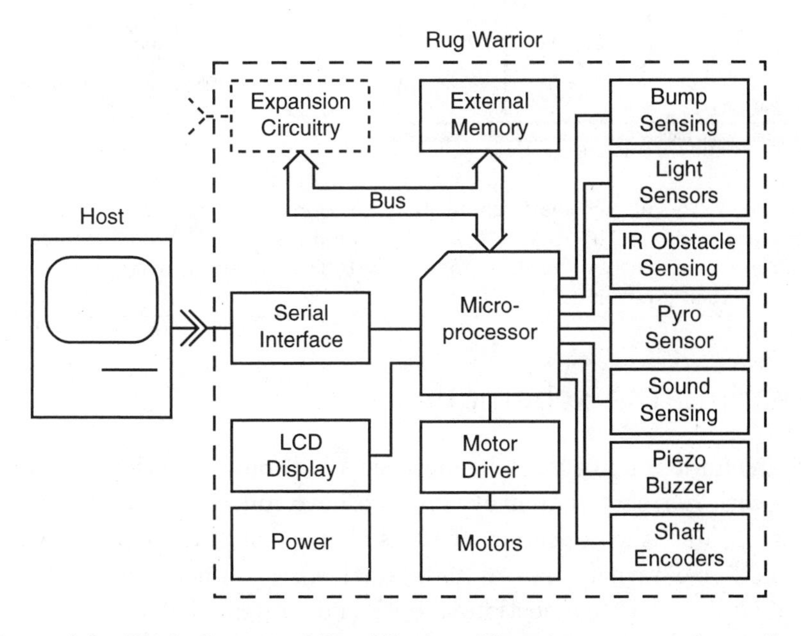

Figure A.1. Block diagram of Rug Warrior. The microprocessor has a direct connection to all onboard sensors and actuators. Additional sensors and actuators can be interfaced via optional expansion circuitry.

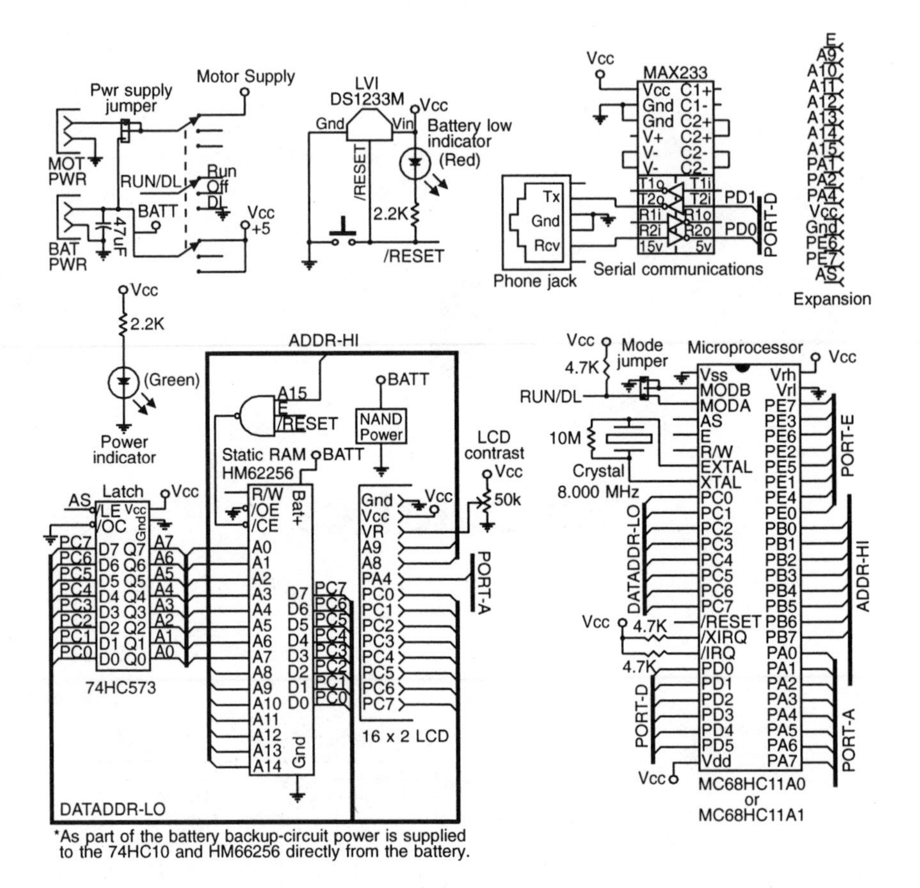

Figure A.2. This schematic shows Rug Warrior's logic, power and serial line interface circuits. The RCV, TX, and GND lines from the MAX233 chip can be connected to the user's host computer. Note that external memory is battery backed when power is off.

Reading the schematic

In an effort to make the schematic clear without cluttering the drawing with a great number of wires we have followed certain conventions. First, the collection of wires that forms the bus is shown as a bold line rather than as individual wires. There should be no ambiguity in the connection as both ends are labeled.

One can determine whether crossing wires are connected or not by examining Figure A.4.

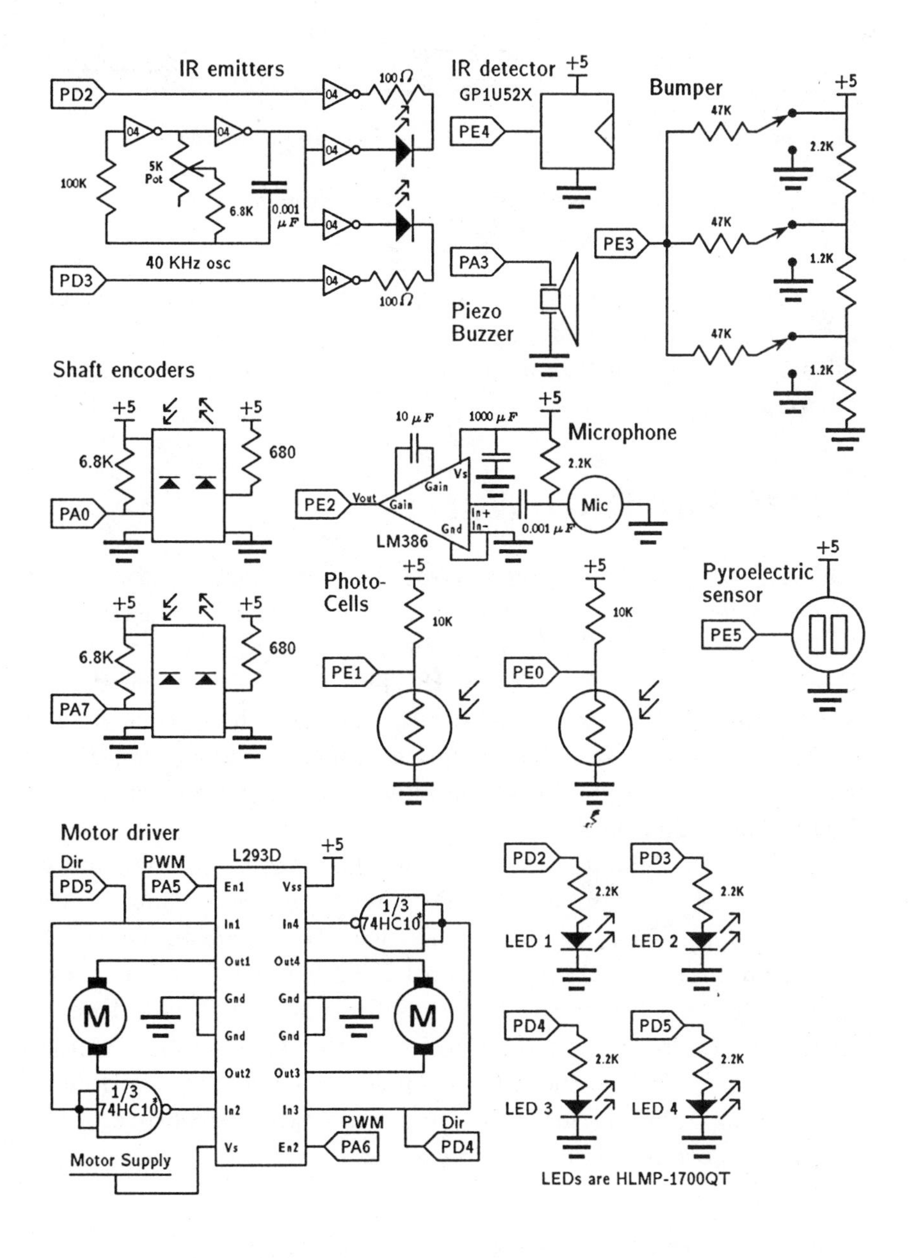

Figure A.3. Rug Warrior's sensor and actuator circuits.

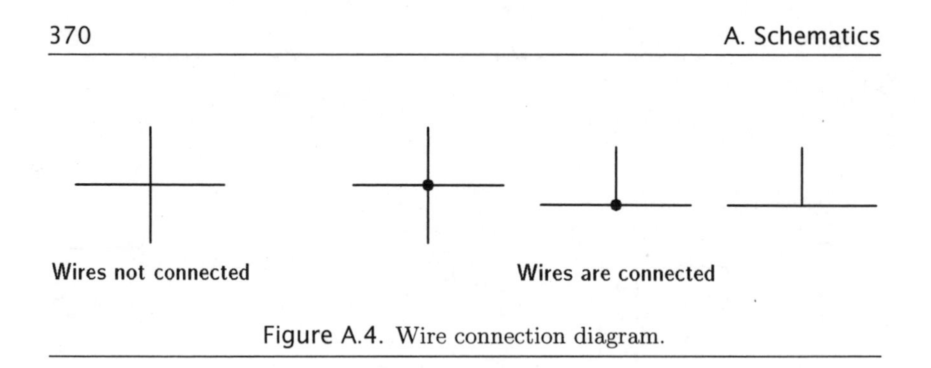

Figure A.4. Wire connection diagram.

A general description of how Rug Warrior's microprocessor, memory, and serial line circuits work can be found in Section 3.6 starting on page 61. The functioning of its sensors is described throughout Chapter 5.

Parts for Rug Warrior

Building a robot from scratch is not easy or quick. Neither is it inexpensive; this is due mostly to the fact that often distributors have hefty minimum order requirements. Indeed, the reason Rug Warrior was offered as a kit in the first place was to satisfy the many readers who complained about having a hard time finding the components we so painstakingly compiled. Nevertheless, for the diehard enthusiast who insists on building a robot from scratch, we include the following list of parts.

We list the manufacturer and part number for components we have used, but any pin-compatible parts from other manufacturers are fine. The Texas Instruments SN754410NE, for example, can be substituted for the SGS-Thompson L293D motor-driver chip. Some manufacturers sell directly to the public, others work only through distributors. Call or visit the web site of the manufacturer (contact information is in Appendix C or Appendix E) to find distributors nearest you. Rug Warrior can be built using either MC68HC11A0FN or the MC68HC11A1FN microprocessor.

Microprocessor and related components

1	Rug Warrior brain	Printed circuit board
1	Motorola MC68HC11A1FN	Microcontroller
1	52-pin plastic leaded chip carrier	PLCC 6811 socket

1 MS62256L-10P	32K×8 static RAM
1 74HC573AN	Latch
1 74HC10E	Triple input NAND
1 DS1233M	Low voltage inhibit
1 Augat MSS-3350	3P3T power switch
1 Panasonic P8037S	SPST reset switch
1 Optrex DMC-16249	16 × 2 LCD Screen
Clock circuit	
1 10 MΩ resistor	$\frac{1}{4}$ W resistor
1 8.000 MHz crystal	Clock
Battery backup capacitor	
1 47 μF capacitor	Tantalum capacitor
Power supply	
1 1000 μF capacitor	Electrolytic capacitor
4 4.7 KΩ resistors	Use one 5-resistor SIP
Serial line interface	
1 Maxim MAX233ACPP	Serial port driver
1 Hirose H9072	Phone jack 6-4 connector
Motor Driver Chip	
1 SGS Thompson L293D or TI SN754410NE	H-bridge
Debugging LEDs	
5 Quality Tech. Corp. HLMP-1700-QT-ND	High efficiency red LEDs
1 Quality Tech. Corp. HLMP-1790-QT-ND	High efficiency green LED
6 2.2 KΩ resistors	1 4-res. SIP, 2 $\frac{1}{4}$ W res.
Pyroelectric	
1 Eltec 442-3	Pyroelectric sensor
Shaft encoders	
2 Hamamatsu P5587 Photo IC	IR emitter/detectors
2 6.8 KΩ resistors	$\frac{1}{4}$ W resistors
2 680 Ω resistors	$\frac{1}{4}$ W resistors
Photocell circuit	
2 EG&G VT801	CdS photocells
2 10 KΩ resistors	$\frac{1}{4}$ W resistors
Piezo buzzer	
1 Panasonic ERB-RD24C411	Piezo buzzer

Bumper circuit

3	Omron SS5GLT	Lever switches
3	47 KΩ resistors	Use one 3-resistor SIP
2	1.2 KΩ resistors	$\frac{1}{4}$ W resistors
1	2.2 KΩ resistor	$\frac{1}{4}$ W resistors

Microphone circuit

1	Panasonic WM-034CY195	Microphone
1	LM386N-1	Op-amp
1	10 μF capacitor	Tantalum capacitor
1	0.001 μF capacitor	Disc capacitor
1	2.2 KΩ resistor	$\frac{1}{4}$ W resistor

IR emitters/detectors

1	Sharp GP1U52X	IR receiver unit
2	Siemens SFH486	IR LEDs
1	74HC04N	Inverter
1	100 KΩ resistor	$\frac{1}{4}$ W resistor
2	100 Ω resistors	$\frac{1}{4}$ W resistor
1	5 KΩ potentiometer	Cermet laydown style
1	6.8 KΩ resistor	$\frac{1}{4}$ W resistor
1	0.001 μF capacitor	Disc capacitor

Connectors

5	Samtec IDMD2S12	Cable plug strips
2	Samtec IDSD2S12	Cable socket strips
4	Samtec SS132T2	32-socket socket strips
2	Samtec TS132T-AA	32-plug terminal strips
1	Amp 2-640463-3	8-pin IC socket
2	Amp 2-640357-3	14-pin IC socket
1	Amp 2-640358-3	16-pin IC socket
2	Amp 2-640464-3	20-pin IC socket

Distributors for Rug Warrior Components

Manufacturers usually work through a number of local distributors. If the distributors do not have parts in stock and the lead times are long, you can call the manufacturer and have them locate distributors for you who do have the parts in stock. Alternatively, the manufacturer may be willing to give you samples. Motorola has a special division, Motorola University Support, which assists universities and schools in acquiring parts which may have long lead times, such as the MC68HC11A0FN microcontroller[1]. Here we list a number of distributors or manufacturers who sell components used in Rug Warrior. The list is not meant to be exclusive, just helpful in getting started. Distributors of semiconductors usually carry whole lines of semiconductor companies' products and specific chips listed below under one distributor can likely also be bought from another distributor. Call around to comparison shop and find out who has what in stock.

In the list below, note that when discrete components such as resistors are necessary, while the schematic will show separate resistors, it is often easier to incorporate resistors into a design using resistor single-in-line-packages (SIP). Resistor SIPs come in two varieties, bussed and isolated. Isolated resistor SIPs have separate pins for each end of the resistor while bussed resistor SIPs have one end of each resistor tied to a common pin. Bussed resistor SIPs are convenient for pullup resistors tied to the positive voltage supply.

Motorola University Support	(602) 952-3855
Newark Electronics	(508) 683-0913
1 MC68HC11A0FN	Microcontroller
1 10F7807	Augat MSSA-3350 3P3T switch
1 50F066	1000 μF capacitor
1 44F7982	Amp 2-640463-3 8-pin IC socket
2 44F983	Amp 2-640357-3 14-pin IC socket
1 44F7984	Amp 2-640358-3 16-pin IC socket
2 44F7986	Amp 2-640464-3 20-pin IC socket
Wyle Laboratories	(800) 444-9953
1 SN754410NE	H-bridge (replacement for L293D)
Maxim Small Orders Desk	(408) 737-7600
1 MAX233ACPP	Serial port driver
Hamamatsu	(908) 231-0960
2 P5587	IR emitter/detector pairs
Allied Electronics	(800) 433-5700
2 980-2500	EG&G VT801 CdS photocells

[1] Rug Warrior's brain can also be constructed using a MC68HC11A1FN microprocessor.

Samtec	(812) 944-6733
5 IDMD2S12	4-pin plug cable assemblies
2 IDSD2S12	4-pin socket cable assemblies
4 SS132T2	32-socket socket strips
2 TS132T-AA	32-pin terminal strips
Hamilton-Hallmark	(800) 272-9255
1 74HC10E	Triple NAND
1 74HC573AN	Latch
1 74HC04N	Inverter
1 MS62256L-10PC	32K×8 static RAM
2 SFH485	IR LEDs
Sterling Electronics	(617) 938-6200
1 Sharp GP1U52X	IR detector module
Dallas Semiconductor	(214) 450-0400
1 DS1233M	Low voltage inhibit
Digi-Key	(800) 344-4539
1 H9072	Phone jack 6-4 socket
1 LM386N-1	Op-amp
1 P8037S	SPST reset switch
3 SW143-ND	Omron SS5GLT bump switch
1 X056	8.000 MHz crystal
1 750-83-R-2.2K	2.2 KΩ (isolated) resistor SIP
1 750-61-R-4.7K	4.7 KΩ (bussed) resistor SIP
1 750-63-R-47K	47 KΩ (isolated) resistor SIP
5 HLMP-1700-QT-ND	2.0 mA red LEDs
1 HLMP-1790-QT-ND	2.0 mA green LED
1 P9924	ERB-RD24C411 piezo buzzer
1 P9962	Panasonic WM-34CY195 microphone
1 100Q	Pack of 5 100 Ω resistors
1 680Q	Pack of 5 680 Ω resistors
1 1.2KQ	Pack of 5 1.2 KΩ resistors
1 2.2KQ	Pack of 5 2.2 KΩ resistors
1 6.8KQ	Pack of 5 6.8 KΩ resistors
1 10KQ	Pack of 5 10 KΩ resistors
1 100KQ	Pack of 5 100 KΩ resistors
1 10MQ	Pack of 5 10 MΩ resistors
1 P4200	Pack of 10 0.001 μF disc capacitors
1 P2026	10 μF tantalum capacitor
1 P2030	47 μF tantalum cap
1 36C53	5 KΩ potentiometer
1 36C54	50 KΩ potentiometer

Accessories and Alternatives

To download code to Rug Warrior from your personal computer, you will need a serial port cable to connect to the Rug Warrior board's modular phone jack connector which is similar to a telephone handset's socket. You will need a cable which has a matching modular phone jack plug on one end and the proper connector for your off-board computer on the other end. Oftentimes, a computer's serial port connector is a D-shaped female DB-25 connector. The easiest thing to do is to buy a special male DB-25 connector which has an attached modular phone socket on the back of the case with wires and plugs in between, which allows you to configure the pin-outs in whichever way you like. Then you can just use a normal phone cord for the cable running from this connector, which you plug onto the back of your workstation, to Rug Warrior. Digi-Key sells the necessary parts.

Another suggestion is to use a non-volatile RAM in place of the static RAM listed above. While the battery-backup circuit on Rug Warrior will keep the program resident in RAM as long as the batteries are plugged in, you may find that in working on Rug Warrior, you tend to take it apart fairly often and disconnect the batteries. Non-volatile RAMs are more expensive than static RAMs, but have a battery inside the chip's package which keeps the memory backed up even when you remove it from Rug Warrior's board. Dallas Semiconductor and Greenwich Electronics both sell 32K×8 non-volatile RAMs. Replacements for other components such as sensors and discrete electronics can also be found at Radio Shack as listed below.[2]

The pyroelectric sensor that we specified (which has an on-chip amplifier) for Rug Warrior is fairly expensive. You can add it as an accessory or buy a discrete pyroelectric sensor and add your own amplifier.

[2]If you use the non-volatile RAM you should disable the battery backup circuit. The quiescent current drawn by the non-volatile RAM is large compared to the current needed by the static RAM and will drain the robot's batteries in a few days.

Digi-Key	(800) 344-4539
1 H164107ND	Phone cable assembly
Dallas Semiconductor	(214) 450-0400
1 DS1230AB-120	32K Non-volatile static RAM
Greenwich Electronics	(800) 476-4070
1 GR3281-100	32K Non-volatile static RAM
Eltec	(800) 874-7780
1 442-3	Pyroelectric sensor w/amplifier
Vero Electronics	(800) 242-2863
1 244-26221G	Reel of 250 Speedwire pins
1 244-26213E	Speedwire wiring pen
Samtec	(812) 944-6733
1 ESW-136-34-T-S	Spacer connector socket for LCD
1 TSW-136-34-T-S	Spacer connector plug for LCD
Radio Shack	(Consult local directory)
2 276-143	SYIR53L IR LED
1 276-137	GP1U52X IR detector module
1 276-099	Infrared sensor display card
1 276-1657	Pack of 5 Cds photocell
1 270-090	Microphone

B

Rug Warrior Programs

The following sections implement a number of example Rug Warrior programs. We invite you to build on these examples to create programs for you own robot. All programs except the velocity control program rely on standard library functions. (The library is distributed with **IC**. **IC** plus the example code in this appendix is included with the new Rug Warrior Pro^{TM} robot kit.)

Important library functions include:

`analog(chan)` Perform an A/D conversion on analog channel `chan`. Wait until conversion has completed before returning.

`bumper()` Return a number identifying which bump switches are closed.

`defer()` This function is used to tell the scheduler that the current process is ready to be suspended. `defer` has no effect other than to increase the efficiency of the code.

`msleep(msec)` Sleep for `msec` milliseconds.

`mseconds()` Return the number of milliseconds since the last reset.

`peek(addr)` Return the 8-bit byte stored at `addr`.

`poke(addr, value)` Load an 8-bit `value` into the memory location specified by `addr`.

`sleep(sec)` Wait for `sec` seconds before returning.

`start_process(proc-name)` Begin a process that will run in the background. (See Chapter 9 for more details.)

`tone(freq, duration)` Activate the piezoelectric buzzer at a frequency of `freq` for `duration` seconds.

`set_beeper_pitch(freq)` Set the frequency the beeper will make when it is activated.

`beeper_on()` Turn beeper on continuously.

`beeper_off()` Turn beeper off.

`init_velocity()` Initialize open loop PWM velocity control.

`get_left_clicks()` Count the number of clicks made by the left encoder since the last call.

`get_right_clicks()` Count the number of clicks made by the right encoder since the last call.

`motor(index, vel)` Low level open loop motor control command. The motor identified by `index` (0 or 1) is commanded to move at velocity `vel`. `vel` is a percentage of maximum velocity in the range -100 to +100.

`drive(trans, rot)` Fundamental open loop robot motion command. `drive` causes the robot to translate at a velocity of `trans`, rotate at a velocity of `rot`. `trans` and `rot` are given as a percentage of maximum. Each argument can take on integer values from -100 to +100. Note that a few examples in the book rely on an earlier version of `driver` (and `motor`) that had float arguments. You can use type conversion to make these examples work with newer libraries.

Each of the example programs below defines a function of the form `start_name`. To run the example from the keyboard, you must type in this function. The first example can thus be started by typing: `start_bugle();`.

B.1 Bugle

Bugle demonstrates one of the simplest possible connections of sensing to actuation. The bumper is used to control the frequency of the tone emitted by the piezo buzzer. The function named `bugle` does just two things: it prints "Bugle" on the LCD when it begins, and it loops endlessly on the function `select_bumper`.

The function `select_bumper` is also fairly simple. It checks to see if the bumper is currently depressed. If the bumper has been pressed (bpr $\neq 0$) then `select_bumper` turns the beeper on; if the bumper is not pressed `select_bumper` turns the beeper off. The function `select_bumper` further sets the frequency of the piezo buzzer depending on where the bumper was pressed. The possible notes are C, F and A; the C, F, and A in the octave above middle C; and the C two octaves above middle C.

With practice, you can play reveille, taps, and other bugle favorites on your robot!

```
/* Bugle -- Connect bump switches to tones */

/* There are 12 steps in an octave.  Step 0 corresponds to A, 1 to A
   sharp, 2 to B, and so on.  A common tuning is to let A = 440 Hz.
   To compute the frequency of a given note we use: freq = 440 * (2 ^
   (s / 12)), where s is the step.  Step 0 is the A below middle C,
   step 12 is the A above middle C.
 */

float octave = 440.0;

float  c_note = octave * (2.0 ^  3.0 / 12.0);
float  f_note = octave * (2.0 ^  8.0 / 12.0);
float  a_note = octave * (2.0 ^ 12.0 / 12.0);
float c1_note = octave * (2.0 ^ 15.0 / 12.0);
float f1_note = octave * (2.0 ^ 20.0 / 12.0);
float a1_note = octave * (2.0 ^ 24.0 / 12.0);
float c2_note = octave * (2.0 ^ 27.0 / 12.0);

/* Select an action based on which bumper switch is closed
Note played       --> none  C   F   A   C1  F1   A1  C2
Switches closed   --> none  B  R&B  R  L&R   L  L&B LRB
bumper() returns -->   000 100 101 001 011 010  110 111
 */
```

```
void select_bumper()
{ int bpr = bumper();           /* Local Var for bumper contents */
  if      (bpr == 0b100)
    set_beeper_pitch(c_note);
  else if (bpr == 0b101)
    set_beeper_pitch(f_note);
  else if (bpr == 0b001)
    set_beeper_pitch(a_note);
  else if (bpr == 0b011)
    set_beeper_pitch(c1_note);
  else if (bpr == 0b010)
    set_beeper_pitch(f1_note);
  else if (bpr == 0b110)
    set_beeper_pitch(a1_note);
  else if (bpr == 0b111)
    set_beeper_pitch(c2_note);

  if (bpr !=0)        /* Turn beeper on if any switch is pressed */
    beeper_on();
  else                   /* Turn beeper off otherwise */
    beeper_off();
}

void bugle()
{ printf("Bugle \n");
  while(1)
    { select_bumper();
    }
}

void start_bugle()
{ start_process(bugle());
}
```

B.2 Theremin

The *Theremin* (named for Russian inventor Leo Theremin) is a musical instrument whose pitch and volume is controlled by capacitive coupling between the instrument's two antenna and the performer's hands. Rug Warrior simulates this control using the amount of light falling on the photocells rather than capacitive coupling. Also, rather than controlling the volume, we select how often the tone is played.

```
/* Theramin */

/* Compute beeper frequency from the difference of the photo cells */

float freq(int left, int right)
{ int delta;
  float frq = 100.0;
  delta = left - right;
  frq = 2500.0 * (1.0 + ((float) delta) /((float) max(left, right)));
  return frq;
}

float period_gain = 0.0005;

/* Compute the length of the pause between beeps from the sum of the
   photo cells */

float period(int left, int right)
  /* Protect against negative periods */
{ return (period_gain * (float) (max(0, (512 - (left + right)))));
}

void theremin()
{
 int left = 0;
 int right = 0;
 while (1)
   {
     left  = analog(photo_left);
     right = analog(photo_right);
     tone(freq(left, right),0.1);/*Play a tone of a certain freq */
     sleep(period(left, right));   /*  then wait a while */
  }
}

void start_theremin()
{ printf("Theremin\n");
  start_process(theremin());
}
```

B.3 Yo-Yo

The yo-yo program exercises the robot's skirt and shaft encoders for input and its motors for output. When you press the skirt at the back, yo-yo directs the robot to drive forward a preset distance, then stop and backup until the robot returns to the starting point.

```
/* Yoyo -- when the back bumper is pressed the robot goes out a
   measured distance.  Then it comes back.
   Requires: track.c
 */

int yo_inches = 30;          /* Go out this many inches */

int yoyo(int yo_inches)
{
  int yo_clicks =  yo_inches * 2; /* Aprx clicks to in. conversion*/
  while(1)
        { if (bumper() == 0b100) /* Bumper was hit from the back */
              {
                    track(90,0,yo_inches);  /* Go out */
                    track(-90,0,yo_inches); /* Come back */
                    sleep(0.5); /* Wait in case bpr is vibrating */
        }
    }
}

void yo_tune()
{ tone(1000.0 * (2. ^ 12./12.) ,0.15);
  tone(1000.0 * (2. ^  8./12.) ,0.15);
  tone(1000.0 * (2. ^  5./12.) ,0.15);
  tone(1000.0 * (2. ^  8./12.) ,0.15);
  tone(1000.0 * (2. ^ 12./12.) ,0.15); }

void start_yoyo()
{   printf("Yo-Yo\n");
    yo_tune();
    init_velocity();                /* Needed by track */
    sleep(1.0);
    start_process(yoyo(yo_inches));
}
```

Yo-yo and other programs depend on the Track function:

```
/* Track - Move a specified number of encoder clicks */
/* init_velocity must be called before calling track */

float stop_time = 0.035;    /* Approximately, one servo cycle */

/* Track - drive at selected velocity until CLICKS encoder clicks */
/* have occurred, then stop.  Track works best with either */
/* trans_vel = 0, or rot_vel = 0. */

int track(int trans_vel, int rot_vel, int clicks)
{   long time_out = mseconds() + 5000L;  /* Time out after 5 sec */
    int l_tot_clks = 0;     /* Total clicks */
    int r_tot_clks = 0;
    int ave_clks = 0;
    get_left_clicks();      /* Reset clicks */
    get_right_clicks();     /* Reset clicks */
    driveb(trans_vel, rot_vel);   /* Turn motors on */
    while(1)
        {
           l_tot_clks = l_tot_clks + get_left_clicks();/* Sum clks*/
           r_tot_clks = r_tot_clks + get_right_clicks();
           ave_clks = min(l_tot_clks, r_tot_clks); /*Trig ave_clks*/
           if ((ave_clks >= clicks) || (mseconds() > time_out))
                { driveb((- trans_vel), (- rot_vel));/*Kill motion*/
                  sleep(stop_time);
                  stop();
                  return (clicks - ave_clks);
                }
        }
}
```

B.4 Wimp

The Wimp program just tries to escape annoyances. Press on the skirt at any point. The robot will beep and try to move away from the point where it was touched.

```
/* Wimp.c -- move away from any touch
   Requires: track.c
 */


float oct = 440.0;
```

```
float wfreq(int step)   /* Freq. of note */
{ if (step == 0)
    return oct;
  else
    return (oct * (2.0 ^ ((float) step / 12.0)));
}

int wimp_active = 0;/* Don't start process when already running */

void start_wimp_tune()
{ if (! wimp_active)
    start_process(wimp_tune());
}

void wimp_tune()
{ wimp_active = 1;
  oct = 1760.0;
  tone(wfreq(5),0.1);
  tone(wfreq(4),0.1);
  tone(wfreq(5),0.1);
  tone(wfreq(4),0.1);
  wimp_active = 0;
}

/* Move away from any bump -- play an annoyed tune */

void wimp()
{ int bmpr;
  printf("Wimp\n");
  while(1)
    { bmpr = bumper();
      if (bmpr != 0)        /* Somehow, we were bumped */
        { start_wimp_tune();       /* Play an-annoyed-wimp tune */
          if (bmpr == 0b110)       /* Back and Left bump */
            track(0,-80,5);        /* Spin right ~60 deg */
          else if (bmpr == 0b010)        /* Left Bump */
            track(0,-80,10);       /* Spin Right 120 deg */
          else if (bmpr == 0b011)   /* Left and Right => Front */
            { track(-80,0,4);      /* Backup then turn away */
              track(0,80,17); }
          else if (bmpr == 0b001)        /* Right Bump */
            track(0,80,10);        /* Spin Left 120 deg */
          else if (bmpr == 0b101)        /* Right and Back bump */
             track(0,80,5);        /* Spin Left 60 deg */
          track(80,0,8);           /* Go forward a bit */
```

```
            sleep(0.75);             /* Let bumper damp out */
        }
    }
}

void start_wimp()
{ tone(1000.0,.1);
  wimp_tune();
  sleep(1.0);
  init_velocity();          /* Needed by track */
  start_process(wimp());
}
```

B.5 Follow

With the Follow program loaded, put your hand in front of the robot. The robot will move toward your hand and stop pressed against it. Follow uses the IR sensors to find and follow the nearest object. A collision with that object causes the robot to turn off its motors.

If you have two robots, an interesting "cooperative behavior" can be exhibited by loading Wimp on one and Follow on the other. The robot executing Follow will move toward the Wimp robot until it collides. The Wimp robot will then move away. Follow will again move toward its target. The robot will thus move across the floor in a dance of apparent longing and annoyance.

```
/* Follow.c -- Try to crowd others.  Use the IRs to find other robots
 * and follow them.
 */

int fol_trans_def = 90; /* Default translational velocity */
int fol_rot_def = 35;          /* Default rotational velocity */

/* Use the IRs to follow closest target */

void follow()
{ int ir = 0;            /* Local var for obstacle infrared data */
  int bmp = 0;           /* Local var for bumper data */
  int old_bmp = 0;       /* Local var for bmp on previous iteration*/
  printf("Follow\n");    /* Print message on LCD screen */
  while(1)
```

```
    { ir = ir_detect();  /* Record IR obstacle detection sensor */
      bmp = bumper() && 0b011;/* Only consider Left or Right col. */
      if (old_bmp & (! bmp))    /* Collision just ended */
        sleep(0.5);          /* Wait a bit before following */
      else if (bmp)          /* If bumper is pressed... */
        stop();              /* Stay right here */
      else if (ir == 0)      /* Don't know which way to go, so stop */
        stop();              /* Stay right here */
      else if (ir == 0b11)        /* Object straight ahead */
        driveb(fol_trans_def,0); /* Race forward */
      else if (ir == 0b01)        /* Object on right */
        driveb(fol_trans_def,(- fol_rot_def));/* Arc to the right */
      else if (ir == 0b10)      /* Object on left */
        driveb(fol_trans_def,fol_rot_def);      /* Arc to the left */
      sleep(0.1);                   /* Debounce bumper */
      old_bmp = bmp;
    }
}

void start_follow()
{ start_process(follow());
}
```

B.6 Echo

With Echo running, whistle or clap loudly a few times in quick succession. The robot will repeat the number of loud sounds it hears by beeping and rocking back and forth. (If you operate the robot in a noisy environment, you may need to change the value of `sound_delta`.)

Echo is an interesting behavior for two robots to execute simultaneously. Have the first robot begin by making some number of beeps. The second robot will echo the beeps, the first will echo the echo, and so on. The robots seem to talk back and forth to each other.

```
/* Echo -- Repeat the number of loud sounds heard */

int sound_delta = 50;/* Must be low for robots to hear each other */
long hold_time = 175L;/* Number of milliseconds to hold high time */
int loud_p = 0;          /* 0 => quiet, 1 => loud */
```

```
int sound_count = 0;

int listen_p = 1;        /* Make it 0 during echo execution */

int s_level_max = 0;

float warb_low = 1500.0;
float warb_high = 1600.0;
float warb_dur = 0.01;

void warb()
{ float freq = warb_low;
  twitch();                      /* Motion accompanies sound */
  beeper_on();
  for (freq = warb_low; freq < warb_high; freq = freq + 10.0)
    { set_beeper_pitch(freq);
      sleep(warb_dur); }
  beeper_off();
}

/* Sample the microphone to see how much noise there is */

void sample_sound()
{ int s_level = 0;               /* Instantaneous sound level */
  long go_low_time = 0L;
  long current_time = 0L;
  while (1)
    { while(listen_p)
        { current_time = mseconds();
          s_level = abs(analog(microphone) - 128);
                                 /*Abs diff from 128*/
          if (s_level > s_level_max)
            s_level_max = s_level;
          if ( s_level > sound_delta)
            go_low_time = current_time + hold_time;
                                 /*New go low time*/
          if (current_time > go_low_time)
            loud_p = 0;
          else
            { if (loud_p == 0)
                    /* Was 0, now 1 => add to trans count */
                sound_count++;
              loud_p = 1; }
        }}}
```

```
/* Count the number of low high transitions during some interval.
   Start a timer at the first low high transition, then keep
   checking for more low high transitions until time runs out */

int echo_cmd = 0;
float cmd_period = 1.2; /* Gather sound for this long */

void capture_command()
{ int old_count = 0;
  while(1)
    { if (old_count != sound_count)
        { sleep(cmd_period); /* Wait for the sounds to happen */
          echo_cmd = sound_count;
          sound_count = 0; }
    }}

void twitch_aux()
{ drive(100,0);
  sleep(.07);
  drive(-100,0);
  sleep(0.07);
  drive(0,0); }

void twitch()
{ start_process(twitch_aux(),1);
}

void echo_control()
{ int i = 0;
  while(1)
    { if (echo_cmd != 0)
        { listen_p = 0;
          for (i = 0; i < echo_cmd; i++)
            { warb();
              sleep(0.2);
            }
          sleep(0.15);  /* Wait longer before listening again */
          echo_cmd = 0;
          listen_p = 1;
        }
    }
}
```

```
void rpt()
{ while(1)
    {
      printf("Echo  Cnt: %d Max sound: %d\n",
                                  sound_count, s_level_max);
      s_level_max = 0;
      sleep(0.4);
    }}

void start_echo()
{ start_process(sample_sound());
  start_process(capture_command(),1);
  start_process(echo_control(),1);
  start_process(rpt(),1);
}
```

B.7 Sonic Commander

Three loud sounds in rapid succession make the robot go forward. Two sounds cause it to spin in place and one sound makes the robot stop. Sonic Commander lets you drive the robot around by clapping, whistling, or even speaking loudly. For example, it will seem that your robot understands speech if you command it by shouting "Stop!," "Turn now!," and "Go forward now!"

```
/* Sonic commander -- steer the robot with sound */

/* The value for snc_sound_delta must be chosen carefully because
   the microphone tends to hear the sound of the motors. */

int snc_sound_delta = 80; /* Pick large number,  make  LOUD sound*/
long snc_hold_time = 100L;/* Number of ms to hold high time */
int snc_loud_p = 0;         /* 0 => quiet, 1 => loud */
int snc_sound_count = 0;

int snc_bmpr = 0;         /* Escape process/analog conflict problem */

void snc_sample_sound()
{
  int s_level = 0;          /* Instantaneous sound level */
  long go_low_time = 0L;
  long current_time = 0L;
```

```
  while (1)
    { snc_bmpr = bumper();
                  /*Must have all analog calls in one process*/
      current_time = mseconds();
      s_level = abs(analog(microphone) - 128);
                                         /*Abs diff from 128*/
      if ( s_level > snc_sound_delta)
        go_low_time = current_time + snc_hold_time;
                                         /* New go low time*/
      if (current_time > go_low_time)
        snc_loud_p = 0;
      else
        { if (snc_loud_p == 0)
                    /* Was 0, now 1 => add to trans. count*/
            snc_sound_count++;
          snc_loud_p = 1; }
    }}

/* Count the number of low high transitions during some interval.
   Start a timer at the first low high transition, then keep
   checking for more low high transitions until time runs out */

int sonic_cmd = 0;
float snc_cmd_period = 1.2; /* Gather the sound for this long */

void snc_capture_command()
{ int old_count = 0;
  while(1)
    { if (old_count != snc_sound_count)
        { sonic_cmd = 1; /* Stop immediately, wait for next cmd */
          sleep(snc_cmd_period);/* Wait for the sounds to happen */
          sonic_cmd = snc_sound_count;
          snc_sound_count = 0; }
    }}

/* Obey the sonic commands:
      1 - Stop
      2 - Turn
      3 - Forward
   Stop in the event of a collision */
void sonic_control()
{ while(1)
    { if (snc_bmpr & 0b011)       /* Bump L or R */
        { drive(-75,0);           /* Backup */
          sleep(0.5);
          sonic_cmd = 0; }
```

```
      else if (sonic_cmd == 2)  /* Spin in place */
        drive(0,75);
      else if (sonic_cmd == 3)  /* Forward */
        drive(75,0);
      else
        drive(0,0);             /* Stop by default */
    }}

void snc_rpt()
{ while(1)                      /* Show what the robot is doing */
    { printf("Sonic Commander Cmd: ");
      if (snc_bmpr & 0b011)
        printf("Backup");
      else if (sonic_cmd == 2)
        printf("Turn  ");
      else if (sonic_cmd == 3)
        printf("Forward");
      else
        printf("Stop  ");
      printf("\n");
      sleep(0.5);
    }}

void start_sonic()
{ start_process(snc_sample_sound());
  start_process(snc_capture_command(),1);
  start_process(sonic_control(),1);
  start_process(snc_rpt(),1);
}
```

The cryptic comment "`Must have all analog calls in one process`" in the above program refers to a peculiarity of **IC**. Calls to the `analog` function occuring in two different processes can, on occasion, cause a problem. If the scheduler interrupts one process and starts another process at just the right moment, a call to `analog` started in the first process can be intercepted by the second process. This leads to an erroneous value being reported to the second process. This potential problem can be eliminated by placing all calls to `analog` in a single process.

B.8 Auxiliary Code

The following functions do not appear in the standard library but are needed by one or more of the examples.

```
/* Common.c */

/* No two drive motors respond in exactly the same way to the same
   applied voltage.  Use the drive_bias term to correct for biases
   in your robot.  If your robot arcs to the right, make drive_bias
   positive, arcs to the left require a negative correction. */

int drive_bias = 0 /* Open loop correction term for drive motors */

/* Common Utilities */

/* Absolute value function for integers */

int abs(int val)
{
  if (val < 0)
    return (- val);
  else
    return val;
}

int min(int a, int b)/* Find the minimum of two arguments */
{  if (a < b)
        return a;
    else
        return b;
}

int max(int x, int y) /* Find the maximum of two arguments */
{ if (x > y)
   return x;
  else
   return y;
}

void driveb(int trans, int rot) /* Correct for motor bias */
{ int rot_bias = (drive_bias * trans) / 100;
  motor(0,trans - (rot + rot_bias));
  motor(1,trans + (rot + rot_bias));
}
```

B.9 Velocity Control Code

Here is the velocity control code described earlier in the text. This code, for the most part, does not use the standard library definitions.

```
/* Components of robot velocity control:
     Velocity monitoring
     Open loop PWM
     Velocity control loop
*/

/*                    VELOCITY MONITORING                    */

int TCTL2 = 0x1021;      /* Timer Control 2, interrupt edge */
int TMSK1 = 0x1022;      /* Timer Interrupt Masks, 8-bit reg */
int TFLG1 = 0x1023;      /* Timer Flags, 8-bit reg */
int PACTL = 0x1026;      /* Pulse accumulator control, 8-bit reg */
int PACNT = 0x1027;      /* Pulse accumulator counter, 8-bit reg */


void init_velocity ()
{ poke(PACTL, 0b01010000); /* DDRA7 in, pulse acc rising edges */
  poke(PACNT,0);          /* Start off with 0 measured velocity */
  bit_set(TCTL2,0b00000001); /* Make IC3 interrupt on rising edges */
  bit_set(TMSK1,0b00000001); /* Enable IC3 interrupts */
}

/* Call get_left_vel and get_right_vel
   at regular intervals to get velocity */

float get_left_vel()     /* Left vel from PA7 using pulse counter */
{ float vel;
  vel = (float) peek(PACNT);
  poke(PACNT,0);           /* Reset for next time */
  return(vel); }

float get_right_vel()    /* Right vel from PA0, interrupt routine */
{ float vel;
  vel = (float) right_clicks;
  right_clicks = 0;        /* Reset for next time */
  return (vel); }

/*                    OPEN LOOP PWM                    */

int DDRD = 0x1009;       /* Port D data direction */
int OC1M = 0x100C;       /* Output Compare 1 Mask */
```

```
int OC1D = 0x100D;        /* Output Compare 1 Data */
int TOC1 = 0x1016;        /* Output Compare Tmr 1, */
int TOC2 = 0x1018;        /* Output Compare Tmr 2, (left motor) */
int TOC3 = 0x101A;        /* Output Compare Tmr 3, (right motor) */
int TCTL1 = 0x1020;       /* Timer Control 1, 8-bit reg */

/* motor_index:  0 => Left motor, 1 => Right motor */
/* int TOCx[2] = {TOC2,TOC3};    /* Index for timer register */
int TOCx[2] = {0x1018, 0x101A}; /* Index for timer register */
int sign[2] = {1,1};            /* Sign of rotation of motor */
int dir_mask[2] = {0b010000, 0b100000}; /* Port D direction bit */

/* Utility functions */

float abs(float arg)            /* Absolute value function */
{ if (arg < 0.0)
    return (- arg); else return arg; }

int get_sign(float val) /* Find the sign of the argument */
{ if (val > 0.0)
    return 1; else return -1; }

        /* Limit range of val */
float limit_range(float val, float low, float high)
{  if (val < low) return low;
   else if (val > high) return high;
   else return val; }

void init_pwm()           /* Initialize Pulse Width Modulation */
{ poke(DDRD,0b110010);     /* Port D dir: OUT 5,4,3,1; IN 0 */
  poke(OC1M,0b01100000);  /* Output Compare 1 affects PA5 and PA6 */
  poke(OC1D,0b01100000);  /* OC1 compare turns on PA5 and PA6 */
  bit_set(TCTL1,0b10100000); /*OC3 turns off PA5, OC2 turns off PA6*/
  pokeword(TOC1,0);        /* When timer (TCNT) = 0, OC1 successful */
  pokeword(TOC2,1);        /* Minimum on time for OC2 */
  pokeword(TOC3,1); }      /* Minimum on time for OC3 */

/* The sign is handled in a special way --
    we have only a 1 channel encoder */
float pwm_motor(float vel, int motor_index)
{ float vel_1;
  if (sign[motor_index] > 0)     /* Choose the dir of rotation */
    bit_set(port_d, dir_mask[motor_index]);
  else
    bit_clear(port_d, dir_mask[motor_index]);
```

```
  vel_1 = limit_range(vel, 1.0, 99.0);/* 1 < PWM-duty-factor 100 */
  pokeword(TOCx[motor_index], (int) (655.36 * vel_1));
  return vel_1;}

/* Top level open loop PWM command */
void move(float l_vel, float r_vel) /* R, L vel: [-100.0, 100.0] */
{ sign[0] = get_sign(l_vel);    /* Desired direction of rotation */
  sign[1] = get_sign(r_vel);
  pwm_motor(abs(l_vel), 0);     /* Set PWM constant */
  pwm_motor(abs(r_vel), 1); }

/*                      CONTROL LOOP                          */

float control_interval = 1.0; /* How often to run the servo loop */
float des_vel_clicks = 0.0;   /* Des vel in clicks per interval */
float des_bias_clicks = 0.0;  /* Des bias in clicks per interval */
float power[2] = {0.0,0.0};   /* Positive power command to motor */
float integral = 0.0;         /* Integral of velocity difference */
float k_integral = 0.10;      /* Integral error gain */
float k_pro = 1.0;            /* Proportional gain */

void alter_power(float error, int motor_index) /* Set, save power */
{  power[motor_index] = limit_range(power[motor_index]
                        + error, 0.0, 100.0);
   pwm_motor(power[motor_index], motor_index); }

float integrate(float left_vel, float right_vel, float bias)
{  integral = limit_range((integral + left_vel + bias - right_vel),
                          -1000.0, 1000.0);
   return integral; }

void speed_control()
{float left_vel, right_vel, integral_error, left_error, right_error;
  while (1)
    {left_vel = get_left_vel();
     right_vel = get_right_vel();
     integral_error =
       k_integral * integrate(left_vel, right_vel, des_bias_clicks);
     left_error  =
       k_pro * (des_vel_clicks - left_vel  - integral_error);
     right_error =
       k_pro * (des_vel_clicks - right_vel + integral_error);
     alter_power(left_error, 0);
     alter_power(right_error, 1);
     sleep(control_interval);
     }}
```

```
float k_clicks = 8.0 / 100.0;

void set_velocity(float vel, float bias)
{ des_vel_clicks = k_clicks * vel;
  des_bias_clicks = k_clicks * bias;
  sign[0] = get_sign(vel - bias);
  sign[1] = get_sign(vel + bias); }

void start_speed_control()
{ init_velocity();
  init_pwm();
  get_left_vel();
  get_right_vel();
  start_process(speed_control()); }

void vel()
{ while (1)
    {
      left_vel = get_left_vel();
      right_vel = get_right_vel();
      sleep(control_interval);
    }
}
```

Yellow Pages

One of the major roadblocks in building robots is not knowing where to get parts. Sensors, motors, electronics, batteries, prototyping equipment, connectors, and tools all come from a variety of vendors. After years of tracking things down, we have compiled a database of suppliers we commonly turn to for interesting robot parts. After the alphabetical listing of suppliers, Section C.1, is a cross-reference list by component category, Section C.2. The best thing to do is start calling these companies, consulting their web pages, and collecting catalogs. Most suppliers will gladly send catalogs, free of charge.

If you are searching for a type of component and have no idea how to find a supplier (i.e., none exist in our list below), the place to start is the *Thomas Register*. This is an index to the world. The *Thomas Register* is a set of over two-dozen very large books that lists manufacturers and suppliers of every type of product that you can imagine. The *Thomas Register* is also available on CD ROM and can be accessed over the web (www.thomasregister.com). To learn more, contact Thomas Publishing Company, Attn: Circulation Department, One Penn Plaza, New York, NY 10117-0138.

C.1 Suppliers

3M Electronic Products
225-1N 3M Center
St. Paul, MN 55144
(800) 328-SPEC
Fax: (651) 737-7117
innovation@mmm.com
www.mmm.com/

Scotchflex prototype wiring technology; distributed by Wallace Electronics Sales

A K Peters, Ltd.
63 South Ave.
Natick, MA 01760
(508) 655-9933
Fax: (508) 655-5847
service@akpeters.com
www.akpeters.com

Publisher of *Mobile Robots: Inspiration to Implementation* and numerous other robotics texts. A K Peters also distributes Rug Warrior Pro kits, expansion modules for Rug Warrior Pro, and Interactive C

A. Cohen Company
353 Washington Street
Boston, MA 02108
(617) 523-7440
Fax: (617) 523-8723

Vigor watchmakers' tools

Acroname
PO Box 1894
Nederland, CO 80466
(303) 258-3161
Fax:(303) 247-1892
www.acroname.com/

Acroname, Inc. makes a Pyro sensor designed for Rug Warrior, sensor is complete with Fresnel lens

Active Electronics
11 Cummings Park
Woburn, MA 01801
(781) 932-0500
Fax: (781) 933-8884

Retail dealer for electronic components

Advanced Design
1101 East Rudsill Road
Tucson, AZ 85718
(602) 544-2390
Fax: (602) 575-0703

This company makes a clever and inexpensive robot arm using airplane servos

Airtronics
1185 Stanford Court
Anaheim, CA 92805
(800) 567-6867
Fax: (714) 928-1540

Motors

Alarm Supply
48 Mechanic Street
Newton Upper Falls, MA 02494
(617) 527-4931
Fax: (617) 527-4534

Pyroelectric sensors

Alberta Printed Circuits Unit 3, 1112 40th Ave. NE Calgary, AB, T2E 5T8, Canada (403) 250-3406 www.apcircuits.com/	Low cost, rapid turnaround circuit board fabrication house
All Electronics Corporation PO Box 567 Van Nuys, CA 91408 (800) 826-5432 Fax: (818) 781-2653 allcorp@allcorp.com www.allcorp.com	Surplus dealer, surplus boards, components, and assemblies
Allied Electronics 10-L Centennial Drive Peabody, MA 01960 (800) 433-5700 www.allied.avnet.com	Electronic components
America's Hobby Center 146 West 22nd Street New York, NY 10001-2466 (212) 675-8922 Fax: (212) 675-0060 www.ahc1931.com	Radio-control products, servos, motors
American Control Technology 850 Church Road Elgin, IL 60123 (847) 468-6000 Fax: (847) 468-8959	LCD thumbwheel switches
American Design Components 400 County Ave. Secaucus, NJ 07094 (800) 776-3800	Surplus dealer, computer equipment, power supplies, motors, batteries, MOVIT robot kits
American Science and Surplus 3605 W. Howard Street Skokie, IL 60076 (847) 982-0874 Fax: (800) 934-0722 www.sciplus.com	Surplus dealer, wide assortment of electronic components
AMP Sensors PO Box 799 Valley Forge, PA 19482 (610) 650-1500 Fax: (610) 650-1509	Thin film piezoelectric/pyroelectric material supplied by this company can be used to build custom-designed sensors
Animate Systems 390 Wakara Way, Suite 56 Salt Lake City, UT 84108 (801) 581-0155 Fax:(801) 581-1151 f.smith@sarcos.com	Entertainment robots, small servo valves

Arrick Robotics PO Box 1574 Hurst, TX 76053 (817) 571-4528 Fax:(817) 571-2317 info@robotics.com www.robotics.com	Stepper motor control system and robot components
AVNET 10M Centennial Drive Peabody, MA 01960 (800) 272-9255 Fax: (978) 532-9802 www.avnet.com	Distributor for many semiconductor manufacturers
Banner Engineering 97114 Tenth Avenue North Minneapolis, MN 55440 (612) 545-0813 Fax: (612) 544-3213 www.baneng.com	Infrared sensors
BEI 7230 Hollister Avenue Goleta, CA 93117-2891 (805) 968-0782 Fax: (800) 960-2726	Encoders
Berg 499 Ocean Avenue E. Rockaway, NY 11518 (516) 599-5010 Fax: (516) 599-3274	Gears, linkages, pulleys, etc
Binsfeld Engineering 4571 W. MacFarlane Maple City, MI 49664 (616) 334-4383 Fax: (616) 334-4903 www.binsfeld.com	Strain gage telemetry system
BNF Enterprises 134 Newbury Street R Peabody, MA 01960 (978) 536-2000 Fax: (978) 536-7400 peterb@bnfe.com www.bnfe.com	Dealer in computer components
Bourns **Sensors and Controls Division** 2533 N. 1500 West Ogden, UT 84404 (801) 786-6200 Fax: (801) 786-6228 Dennis Snavely@bourns.com www.bourns.com	Encoders, potentiometers

Burden's Surplus Center 1015 West O Street PO Box 82209 Lincoln, NE 68501 (800) 228-3407 Fax: (402) 474-5198	Mechanical parts
Canon One Canon Plaza Lake Success, NY 11042 (516) 488-6700 Fax: (516) 328-4609 RonTravis@cusa.canon.com	Encoders, motors
Capsella See MIT Museum Shop	
Centro Vision 2088 Anchor Court Newbury Park, CA 91320-1601 (805) 499-5902 Fax: (805) 499-7770 www.centrovision.com	Silicon photodetectors and linear arrays
Chinon America Industrial Products Division PO Box 1248 1065 Bristol Road Mountainside, NJ 07092-1248 (908) 654-0404	Small cameras
Circuit Board Fabrications 179 Bear Hill Road Waltham, MA 02254 (617) 890-1878 Fax: (781) 890-7098 sales@circuitfab-co.com www.circuitfab-co.com	Printed circuit board for Rug Warrior
Circuit-Wise 400 Sackett Point Road North Haven, CT 06473 (203) 281-6511 Fax: (203) 287-8409	Molded boards for mechanical/electrical integration
Detection Systems 130 Perinton Parkway Fairport, NY 14450 (716) 223-4060 Fax: (716) 223-9180 www.detectionsys.com	Pyroelectric sensors

Digi-Key
701 Brooks Avenue South
PO Box 677
Thief River Falls, MN 56701-0677
(800) 344-4539
Fax: (218) 681-3380
sales@digikey.com
www.digikey.com

Digi-Key is a “hobbyist friendly” business; they accept small orders, ship products promptly, and have huge assortment of products in stock

Direct Imaging
PO Box 820
Wilder, VT 05088
(802) 295-3770
Fax: (802) 295-3862
directim@aol.com
www.uppervalleydirectory.com/

Circuit boards

Dunfield Development Systems
PO Box 31044
Nepean, Ontario
K2B 8S8 Canada
(613) 256-5820
Fax: (613) 256-5821
info@dunfield.com
www.dunfield.com

Markets inexpensive C compiler compatible with the MC68HC11 as well as several other popular microprocessors

Duracell
Berkshire Industrial Park
Bethel, CT 06801
(800) 431-2656
Fax: (203) 791-3021
www.duracell.com

Batteries

Edlie Electronics
2700 Hempstead Turnpike
Levittown, NY 11756-1443
(516) 735-3330
Fax: (516) 731-5125

Surplus assortment of tools, test equipment, parts

Edmund Scientific
101 E. Gloucester Pike
Barrington, NJ 08007
(609) 547-3488
Fax: (609) 573-6295
info@edsci.com
www.edsci.com

Optical components, science kits, surplus motors

EDO Corporation
Barnes Engineering Division
88 Long Hill Cross Road
PO Box 867
Shelton, CT 06484-0867
(203) 926-1777
Fax: (203) 926-1030
www.edocorp.com

Temperature sensors

Electronic Goldmine PO Box 5408 Scottsdale, AZ 85261 (602) 451-7454	Electronic components
Elmec 4127 Avenida De La Plata Oceanside, CA 92056 (760) 631-0202 Fax: (760) 631-0237 sales@elmecmfg.com www.elmecmfg.com	Flexible circuit design and fabrication
Eltec Instruments PO Box 9610 Central Business Park Daytona Beach, FL 32020 (800) 874-7780	Pyroelectric sensors, Fresnel lenses
Entran Devices 10 Washington Avenue Fairfield, NJ 07004 (800) 635-0650 Fax: (973) 227-6865 sales@entran.com www.entran.com	Accelerometers, Force sensors, Pressure sensors, Strain gauges
EP Circuits 5468 Highroad Crescent Chilliwack, BC, V2R 3Y1, Canada (604) 824-1238 fax: (604)858-7663 epproto@uniserve.com www.uniserve.com/epicircuits	Circuit boards
Erector Set See MIT Museum Shop	
ETAK 1605 Adams Drive Menlo Park, CA 94025 (650) 328-3825 Fax: (650) 328-3148 info@etak.com www.etak.com	Navigation systems for cars
Fischer-Technik See MIT Museum Shop	
Fresnel Technologies 101 West Morningside Drive Fort Worth, TX 76110 (817) 926-7474 Fax: (817) 926-7146 info@fresneltech.com www.fresneltech.com	Fresnel lenses

Futaba Corporation 1605 Perry Lane Schaumberg, IL 60173 (8437) 884-1444 www.futaba-na.com	Accessories for radio-controlled toys
Gerber Electronics 128 Carnegie Row Norwood, MA 02062 (800) 225-1800 Fax: (781) 762-8931 Postmaster@gerberelec.com www.gerberelec.com	Electronic components
Gleason Research PO Box 1247 Arlington, MA 02174 (781) 641-2551 Fax: (781) 641-2551 info@gleasonresearch.com www.gleasonresearch.com/	Supplier of the Handy board
Globe Motors 2275 Stanley Avenue Dayton, OH 45404 (937) 228-3171 Fax: (937) 461-1017	Motors
Graymark International PO Box 2015 Tustin, CA 92781 (800) 854-7393 Fax: (714) 544-2323 www.labvolt.com	Robots
Gurley Precision Instruments 514 Fulton Street Troy, NY 12181-0088 (800) 759-1844 Fax: (518) 274-0336 Info@Gurley.com www.Gurley/com	Optical encoders
Hamamatsu Photonics 360 Foothill Road Bridgewater, NJ 08807-0910 (908) 231-0960 Fax: (908) 231-1218 www.hamamatsu.com	Photoresistors, infrared detectors, rangers, color sensors, shaft encoder sensors
Harbor Tool 20 Southwest Park Westwood, MA 02090 (617) 329-4432 info@harbortool.com www.harbortool.com	Machine tools, hardware

HDS
12310 Pinecrest Road
Reston, VA 20191
(703) 620-6200
hds@hdscorp.com
www.hdscorp.com

Small cameras

Heathkit Company
455 Riverview Drive
Benton Harbor, MI 49022
(800) 253-0570
Fax: (616) 925-2898
heathkit@heathkit.com
www.heathkit.com

Many electronic products, including test equipment

Herbach and Rademan
16 Roland Avenue
Mt. Laurel, NJ 08054
(800) 848-8001
Fax: (609) 802-0465
sales@herbach.com
www.herbach.com

Surplus dealer

Hohner Corp.
5536 Regional Road No. 81
Beamsville, Ontario, L0R 1B3, Canada
(800) 295-5693
Fax: (905) 563-4924
hohner@hohner.com
www.hohner.com

Encoders

Humphrey
9212 Balboa Avenue
San Diego, CA 92123
(619) 565-6631
Fax: (619) 565-6873
www.remec.com

Gyros

IC Sensors
1701 McCarthy Blvd.
Milpitas, CA 905035-7416
(800) 767-1888
Fax: (408) 432-7322

Micromachined accelerometers, pressure sensors

Images Company
PO Box 140742
Staten Island NY 10314
(718) 698-8305
Fax: (718) 982-6145
www.imagesco.com

Source for bend sensors

Integrated Circuit Systems
2435 Boulevard of the Generals
PO Box 968
Valley Forge, PA 19482
(610) 630-5300
Fax: (610) 630-5399
www.icst.com

Battery-charging ICs

Interlink Electronics
546 Flynn Road
Camarillo, CA 93012
(805) 484-8855
Fax: (805) 44-8989
www.interlinkelec.com

Force-sensing resistors

IS Robotics
Suite 6
22 McGrath Highway
Somerville, MA 02143
(617) 629-0055
Fax: (617) 629-0126
www.isr.com

Research robots, commercial robots, and sensor systems

ITT Cannon
666 E. Dyer Road
Santa Ana, CA 92705
(714) 557-4700
Fax: (714) 654-2142
www.ittcanon.com

Microminiature connectors

Jameco
1355 Shoreway Road
Belmont, CA 94002
(650) 592-8097
Fax: (650) 592-2503
info@jameco.com
www.jameco.com

Electronic components, Bend sensors, Compasses

Jenson Tools
7815 S. 46th Street
Phoenix, AZ 85044-5399
(800) 426-1194
Fax: (800) 366-9662
sales@jensentools.com
www.jensentools.com

Hand tools, test equipment

Johuco Ltd.
PO Box 385
Vernon, CT
johuco@pcrealm.net
www.pcrealm.net/ johuco/index.html

Robots

Joker Robotics Muenchinger Str. 8 71282 Hemmingen, Germany +49 (172) 711-3633 Fax: +49 (7150) 970 850 joker@joker-robotics.com www.joker-robotics.com/	Distributes Rug Warrior Pro™ and accessories in Europe
K-Team Ch. de Vuasset CP 111 CH 1028 Préverenges, Switzerland +41 (21) 802-5472 Fax: +41 (21) 802-5471 info@k-team.com www.k-team.com	Khepera miniature mobile robots
Kaufman Tools 110 Second Street Cambridge, MA 02141 (800) 338-8023 Fax: (800) 638-8805 sales@kaufmanco.com www.kaufmanco.com	Machine tools, hand tools
Laser Services 123 Oak Hill Road Westford, MA 01886 (508) 692-6180	Laser job shop
LEGO LEGO Educational Dept. PO Box 39 Enfield, CT 06082 (800)527-8339 www.lego.com	All components needed for quickly building robot prototypes; LEGO Mindstorms^{TM}, educational department sells primarily to schools
Lucas Control Systems **Shaevitz Sensors** 100 Lucas Way Hampton, VA 23666 (757) 766-1500 Fax: (757) 766-4297 sales@schaevitz.com www.schaevitz.com	Gyros, Accelerometers
Lucas Ledex 801 Scholz Drive PO Box 427 Vandalia, OH 45377-0427 (937) 898-3621 Fax: (937) 898-8624 www.golucas.com	Encoders

Lucas Novasensor
1055 Mission Court
Fremont, CA 94539
(510) 490-9100
www.golucas.com

Micromachined pressure sensors and accelerometers

Lucas Schaevitz
7905 N. Route 130
Pennsauken, NJ 08110-1489
(609) 662-8000
www.golucas.com

Force sensors, displacement sensors

Mabuchi Motors America
3001 W. Big Beaver Road
Suite 520
Troy, MI 48084
(248) 816-3100
Fax: (248) 816-3242

Mfrs. of a full range of motors

Marshall Electronics
33 Upton Drive
Wilmington, MA 01887
(978) 658-0810
Fax: (978) 657-5931
www.marshall.com

Electronic components

Maxon Precision Motors
838 Mitten Road
Burlingame, CA 94010
(650) 697-9614
Fax: (650) 697-2887
www.mpm.maxonmotor.com

Small high-quality motors

MCM Electronics
650 Congress Park Drive
Centerville, OH 45459-4072
(800) 543-4330
Fax: (937) 434-6959
www.mcmelectronics.com

Tools, connectors, transistors

McMaster-Carr
PO Box 440
New Brunswick, NJ 08903-0440
(732) 329-3200
Fax: (732) 329-3772
nj.sales@mcmaster.com
www.mcmaster.com

Machine tools, hardware, materials, everything you'd find in a factory

Meccano
See MIT Museum Shop

Mendelson Electronics
340 E. First Street
Dayton, OH 45402
(800) 422-3525
Fax: (937) 461-3391
meci@meci.com
www.meci.com

Subassemblies of discontinued Heathkit HERO 2000 robot

Methode Electronics 1700 Hicks Road Rolling Meadows, IL 60656 (800) 323-6864 Fax: (847) 392-9404	Sockets and connectors
Micro Gage 9537 Telstar Avenue El Monte, CA 91731 (626) 443-1741 Fax: (626) 443-7290	Force sensors
Micro Measurements PO Box 27777 Raleigh, NC 27611 (919) 365-3800 Fax: (919) 365-5945 email@measurementsgroup.com www.measurementsgroup.com	Strain gauges
Micro Miniature Bearing 7 Jocama Boulevard Old Bridge, NJ 08857 (800) 526-2353 Fax: (732) 591-1890 www.mmbearco.com	Bearings
Micro Mo Electronics 14881 Evergreen Ave. Clearwater, FL 33762 (813) 572-0131 Fax: (813) 573-5981 www.micromo.com	Small motors
Mikron Instrument Company 16 Thornton Road Oakland, NJ 07436 (201) 891-7330 Fax: (201) 405-0900 sales@mikroninst.com www.mikroninst.com	Pyroelectric sensors
Minco Products 7300 Commerce Lane Minneapolis, MN 55432-3177 (612) 571-3120 Fax: (612) 571-0927 info@minco.com www.minco.com	Flexible coils
MIT Media Laboratory 20 Ames Street Room E15-315 Cambridge, MA 02139 (617) 253-0300 Fax: (617) 358-6264 www.media.mit.edu	Developed IC, performs research on robots in education

MIT Museum Shop Building N52 MIT Student Center Cambridge, MA 02139 (617) 253-4462	Sells Fischer-Technik, Meccano, Capsella, LEGO and Erector Set construction kits
Model-A Technology Fischer-Technik 2420 Van Layden Way Modesto, CA 95356 (209) 575-3445 Fax: (209) 527-6016	Construction kits
Mondo-tronics 4186 Redwood Highway No. 226 San Rafael, CA 94903 (800) 374-5764 Fax: (415) 455-9333 www.robotstore.com	Shape memory metal, robots, robot books, robot videos
Mouser Electronics 958 North Main St. Mansfield, TX 76063-4287 (800) 34-MOUSER Fax: (817) 483-6899 sales@mouser.com www.mouser.com	Wide selection of electronic components; will fax detailed specs; regional distribution centers; accepts small orders
MTI Instruments Division 968 Albany-Shaker Road Latham, NY 12110 (800) 828-8210 Fax: (518) 785-2127 www.mechtech.com	Fotonic sensor for displacement
Murata 2200 Lake Park Drive Smyrna, GA 30080 (404) 436-1300 www.murata.com	Temperature sensors
Namiki 201 West Passaic Street Rochelle Park, NJ 07662 (201) 368-0123 Fax: (201) 368-2244 motor@namiki.co.jp www.namiki.co.jp	Very small motors
New Micros 1601 Chalk Hill Road Dallas, TX 75212 (214) 339-2204 general@newmicros.com www.newmicros.com	Single-board computer uses MC68HC11 chip; Forth language in ROM

Newark Electronics 59 Composite Way Lowell, MA 01851 (800) 463-9275 Fax: (508) 229-2222 www.newark.com	Distributor of electronic components
Newton Research Labs 4140 Lind Ave SW Renton, WA 98055 (425) 251-9600 Fax: (425) 251-8900 sales@newtonlabs.com www.newtonlabs.com/	Newton Labs offers a commercial (supported) version of IC
Nomadic Technologies 2133 Leghorn St. Mountain View, CA 94043-1603 (650) 988-7200 Fax: (650) 988-7201 nomad@robots.com www.robots.com	Robots
Omron Electronics 1 East Commerce Drive Schaumburg, IL 60173 (847) 843-7900 Fax: (847) 843-8568 www.omron.com	Photomicrosensors, relays, bump switches
Optima Batteries 17500 East 22nd Avenue Aurora, CO 80011 (888) 867=8462 Fax: (303) 340-7474 cdouglass@optimabatteries.com www.optima.com	Manufacturer of batteries, especially lead acid
Optoelectronic Center Lincoln Loop Sauk Centre, MN 56378 (320) 352-6556 Fax: (320) 352-3617 oci@sockherald.com www.optoelectronics-oci.com	Optical switches
Pace Electronics 34 Foley Drive Solus, NY 14551-0067 (315) 483-9122 Fax: (315) 483-0480 pace@PaceElectronics.com www.PaceElectronics.com	Distributor for Nippon Ceramics; cheap pyros

Pacer Electronics
112 Commerce Way
Woburn, MA 01801
(781) 935-8330
Fax: (781) 938-7881
pacerelect@msn.com
www.pacerelc.com

Electronic components

Parallax
3805 Atherton Road
Suite 102
Rocklin, California 95765
(888) 512-1024
Fax: (916) 624-8003
info@parallaxinc.com
www.parallaxinc.com

Parallax offers the very popular Basic Stamp computer for embedded systems

Pico Electronics
143 Sparks Avenue
Pelham, NY 10803
(800) 431-1064
Fax: (914) 738-8225
info@picoelectronics.com
www.picoelectronics.com

DC-DC converters

Piezo Systems
186 Massachusetts Avenue
Cambridge, MA 02139
(617) 547-1777
Fax: (617) 354-2220
sales@piezo.com
www.piezo.com

Piezoelectric ceramics, sensors, actuators

Pioneer Electronics
44 Hartwell Avenue
Lexington, MA 02173
(781) 861-9200
Fax: (781) 863-1547
www.pios.com

Distributor of many semiconductor manufacturers' lines.

Pittman
P.O. Box 3
Harleysville, PA 19438-0003
(215) 256-6601
Fax: (215) 256-6601
info@pittmannet.com
www.pittmannet.com

Motors

Polaroid Corporation
OEM Components Group
153 Needham Street
PO Box 9122
Newton, MA 02464-9122
(781) 386-3964
Fax: (781) 386-3966
ultrason@polaroid.com
www.polaroid-oem.com

Sonar transducers, sonar drive boards

Portescap US 110 Westtown Road Westchester, PA 19382 (610) 692-2700 Fax: (610) 696-4598 pub@portescap.com www.portescap.com	High-quality DC gearhead motors
Precision Navigation 1235 Pear Ave. Suite 111 Mountain View, CA 94043 (650) 962-8777 Fax: (650) 962-8776 sales@precisionNav.com www.PrecisionNav.com/	Precision Navigation offers several versions of electronic compasses
Radio Shack *National chain – consult telephone directory for nearest distributor* *(800) THE-SHACK* *www.radioshack.com*	Offers a variety of electronic components from local distributors; to mail order, see **Tech America**
Ramtron 1850 Ramtron Drive Colorado Springs, CO 80921 (719) 481-7000 Fax: (719) 481-9294 www.ramtron.com	RAM and DRAM, Discrete semiconductors, LEDs, LCDs
RC Systems 1609 England Avenue Everett, WA 98203-2627 (425) 355-3800 Fax: (425) 355-1098 www.rcsys.com	High-quality, inexpensive speech boards
RCD Components 520 East Industrial Park Drive Manchester, NH 03109 (603) 669-0054 Fax: (603) 669-5455 info@rcd-comp.com www.rcd-comp.com	Temperature sensors
Reactive Technologies P.O. Box 2095 Merrimack, NH 03054 info@reactivetechnologies.com www.reactivetechnologies.com	Reactive Technologies supplies Rug Warrior Pro compatible modules and robotics hardware and software development services.

Real World Interface 32 Fitzgerald Drive PO Box 375 Jaffrey, NH 03452 (603) 532-6900 www.rwii.com	Robot bases, sensory systems; RWI is now a division of IS Robotics
Redwood Microsystems 959 Hamilton Ave. Menlo Park, CA 94025 (650) 326-1896 Fax: (650) 326-1899 www.redwoodmicro,com	Micromachined miniature valves
Redzone Robotics 2425 Liberty Avenue Pittsburgh, PA 15222-4639 (412) 765-3064 Fax: (412) 765-3069 info@redzone.com www.redzone.com	Hazardous waste robots, applications in nuclear energy, and mobile robots
Reptron 20 Blanchard Road Burlington, MA 01803 (800) 345-2921 Fax: (412) 765-3064 info@reptron.com www.reptron.com	Distributor of electronic components
Richards Micro Tool 250 Nicks Rock Road Plymouth, MA 02360 (508) 746-6900 Fax: (508) 747-4339	Small tools
RMB Miniature Bearings 29 Executive Parkway Ringwood, NJ 07456 (973) 962-1111 RMB Tech@comopuserve.com www.rmb-ch.com	Bearings
Rogers Corporation One Technology Drive Rogers, CT 06263 (860) 774-9605 Fax: (860) 779-5509 info@rogers-corp.com www.rogers-corp.com	Bendflex flexible printed circuit boards
Royal Products Corporation 790 W. Tennessee Avenue Denver, CO 80223 (303) 778-7711 Fax: (303) 778-7721	Inexpensive model airplane servo motors

Samtec PO Box 1147 New Albany, IN 47150-1147 (812) 944-6733 Fax: (812) 948-5047 tammy.rudy@samtech.com www.samtech.com	Distributor of electronic components
Sanyo Electric 2055 Sanyo Avenue San Diego, CA 92173 (619) 661-6620 Fax: (619) 661-6743 www.sanyo.com/	Batteries
Sarcos Microsystems 390 Wakara Way, Suite 65C Salt Lake City, UT 84108 (801) 581-0155 Fax: (801) 581-1151 info@sarcos.com www.sarcos.com	Multi-axis strain sensors, rotary displacement transducers
Sharp Electronics Corporation Sharp Plaza Mahwah, NJ 07430-2135 (201) 529-8200 Fax: (201) 529-8425 www.sharpelectronics.com	Sharp makes many useful types of photosensors
Sheldon's Hobbies 2135 Old Oakland Road San Jose, CA 95131 (800) 822-1688 Fax: (408) 943-0904 www.btown.com/sheldons/	Radio-control products, servos, motors, gyros
Shinkawa Electric Co. Ltd. Shinkojimachi Building 3F 3-3 Kojimachi 4-chome Chiyoda-ku, Tokyo 102, Japan +81-332-62-4417 Fax: +81-332-62-2171 soishi@shinkawa.co.jp	Shinkawa distributes Rug Warrior Pro™ kits and accessories in Japan
Small Parts 13980 NW 58th Court PO Box 4650 Miami Lakes, FL 33014-0650 (305) 557-8222 Fax: (305) 558-0509 smlparts@smallparts.com www.smallparts.com	Supply of metal, plastics, tools, and hardware

Southco 210 North Brinton Lake Road Concordville, PA 19331 (610) 459-4000 Fax: (610) 459-4012 www.southco.com	Mechanical fasteners
Spectron 595 Old Willets Path PO Box 13368 Hauppauge, NY 11788 (516) 582-5600 Fax: (516) 582-5671 info@spectronsensors.com www.spectronsensors.com	Inclinometers, mercury switches
Spiricon 2600 North Main Logan, UT 84321 (435) 753-3729 Fax: (435) 753-5231 sales@spiricon.com www.spiricon.com	Sensors for laser systems
Sterling Electronics 15D Constitution Way Woburn, MA 01801 (781) 938-6200 Fax: (781) 933-5468	Distributor of electronic components
Stock Drive Products 55 South Denton Avenue New Hyde Park, NY 11040 (516) 328-0200 Fax: (516) 326-8827 www.sdp-si.com	Assortment of small parts
Strataflex Corporation 11 Dohme Avenue Toronto, Ont. M4B 1Y7, Canada (416) 752-2224 Fax: (416) 752-6719 keirstead@strataflex.com	Laser machining of flexible circuits
Supercircuits One Supercircuits Plaza Leander, TX 78641 (512) 260-0333 Fax: (512) 260-0444 www.supercircuits.com	Microvideo cameras and transmitters
T-Tech 5591-B New Peachtree Road Atlanta, GA 30341 (404) 455-0676 www.T-Tech.com	Inhouse milling machine for fabbing PC boards

Tech America PO Box 1981 Fort Worth, TX 76101 (800) 877-0072 Fax: (800) 813-0087 tacusrel01@tandy.com www.techamerica.com	Mail order source for a wide variety of electronic components, no minimum order
TestEquity 2450 Turquoise Circle Thousands Oaks, CA 91320 (800) 732-3457 Fax: (800) 272-4329 www.testequity.com	Used equipment, scopes, and meters
Tower Hobbies PO Box 9078 Champaign, IL 61826 (800) 637-4989 Fax: (800) 637-7303 www.towerhobbies.com	Radio-control products, servos, motors, gyros
Trilogy Linear Motors 141 Bay Area Boulevard Webster, TX 77598 (281) 338-2739 Fax: (281) 338-1227 info@trilogsystems.com www.trilogysystems.com	Linear motors
Unitrode Corporation 7 Continental Blvd. Merrimac, NH 03054 (603) 424-2410 Fax: (603) 429-8771 macdonald@unitrode.com www.unitrode.com	Battery charging ICs
Vero Electronics 5 Sterling Drive Wallingford, CT 06492 (800) 242-2863 Fax: (203) 949-1101 vero@vero-usa.com www.vero-usa.com/index.html	Speedwire wiring equipment, pins, sockets
Wallace Electronics Sales 935-K East Mountain Street Kernersville, NC (336) 996-2742 Fax: (336) 996-1630 info@wes-inc.com www.wes-inc.com/	Electronics distributor; carries 3M Scotchflex wiring technology

Watlow 5710 Kenosha Street Richmond, IL 60071 (815) 678-2211 Fax: (800) 537-4644 www.watlow.com	Pyroelectric sensors
Watson Industries 3041 Melby Road Eau Claire, WI 54703 (800) ABC-GYRO Fax: (715) 839-8248 support@watson-gyro.com	Gyros
Wirz Electronics PO Box 457 Littleton, MA 01460-0457 (888) 289-9479 Fax: (978) 448-0196 sales@wirz.com www.wirz.com	Supplies microcontroller prototyping systems, sonar sensor kits, stepper motor controllers and an LCD serial interface
Wyle Laboratories 5 Oak park Drive Bedford, MA 01730 (781) 271-9953 Fax: (781) 275-3809 www.wyle.com	Carries TI replacement, the SN754410NE, for the SGS-Thompson L293D motor-driver chip
Z-World Engineering 2900 Spafford Street Davis, CA 95616 (530) 753-3737 Fax: (916) 753-5141 www.zworld.com	C-based single board computer

C.2 Products

Accelerometers
Entran Devices, IC Sensors,
Lucas Control Systems
Schaevitz Sensors, Lucas Novasensor

Actuators
Lucas Ledex, Piezo Systems

Batteries
Duracell, Optima Batteries,
Sanyo Electric

Battery-charging ICs
Integrated Circuit Systems,
Unitrode Corporation

Bearings
Micro Miniature Bearing,
RMB Miniature Bearings

Bend sensors
Images Company, Jameco

Buzzers, piezoelectric
Digi-Key

C compilers
Dunfield Development Systems

Cameras
All Phase Video Security, Chinon America, HDS, Supercircuits

Circuit boards
Alberta Printed Circuits,
Circuit Board Fabrications,
Circuit-Wise, Direct Imaging, Elmec,
EP Circuits, Rogers Corporation,
Strataflex Corporation, T-Tech

Color sensors
Hamamatsu Photonics

Compasses
ETAK, Jameco, Precision Navigation

Computer components
BNF Enterprises

Connectors
ITT Cannon, MCM Electronics,
Methode Electronics, Samtec

Construction kits
Capsella, Erector Set, Meccano,
MIT Museum Shop,
Model A Technology-Fischer-Technik

DC-DC converters
Pico Electronics

Displacement sensors
Lucas Schaevitz

Electronic components
Active Electronics,
All Electronics Corporation,
Allied Electronics,
American Science and Surplus, AVNET,
Digi-Key, Edlie Electronics,
Electronic Goldmine,
Gerber Electronics, Jameco, Marshall Electronics, Mouser Electronics,
Newark Electronics, Pacer Electronics, Pioneer Electronics,
Radio Shack, Ramtron, Reptron,
Sterling Electronics, Tech America,
Wyle Laboratories

Encoders
BEI,
Bourns-Sensors and Control Division,
Canon, Gurley Precision Instruments,
Hamamatsu Photonics,
Hohner Corporation,
Sarcos Microsystems

Force sensors
Entran Devices, Interlink Electronics,
Lucas Schaevitz, Micro Gage,
Sarcos Microsystems

Fresnel lenses
Eltec Instruments, Fresnel Technologies

Gyros
Futaba Corporation, Humphrey,
Lucas Control Systems
Schaevitz Sensors, Tower Hobbies,
Watson Industries

IC
A K Peters Ltd., Newton Research Labs

Inclinometers
Spectron

Infrared sensors
Banner Engineering,
Hamamatsu Photonics

Laser machining services
Laser Services

LCD thumbwheel switches
American Control Technology

Mechanical parts
Berg, Burden's Surplus Center, Small Parts, Southco, Stock Drive Products

Microphones
Digi-Key

Motors
Canon, Edmund Scientific, Globe Motors, Mabuchi Motors America, Maxon Precision Motors, Namiki, Pittman, Portescap US, Trilogy Linear Motors

Optical components
Edmund Scientific

Photosensors
Centro Vision, Hamamatsu Photonics, Omron Electronics, Optoelectronic Center, Sharp Electronics Corporation

Piezoelectric materials
AMP Sensors

Piezoelectric sensors
Piezo Systems

Pressure sensors
Entran Devices, IC Sensors, Lucas Novasensor

Proximity sensors
MTI Instruments Division

Pyroelectric sensors
Acroname, Alarm Supply, Detection Systems, Eltec Instruments, Mikron Instrument Company, Pace Electronics, Spiricon, Watlow

Radio-control products
Airtronics, America's Hobby Center, Royal Products Corporation, Sheldon's Hobbies, Tower Hobbies

Relays
Omron Electronics

Robots
A K Peters, Ltd., Advanced Design, Animate Systems, Graymark International, IS Robotics, Johuco Ltd., Joker Robotics, K-Team, Mendelson Electronics, Mondo-tronics, Real World Interface, Redzone Robotics

Rug Warrior Pro^{TM} kits/modules
A K Peters, Ltd., Joker Robotics, Mondo-tronics, Reactive TechnologiesShinkawa Electric

Shape memory metal
Mondo-tronics

Single-board computers
Gleason Research, New Micros, Parallax, Z-World Engineering

Sonar sensors
A K Peters, Ltd., Polaroid Corporation, Wirz Electronics

Speech products
RC Systems, Reactive Technologies

Stepper motor controllers
Arrick Robotics, Wirz Electronics

Strain gauges
Binsfeld Engineering, Entran Devices, Micro Measurements,

Surplus dealers
All Electronics Corporation, American Design Components, American Science and Surplus, Herbach and Rademan

Temperature sensors
EDO Corporation/Barnes Engineering Division, Minco Products, Murata, RCD Components

Test equipment
All Phase Video Security,Edlie Electronics, Heathkit Company, Jenson Tools, TestEquity

Tools
Edlie Electronics, Harbor Tool, Jenson Tools, Kaufman Tools, MCM Electronics, McMaster-Carr, Richard's Micro Tool

Valves
Animate Systems, Redwood Microsystems

Vision systems
Joker Robotics, Newton Research Labs

Watchmakers' tools
A. Cohen Company

Wiring products
3M Electronic Products, Vero Electronics, Wallace Electronics Sales

D

Trade Magazines

Technology changes so quickly that a "how to build a robot" book can swiftly become outdated. We recommend that robot enthusiasts and engineers make every effort to stay abreast of technology because a circuit that takes five chips and seven discrete components today might come out tomorrow in single-chip form (and at lower cost).

The best way to remain aware of what new parts are available is to subscribe to the numerous trade magazines that advertise suppliers and their latest products. Most of these publications are free if you qualify when filling out their subscriber forms, either by working in a related profession or by being a student. In this appendix, we list the publications we have found helpful over the years. Again, most are free, but a few listed are of the pay-for-subscription variety.

Other important sources of information are the electronic bulletin boards and online interest groups available through various computer network services. In particular, a number of ideas and suggestions for this book have come from the comp.robotics news group available on the Internet. By using such a network to offer a comment or pose a question, it is literally possible to reach, overnight, a large audience throughout the world who have an interest in the subject.

Circuit Cellar INK: The Computer Applications Journal
Circuit Cellar Incorporated
4 Park Street, Suite 20
Vernon, CT 06066-3233
(800) 269-6301

Design News
Cahners Publishing
275 Washington Street
Newton, MA 02158
(617) 964-3000

Designfax
A Huebcore Publication
PO Box 1151
Skokie, IL 60076-9917

EDN
(Electronic Design News)
Computer Center
PO Box 5563
Denver, CO 80217-5563

EDN News Edition
Computer Center
PO Box 17844
Denver, CO 80217-0844

EE Product News
PO Box 12982
Overland Park, KS 66282-9818

Electrical Manufacturing
Lake Publishing
PO Box 159
Libertyville, IL 60048-9961

Electronic Component News
Box 2011
Radnor, PA 19080-9511

Electronic Engineering Times
Circulation Dept.
Box 2010
Manhasset, NY 11030

Electronic Packaging and Production
PO Box 5690
Denver, CO 80217

Electronics
1100 Superior Avenue
Cleveland, OH 44197-8118

Electronics Now
Gernsback Publications, Inc.
Subscription Dept., Box 55115
Boulder, CO 80321-5115
(516) 293-3000

Embedded Systems
PO Box 41094
Nashville, TN 37204
(800) 950-0523

Evaluation Engineering
2504 North Tamiami Road
Nokomis, FL 34275-9987
(941) 966-9521

Fiber Optic Product News
301 Gibraltar Drive
PO Box 650
Morris Plains, NJ 07950-0650
(973) 292-5100

IEEE Robotics and Automation
345 East 47th Street
New York, NY 10017-2394
(212)705-7900

Instrumentation and Automation News
Box 2005
Radnor, PA 19080-0405

Integrated Device Technology, Inc.
2975 Stender Way
Santa Clara, CA 95054

Journal of Electronic Engineering
Dempa Publications, Inc.
11-15, Higashi Gotanda 1-chome
Shinagawa-ku 141
Tokyo, Japan

Lasers and Optronics
301 Gibraltar Drive
PO Box 601
Morris Plains, NJ 07950-9827

Literature Distribution Services for Motorola
5005 E. McDowell Road
M/D A 201
Phoenix, AZ 85005
(602) 244-6548

Machine Design
Penton Publishing
PO Box 95759
Cleveland, OH 44101

Measurement Science and Technology
Techno House, Redcliffe Way
Bristol BS1 6NX,
United Kingdom
+44-(117)-930-1128

Medical Equipment Designer
Subscriber Services
Huebcore Communications
29100 Aurora Road, Suite 200
Cleveland, OH 44139

Microsensor Research
Tech Trends Associates
PO Box 386
Bel Air, MD 21014

Microwaves and RF
1100 Superior Avenue
Cleveland, OH 44197-8040

Motion Control
Attn: Circulation Dept.
PO Box 7907
Wheaton, IL 60189-9850

NASA Tech Briefs
NASA STI Facility
Manager TU Division
PO Box 8757
Baltimore, MD 21240-9985

Nuts and Volts Magazine
430 Princeland Court
Corona, CA 91719
(800) 783-4624

PCIM – Power Conversion and Intelligent Motion
PO Box 420374
Palm Coast, FL 32142-0374

Personal Engineering and Instrumentation News
Circulation Department
PO Box 430
Rye, NH 03870-0430

Power Transmission Design
1100 Superior Avenue
Cleveland, OH 44197-8038

Printed Circuit Design
200 Powers Ferry Center
Suite 450
Marietta, GA 30067
(888) 847-6177

Product Design and Development
PO Box 2001
Radnor, PA 19080-9501

Research and Development Magazine
Reader Service Dept.
Computer Center
PO Box 5833
Denver, CO 80217-9937

Robot Science and Technology
2351 Sunset Blvd. No. 170-235
Rocklin, CA 95765
(888) 510-7728

Robotics Digest
1700 Washington Ave,
Rocky Ford, CO 81067
(719) 254-4558

Security Magazine
Reader Service Department
Computer Center
PO Box 5500
Denver, CO 80217-9808

Sensor Review
MCB University Press
62 Toller Lane, Bradford
BD8 9BY, United Kingdom

Sensor Technology
Technical Insights
PO Box 1304
Fort Lee, NJ 07024-9967

Sensors
Helmers Publishing 174 Concord Street
PO Box 874
Peterborough, NH 03458-0874

Surface Mount Technology
IHS Publishing Group
17730 West Peterson Road
Libertyville, IL 60048-0159
(847) 362-8711

Data Books

Semiconductor companies publish a series of data books that give the specifications and pinouts of their chips. Often, chapters are included in each book that contain application notes and brief reviews of theory. A set of data books for a large semiconductor company might number a dozen or more volumes, while more specialized or newer companies might have only a single data book. Typically, data books will be sent free if you call the literature department of each manufacturer and ask for copies.

The following list comes from the collection we have acquired over the years. Probably the most important reference to have, however (which is not free), is the first item on the list, the *IC Master*. This multivolume set lists all chips made by all manufacturers in the world and has an index by part number. That is, if you come across a chip marked with some part number but you have no idea what its function is, you can look it up in *IC Master* and find out all the companies that make that chip and what it is. Then you can go to the data book for one of the companies for the pinouts and electrical characteristics. There is also an online version of *IC Master*.

IC Master Hearst Business Communications 645 Stewart Avenue Garden City, NY 11530 (516) 227-1300 Fax: (516) 227-1453 www.icmaster.com	Index of all manufacturers' integrated circuits
Advanced Micro Devices 1 AMD Place PO Box 3453 Sunnyvale, CA 94088 (800) 538-8450 www.amd.com	Memories, microprocessors, analog chips
Allegro Microsystems, Inc. 115 Northeast Cutoff Box 15306 Worcester, MA 01615 (508) 255-3476 Fax: (508) 853-7895 www.allegromicro.com	Hall effect sensors and more
Analog Devices One Technology Way PO Box 9106 Norwood, MA 02062-9106 (781) 329-4700 Fax: (7810 326-8703 www.analog.com	D/A and A/D converters, analog electronics
Apex Microtechnology Corp. 5980 N. Shannon Road Tucson, AZ 85741 (520) 690-860 Fax: (520) 888-3329 support.apexmicrotech.com www.apexmicrotech.com	Power op-amps
Benchmarq Microelectronics 17919 Waterview Parkway Dallas, TX 75252 (800) 966-0011 Fax: (9720 437-9198 www.benchmarq.com	Battery-charging ICs

Burr-Brown Corporation PO Box 11400 Tucson, AZ 85734-1400 (520) 746-1111 Fax: (520) 746-7401 www.burr-brown.com	Instrumentation amplifiers, linear circuits
Cherry Semiconductor 2000 South County Trail East Greenwich, RI 02818 (401) 885-3600 Fax: (888) 427-2328 info@cherry-semi.com www.cherry-semi-com	Telecom circuits, motor control, power and automotive ICs
Cypress Semiconductor 3901 North First Street San Jose, CA 95134 (408) 943-2600 www.cypress.com	Memories
Dallas Semiconductor 4401 S. Beltwood Parkway Dallas, TX 75244 (972) 371-4000 Fax: (972) 371-3715 www.dalsemi.com	Nonvolatile RAM, microprocessor and support circuits
Dense-Pac Microsystems 7321 Lincoln Way Garden Grove, CA 92641-1428 (714) 898-0007 Fax: (714) 897-1772 www.dense-pac.com	Memory modules
EEM – Electronic Engineers Master Catalog Hearst Business Communications 645 Stewart Avenue Garden City, NY 11530 (516) 227-1300 Fax: (516) 227-1901 www.hearstelectroweb.com	Suppliers of electronic components
EG&G Reticon 345 Potrero Avenue Sunnyvale, CA 94086 (408) 738-4266 Fax: (408) 738-3832 www.egginc.com/reticon	Image-sensing products

Elantec 675 Trade Zone Boulevard Milpitas, CA 95035 (408) 945-1323 Fax: (408) 945-9305 www.elantec.com	Operational amplifiers
Electronic Designs Inc. 1 Research Drive Westboro, MA 01581 (508) 366-5151 Fax: (508) 836-4850	Hybrid memory modules
Exar Corporation 48720 Kato Road Fremont, CA 94538 (510) 6668-7000 www.exar.com	Telecommunications ICs
Fujitsu Microelectronics, Inc. 3545 North First Street San Jose, CA 95134 (408) 922-9000 www.fujitsu.com	Memories
Gennum Corporation PO Box 489, Stn A Burlington, Ontario, Canada L7R 3Y3 (905) 632-2996 Fax: (905) 632-2055 www.gennum.com	Video and power supply products
Greenwich Instruments USA 11925 Ramah Church Road Huntersville, NC 28078 (800) 476-4070 Fax: (704) 875-2801 www.greenwichinst.com	Nonvolatile memories
Harris Semiconductor 1025 West NASA Blvd. Melbourne, FL 32919-0001 (407) 727-9207 Fax: (407) 727-9344 www.harris.com	Digital and analog ICs, microprocessors

Hewlett Packard
3175 Scott Blvd.
Santa Clara, CA 95054
(408) 654-8675
Fax: (408) 654-8575
www.hp.com

Optoelectronics, microprocessors, radio-frequency ICs

Hitachi America, Ltd.
2000 Sierra Point Pkwy., MS-080
Brisbane, CA 94005-1897
(800) 285-1601
Fax: (303) 297-0447
www.halsp.hitachi.com

Microcontrollers, peripherals, LCDs, memories

Hyundai Electronics America
3101 N. First Street
San Jose, CA 95134
(408) 232-8000
www.hea.com

Memories, serial EEPROMs

IMP Inc.
2830 N. First Street
San Jose, CA 95134
(408) 432-9100
Fax: (408) 434-0335
www.impweb.com

Communications components

Integrated Device Technology
2975 Stender Way
Santa Clara, CA 95054
(408) 727-6116
Fax: (408) 492-8674
info@idt.com
www.idt.com

Semiconductors and related devices

Intel Corporation
21515 Vanowen Street
Suite 116
Canoga Park 91303
(800) 628-8686
support@intel.com

Microprocessors and peripherals

International CMOS Technology, Inc.
2123 Ringwood Avenue
San Jose, CA 95131
(408) 434-0678
Fax: (408) 434-0688
www.ictpld.com

Electronically erasable PROMs and PLDs

International Rectifier 233 Kansas Street El Segundo, CA 9024 (310) 322-3331 Fax: (310) 252-7175 www.irf.com	Power MOSFETs
IXYS Corporation 3540 Bessett Street Santa Clara, CA 95054-2704 (408) 982-0700 Fax: (408) 748-9788 sales@ixys.com www.ixys.com	Stepper motor controllers, power ICs
Lambda Advanced Analog 2270 Martin Avenue Santa Clara, CA 95050-2781 (408) 988-4930 Fax: (408) 988-2702 www.lambdaaa.com	A/D converters, power supplies
Lambda Electronics 515 Broadhollow Road Melville, NY 11747 (800) 526-2324 Fax: (516) 293-0519 www.lambda.com	Power semiconductors
Marktech Optoelectronics 5 Hemlock Street Latham, NY 12110 (800) 984-5337 Fax: (518) 786-6599 info@marktechopto.com www.marktechopto.com	Optoelectronics
Maxim Integrated Products 120 San Gabriel Drive Sunnyvale, CA 94086 (408) 737-7600 x6380 www.maxim-ic.com	Data converters, RS232 chips, video products, amplifiers
Micrel 1849 Fortune Drive San Jose, CA 95131 (408) 944-0800 www.micrel.com	Smart-power ICs

Micro Linear 2092 Concourse Drive San Jose, CA 95131 (408) 433-5200 Fax: (408) 432-1627 info@mlinear.com www.microlinear.com	Data converters, communications, and power ICs
Micro Semiconductor Inc. 1010 N. Shiloh Road Garland, TX 75042 (972) 272-9811 Fax: (972) 487-0406	Rectifiers
Microchip 2355 W. Chandler Blvd. Chandler, AZ 85224-6199 (602) 963-7373 Fax: (602) 899-9211 www.microchip.com	Memories, microcontrollers, and peripherals
Micron Technology, Inc. 800 South Federal Way PO Box 6 Boise, Idaho 83707=0006 (208) 368-4000 Fax: (208) 368-4435 www.micron.com	Memories
MITEL Semiconductors Sequoia Research Park 1500 Green Hills Road Scotts Valley, CA 95066 (408) 438-2900 Fax: (408) 438-6231 www.mitel.com	Digital signal processors
Mosel Vitelic Corp. 3910 N. First Street San Jose, CA 95134-1501 (408) 433-6000 Fax: (408) 433-0331 www.moselvitelic.com	Memories
Motorola Literature Dist. PO Box 20912 Phoenix, AZ 85036 (800) 544-9497 www.mot.com	Digital, analog, and optical ICs.

National Semiconductor 2900 Semiconductor Drive PO Box 58090 Santa Clara, CA 95052-8090 (408) 721-5000 www.nsc.com	Digital and analog ICs
NEC Electronics, Inc. 2880 Scott Blvd. Santa Clara, CA 95052 (408) 588-6000 Fax: (408) 588-6130 www.nec.com	Linear and digital products
Oki Semiconductor 785 North Mary Avenue Sunnyvale, CA 94086-2909 (408) 720-1900	Memories, telecom and networking ICs
Opto-Diode Corporation 750 Mitchell Rd. Newbury Park, CA 91320 (805) 499-0335 Fax: (805) 499-8108 www.optodiode.com	Optoelectronics
Phillips Semiconductors 811 E. Arques Ave. PO Box 3409 Sunnyvale, CA 94088-3409 (408) 991-2000 www.semiconductors.phillips.com	Digital, linear Ics, microprocessors, peripherals
Power Trends 27715 Diehl Road Warrenville, IL 60555 (800) 531-5782 Fax: (630) 393-6902 marketing@powertrends.com www.powertrends.com	Switching regulators for voltage regulators
Powerex 200 Hillis Street Youngwood, PA 15697-1800 (724) 925-7272	Power semiconductors

Qualcomm
6455 Lusk Blvd.
San Diego, CA 92121-2779
(619) 587-1121
Fax: (619) 658-2100
www.qualcomm.com

Digital frequency synthesizers, signal-processing ICs

Reliability Inc.
PO Box 218370
Houston, TX 772118
(281) 492-0550
d denning@relinc.com

DC-DC converters

Rockwell Semiconductor Systems
9868 Scranton Road
San Diego, CA 92121
(619) 452-7580
Fax: (619) 452-7294
www.rockwell.com

Video and graphics chips

Samsung
3655 N. First Street
San Jose, CA 95134-1708
(408) 544-4000
Fax: (408) 544-4980
www.samsungsemi.com

Digital, analog, optical electronics

Seeq Technology Inc.
47200 Bayside Pkwy.
Fremont, CA 94538
(800) 333-7766
Fax: (510) 657-2837
www.seeq.com

EEPROMs

SenSym
1804 McCarthy Blvd.
Milpitas, CA 95035
(408) 954-1100
Fax: (408) 954-9458
www.sensym.com

Pressure sensors and accelerometers

Sharp Electronics Corp.
Sharp Plaza
Mahwah, NJ 07430-2135
(201) 529-8200
Fax: (201) 529-8425
www.sharpelectronics.com

Photosensors

Siemens Components
19000 Homestead Road
Cupertino, CA 95014
(408) 725-3586
www.smi.siemens.com

Digital, linear electronics, microprocessors, optoelectronics

Siliconix
2201 Laurelwood Road
Santa Clara, CA 95054-1516
(408) 988-8000
Fax: (408) 567-8979
www.siliconix.com

Power MOSFETs, data converters

STMicroelectronics
10 Maguire Road
Building 1, 3rd Floor
Lincoln, MA 02421
(781) 861-2650
Fax: (781) 861-2664
www.st.com

Digital signal processing chips; analog, digital motion control ICs

Stanford Telecom
1221 Crossman Avenue
PO Box 3733
Sunnyvale, CA 94088
(408) 745-0818
Fax: (408) 745-7756
www.stelhg.com

Digital frequency synthesizers

Supertex, Inc.
1235 Bordeaux Drive
Sunnyvale, CA 94089
(408) 744-0100
Fax: (408) 222-4800
www.supertex.com

Power MOSFETs, high-voltage ICs

Telcom Semiconductor
1300 Terra Bella Avenue
PO Box 7267
Mountain View, CA 94039-7267
(650) 968-9241
Fax: (650) 947-1590
www.telcom-semi.com

Data-acquisition ICs

Telephonics Large Scale Integration, Inc.
770 Park Avenue
Huntington, NY 11743
(516) 755-7610
Fax: (516) 755-7626
www.tlsi.com

Strain gauge conditioner ICs

Texas Instruments
Literature Response Center
PO Box 172228
Denver, CO 80217
(800) 477-8924
Fax: (303) 297-0447
www.ti.com

Memories, microprocessors, analog, digital, optoelectronics

Toshiba America, Inc.
9740 Irvine Blvd.
Irvine, CA 92618
(949) 455-2000
Fax: (949) 859-3963
www.toshiba.com

CCD imagers, LCD displays, optoelectronics, memories

Unitrode Integrated Circuits
7 Continental Blvd.
Merrimack, NH 03054
(603) 424-2410
Fax: (603) 429-8771
macdonald@unitrode.com
www.unitrode.com

Power-management ICs

Vishay Sprague
70 Pembroke Road
Concord, NH 03301
(603) 224-1961
Fax: (603) 224-1339
www.vishay.com

Discrete actives

VLSI Technology, Inc.
1109 McKay Drive
San Jose, CA 95131
(408) 434-3000
Fax: (408) 922-5252
www.vlsi.com

RISC microprocessors and peripherals

White Microelectronics 3601 E. university Drive Phoenix, AZ 85034 (602) 437-1520 Fax: (602) 437-9120 www.whitemicro.com	Memories
WSI, Inc. 47280 Kato Road Fremont, CA 94538-7333 (800) 832-6974 Fax: (510) 657-8495 www.wsipsd.com	Programmable system devices
Xicor, Inc. 1511 Buckeye Court Milpitas, CA 95035 (408) 432-8888 Fax: (408) 432-0640 www.xicor.com	Memories, EEPROMs, digital potentiometers
Xilinx, Inc. 2100 Logic Drive San Jose, CA 95124-9920 (408) 559-7778 Fax: (408) 559-7114 www.xilinx.com	Programmable gate arrays
Zilog 910 E. Hamilton Avenue Campbell, CA 95008 (408) 558-8500 Fax: (408) 558-8300 www.zilog.com	Microprocessors and peripherals

F

Robot Contests

This appendix contains a sample of robot contests and information on how to find out more about them. They are presented here to give you an idea of what kind of robotics contests might be available. Many thanks to Steve Rainwater (email: srainwater@ncc.com) who maintains a robot contests and competitions list that is on the Web at www.ncc.com/misc/rcfaq.html. Steve's site is copyrighted ©1997, Steve Rainwater/Network Cybernetics Corp. with all rights reserved. Permission has been received for using his site to gain much of this information.

Another web site to find infomation about robot contests is: www.frc.ri.cmu.edu/robotics-faq/5.2html. Due to the rapid changes on the Internet, the authors suggest you use a search engine and type in "Robotics Contests" to get the latest information.

AAAI Robot Competitions
www.aaai.org

> The American Association of Artificial Intelligence has an annual robotics competition. Rules and locations vary from year to year.

All Japan MicroMouse Contest
www.bekkoame.or.jp/~ntf/mouse/mouse-e.html

Micromouse Contest is a contest in which contestants enter their robots to compete for intelligence and speed while the robots negotiate a specified maze. A robot participating in this contest is termed a micromouse. (email: KYD02036@niftyserve.or.jp)

ANS Remote Material Handling Robot Competition
www.ri.cmu.edu/ans99/

This American Nuclear Society competition has been designed to emulate a teleoperated robotic mission. Both navigation and manipulation tasks will have to be accomplished from outside the "hot room" with no direct line of site to the mobile platform.

AUVS International Aerial Robotics Competition
avdil.gtri.gatech.edu/AUVS/index.html

The Association of Unmanned Vehicle Systems International Aerial Robotics Competition usually has prize money for the "winner."

AUVS Ground Robotics Competition
avdil.gtri.gatech.edu/AUVS/index.html

The objective of this competition is to build a completely autonomous vehicle capable of navigating itself around a grass track outlined with white lines. There are also several obstacles on the track which the robot must avoid.

BEAM Robot Olympics
sst.lanl.gov/robot/

BEAM (Biology, Electronics, Art and Mechanics) organizer Mark Tilden (mwtilden@math.uwaterloo.ca) advocates using the parts from discarded electronics items such as printers, disk drives, radios, etc., to make machines that move. He avoids the use of computers and microcontrollers in his machines. Many BEAM contests are held throughout the world.

Canada FIRST
www.canadafirst.org/

Canada FIRST is patterned after the US FIRST competition (now just FIRST). In CANADA FIRST the students build the robots while the engineers advise. (email: canfirst@inforamp.net; Mailing address: CANADA FIRST Robotics Games, 3 Rice Drive, Suite 100, Whitby, Ontario, L1N 7X1 CANADA)

FIRST Competition
www.usfirst.org/

Given the standard FIRST kit of parts, a set of rules, and a brief robot-building workshop led by Woodie Flowers of MIT, corporate-sponsored teams of high school students work with professional engineers to design and build robots for the contest. (Mailing address: Attn: Susan Howland FIRST 340 Commercial Street Manchester, NH 03101; tel 603-666-3906; fax 603-666-3907)

International Fire-Fighting Home Robot Contest
shakti.trincoll.edu/~jhough/fire_robot/comp.html

This annual contest at Trinity College organized by Jake Mendelssohn has many classes of robots trying to put out a fire (candle) in a house-like maze. (email: JMENDEL141@AOL.COM)

Robot Symposium and Navigation Contest

The Robotics Society of Southern California holds an annual Robot Symposium and Navigation Contest. Detailed rules will provided on request. (email: pir2@aol.com ; Mailing address: Jerry Burton 10471 S. Brookhurst St., Anaheim, CA 92804; tel: (714) 535-8161)

Sumo Robot Competition
www.ncc.com/misc/rcfaq.html

There are Robot Sumo Tournaments throughout the United States and other countries.

Color and ASCII Codes

Resistor Color Code

The value of a resistor may be determined from its color bands. For example, if the bands on the above resistor, running from left to right, are red, yellow, and orange then the resistance would be: $24 \times 10^3 = 24\text{K}$ ohms.

The tolerance band tells how closely the resistance of a given resistor will match its color code. A silver band indicates that the actual resistance will be within 10% of the marked value; a gold band means 5%.

The value of a small capacitor is sometimes indicated by a three-number code stamped on the body of the device. To get the capacitance in pico-

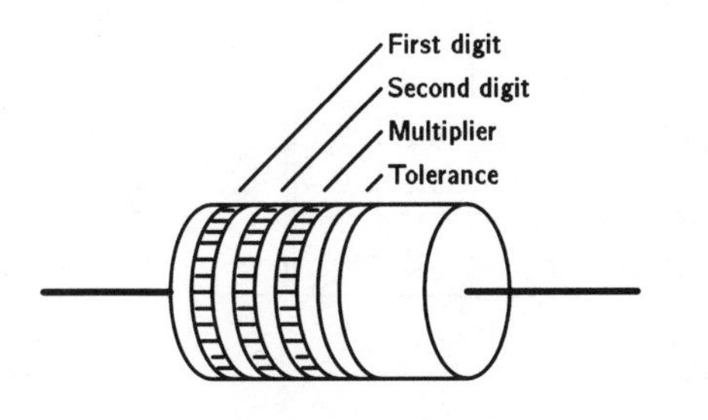

Color	Value	Color	Value
Black	0	Green	5
Brown	1	Blue	6
Red	2	Violet	7
Orange	3	Gray	8
Yellow	4	White	9

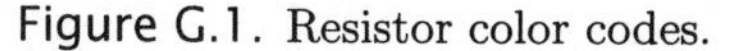

Figure G.1. Resistor color codes.

farads, multiply the first two digits by 10 to the power specified by the third digit. For example, the code 124 would indicate a value of 12×10^4 picofarads, or 0.12 microfarads.

ASCII Code

Dec	Hex	Char	Dec	Hex	Char	Dec	Hex	Char	Dec	Hex	Char
00	00	NUL	32	20		64	40	@	96	60	‘
01	01	SOH	33	21	!	65	41	A	97	61	a
02	02	STX	34	22	“	66	42	B	98	62	b
03	03	ETX	35	23	#	67	43	C	99	63	c
04	04	EOT	36	24	$	68	44	D	100	64	d
05	05	ENQ	37	25	%	69	45	E	101	65	e
06	06	ACK	38	26	&	70	46	F	102	66	f
07	07	BEL	39	27	’	71	47	G	103	67	g
08	08	BS	40	28	(	72	48	H	104	68	h
09	09	HT	41	29	)	73	49	I	105	69	i
10	0A	LF	42	2A	*	74	4A	J	106	6A	j
11	0B	VT	43	2B	+	75	4B	K	107	6B	k
12	0C	FF	44	2C	,	76	4C	L	108	6C	l
13	0D	CR	45	2D	-	77	4D	M	109	6D	m
14	0E	SO	46	2E	.	78	4E	N	110	6E	n
15	0F	SI	47	2F	/	79	4F	O	111	6F	o
16	10	DLE	48	30	0	80	50	P	112	70	p
17	11	DC1	49	31	1	81	51	Q	113	71	q
18	12	DC2	50	32	2	82	52	R	114	72	r
19	13	DC3	51	33	3	83	53	S	115	73	s
20	14	DC4	52	34	4	84	54	T	116	74	t
21	15	NAK	53	35	5	85	55	U	117	75	u
22	16	SYN	54	36	6	86	56	V	118	76	v
23	17	ETB	55	37	7	87	57	W	119	77	w
24	18	CAN	56	38	8	88	58	X	120	78	x
25	19	EM	57	39	9	89	59	Y	121	79	y
26	1A	SUB	58	3A	:	90	5A	Z	122	7A	z
27	1B	ESC	59	3B	;	91	5B	[	123	7B	{
28	1C	FS	60	3C	<	92	5C	\	124	7C	|
29	1D	GS	61	3D	=	93	5D	]	125	7D	}
30	1E	RS	62	3E	>	94	5E	ˆ	126	7E	˜
31	1F	US	63	3F	?	95	5F	_	127	7F	DEL

Bibliography

Some of the references listed here may be hard to find for the general reader. A number of the papers listed came out of work at the Mobile Robot Group at the MIT Artificial Intelligence Laboratory. While often published in journals or conference proceedings, these papers are usually also published internally as *AI Laboratory Memos. AI Memos* can be acquired for a small copying fee by writing or calling the MIT Artificial Intelligence Laboratory Publications Office:

Publications Office
MIT AI Lab, Room 818
545 Technology Square
Cambridge, MA 02139
(617) 253-6773

While journal articles can often be found in a university library, conference proceedings can be difficult to locate. If the journal or conference papers were published by the Institute of Electrical and Electronics Engineers, they can be ordered directly from the IEEE:

IEEE Publishing Services
345 47th St.
New York, NY 10017
(212) 705-7900

Many authors now maintain their papers on line in downloadable form. Searching the web for the author or title is often the fastest way to find the paper you seek.

(AAAI Proceedings) MIT Press, Cambridge, MA.

(Angle and Brooks 90) Colin M. Angle and Rodney A. Brooks. Small Planetary Rovers. *Proceedings of the IEEE International Workshop on Intelligent Robots and Systems.* Tokyo, Japan, July.

(Angle 89) Colin M. Angle. Genghis, A Six Legged Autonomous Walking Robot. *S.B. Thesis, MIT Dept. of Electrical Engineering and Computer Science.* March.

(Angle 91) Colin M. Angle. Design of an Artificial Creature. *Master's Thesis, MIT Electrical Engineering and Computer Science Department.* June.

(Arkin 98) Ronald C. Arkin. *Behavior-Based Robotics.* Bradford Books. 1998.

(Artificial Life) Addison-Wesley Publishing Co, Redwood City, CA.

(Beckwith and Marangoni) Thomas G. Beckwith and Roy D. Marangoni. *Mechanical Measurements.* Addison-Wesley Publishing Co., MA. Reading, MA, 1990.

(Borenstein, Everett, Feng 96) Johann Borenstein, H.R. Everett, Liqiang Feng. *Navigating Mobile Robots* A K Peters, Ltd., Natick, MA. 1996.

(Braitenberg) Valentino Braitenberg. *Vehicles: Experiments in Synthetic Psychology.* MIT Press. Cambridge, MA, 1984.

(Brock 91) David L. Brock. Review of Artificial Muscle Based on Contractile Polymers. *MIT AI Lab Memo 1330.* November.

(Brooks 86) Rodney A. Brooks. A Robust Layered Control System for a Mobile Robot. *IEEE Journal of Robotics and Automation.* RA-2, 14-23 April, also appears as *MIT AI Memo 864*, September, 1985.

(Brooks and Ferrell 98) Rodney A. Brooks and Cynthia Ferrell. *Embodied Intelligence*, MIT Press, Cambridge, MA (in preparation).

(Brooks and Flynn 89) Rodney A. Brooks and Anita M. Flynn. Fast, Cheap and Out of Control: A Robot Invasion of the Solar System. *Journal of the British Interplanetary Society.* Vol. 42, pp. 478-485, also appears as *MIT AI Memo 1182*, December, 1989.

(Brooks 89) Rodney A. Brooks. A Robot that Walks; Emergent Behavior from a Carefully Evolved Network. *Neural Computation 1:2.* pp. 253–262, also appears as *MIT AI Memo 1091*, February, 1989.

(Brooks 91a) Rodney A. Brooks. New Approaches to Robotics. *Science.* Vol. 253, pp. 1227–1232, September 13.

(Brooks 91b) Rodney A. Brooks. Intelligence Without Reason. Prepared for *Computers and Thought, IJCAI-91, MIT AI Laboratory Memo 1293.* April.

(Connell 88) Jonathan H. Connell. The Omni Photovore: How to Build a Robot that Thinks like a Roach. *Omni Magazine.* October.

(Connell) Jonathan H. Connell. *Minimalist Mobile Robotics: A Colony-Style Architecture for an Artificial Creature.* Academic Press. Boston, MA, 1990.

(Connell 91) Jonathan H. Connell. Design Your Own Robot. *Popular Electronics.* August.

(Everett 95) H. R. Everett, *Sensors for Mobile Robots.* A K Peters, Ltd., Natick, MA. 1995.

(Everett, Gilbreath and Tran 90) H.R. Everett, G.A. Gilbreath and T. Tran. Modeling the Environment of a Mobile Security Robot. *Technical Document 1835, Naval Command Control and Ocean Surveillance Center, San Diego, CA, 92152-5000.* June.

(Everett and Stitz 92) H.R. Everett and E.H. Stitz. Survey of Collision Avoidance and Ranging Sensors for Mobile Robots. *Technical Report 1194, Update 1, Naval Command Control and Ocean Surveillance Center, San Diego, CA, 92152-5000.* December.

(Ferrell 92) Cynthia Ferrell. Multiple Sensors, Virtual Sensors and Robustness. *Sensors Expo.* Chicago, IL, September 29–October 1.

(Fitzgerald, Kingsley and Umans) A.E. Fitzgerald, Charles Kingsley and Stephen D. Umans. *Electric Machinery.* McGraw-Hill. New York, NY, 1990.

(Flynn 87) Anita M. Flynn. Gnat Robots (and How They Will Change Robotics). *Proceedings of the IEEE Micro Robots and Teleoperators Workshop.* Hyannis, MA, November. Also appears in *AI Expert*, December 1987.

(Flynn et al. 88) Edited by Anita Flynn, with contributions from Colin Angle, Rodney Brooks, Jon Connell, Anita Flynn, Ian Horswill, Maja Mataric, Henry Minsky, Peter Ning, Paul Viola and William Wells. The Olympic Robot Building Manual. *MIT AI Lab Memo 1230.* December.

(Flynn 89) Anita M. Flynn. The Official Photograph Album of the 1989 Robot Olympics. *MIT AI Lab Manual.* April.

(Flynn, Brooks and Tavrow 89) Anita M. Flynn, Rodney A. Brooks and Lee S. Tavrow. Twilight Zones and Cornerstones: A Gnat Robot Double Feature. *MIT AI Memo 1126.* July.

(Flynn, Brooks, Wells and Barrett 89) Anita M. Flynn, Rodney A. Brooks, William M. Wells III and David S. Barrett. Intelligence for Miniature Robots. *Journal of Sensors and Actuators.* Vol. 20, pp. 187-196, also appears as *Squirt: The Prototypical Mobile Robot for Autonomous Graduate Students*, MIT AI Memo 1120, July, 1989.

(Flynn, et al. 92) Anita M. Flynn, Lee S. Tavrow, Stephen F. Bart, Rodney A. Brooks, Daniel J. Ehrlich, K.R. Udayakumar and L. Eric Cross. Piezoelectric Micromotors for Microrobots. *IEEE Journal of Microelectromechanical Systems,* Vol. 1, No. 1, pp. 44–51. March, also appears as *MIT AI Memo 1269*, February, 1991.

(Foster) Caxton C. Foster. *Real Time Programming - Neglected Topics.* Addison-Wesley. Reading, MA, 1982.

(Gat, et al. 94) Erann Gat, Rajiv Desai, Robert Ivlev, John Loch, and David Miller. Behavior Control for Robotic Exploration of Planetary Surfaces. *IEEE Transactions on Robotics and Automation.* Vol. 10. No. 4, August 1994.

(Grant and Gowar) Duncan A. Grant and John Gowar. *Power MOSFETs.* John Wiley & Sons. New York, NY, 1989.

(Hayes and Horowitz) Thomas C. Hayes and Paul Horowitz. *The Student Manual for the Art of Electronics.* Cambridge University Press. Cambridge, UK, 1989.

(Hennesy and Patterson 96) John L. Hennessy and David A Patterson. *Computer architecture: A Quantitative Approach: Second Edition.* Morgan Kaufman Publishers, Inc. San Francisco, CA 1996.

(Hollerbach, Hunter and Ballantyne) John M. Hollerbach, Ian W. Hunter and John Ballantyne. *A Comparative Analysis of Actuator Technologies for Robotics.* In Robotics Review 2, MIT Press. Edited by Khatib, Craig and Lozano-Pérez, 1991.

(Horn) Berthold K.P. Horn. *Robot Vision.* MIT Press. Cambridge, MA, 1986.

(Horowitz and Hill) Paul Horowitz and Winfeld Hill. *The Art of Electronics.* Cambridge University Press. Cambridge, UK, 1989.

(Hosoe 89) Kazuya Hosoe. An Ultrasonic Motor for Use in Autofocus Lens Assemblies. *Techno.* pp. 36-41, in Japanese.

(Howe, Muller, Gabriel and Trimmer 90) Roger T. Howe, Richard S. Muller, Kaigham J. Gabriel and William S. N. Trimmer. Silicon Micromechanics: Sensors and Actuators on a Chip. *IEEE Spectrum.* July 19.

(Inaba et al. 87) R. Inaba, A. Tokushima, O. Kawasaki, Y. Ise and H. Yoneno. Piezoelectric Ultrasonic Motor. *Proceedings of the IEEE Ultrasonics Symposium.* pp. 747-756.

(IRAS) *IEEE Robotics and Automation Proceedings*, IEEE Computer Society Press, Los Alamitos, CA.

(Jung) Walter G. Jung. *IC Op-Amp Cookbook.* Howard W. Sams & Company. Indianapolis, IN, 1986.

(Kassakian, Schlect and Verghese) John C. Kassakian, Martin F. Schlect and George C. Verghese. *Principles of Power Electronics.* Addison-Wesley. Reading, MA, 1991.

(Kasuga et al. 92) Masao Kasuga, Takashi Satoh, Jun Hirotomi and Masayuki Kawata. Development of Ultrasonic Motor and Application to Silent Alarm Analog Quartz Watch. *4th Congres Europeen de Chronometrie.* Lausanne, Switzerland, 29–30 October, pp. 53–56.

(Kenjo) Takashi Kenjo. *Power Electronics for the Microprocessor Age.* Oxford University Press. New York, NY, 1990.

(Kleinschmidt) Kirk A. Kleinschmidt, Editor. *The ARRL Handbook for the Radio Amateur.* American Radio Relay League. Newington, CT, 1990.

(Lancaster) Don Lancaster. *CMOS Cookbook.* Howard W. Sams & Company. Indianapolis, IN, 1977.

(Latombe) J. C. Latombe. *Robot Motion Planning.* Kluwer Academic Press. Norwell, MA, 1991.

(Lozano-Pérez, Jones, Mazer and O'Donnell) Tomás Lozano-Pérez, Joseph L. Jones, Emmanuel Mazer and Patrick A. O'Donnell. *Handey - A Robot Task Planner.* MIT Press. Cambridge, MA, 1992.

(Maes and Brooks 90) Pattie Maes and Rodney A. Brooks. Learning to Coordinate Behaviors. *AAAI-90.* August.

(Maes) Pattie Maes. *Designing Autonomous Agents: Theory and Practice from Biology to Engineering and Back.* MIT Press. Cambridge, MA, 1991.

(Martin) Fred Martin. *The 6.270 Robot Builder's Guide.* MIT Media Laboratory. Cambridge, MA, 1992.

(Martin 1998) Fred G. Martin, *The Art of Robotics: A Hands-On Introduction to Engineering.* Addison-Wesley. 1998.

(McClelland and Rumelhart) James L. McClelland and David E. Rumelhart. *Parallel Distributed Processing, Vols. I and II.* MIT Press. Cambridge, MA, 1986.

(MEMS) *Proceedings of the IEEE Micro Electro Mechanical Systems Workshops,* IEEE, 47th St., New York, NY.

(Miller, Desai, Gat, Ivlev and Loch 92) D.P.Miller, R.S.Desai, E.Gat, R. Ivlev and J.Loch. Reactive Navigation through Rough Terrain: Experimental Results. *Proceedings of the 1992 AAAI Conference.* pp. 823-828, San Jose CA.

(Miller, Winkless, Bosworth 98) Merl Miller, Helson Winkless, Joe Bosworth. *Personal Robot Navigator* PRT Press, Conifer, CO. 1998.

(Minsky) Marvin Minsky. *The Society of Mind.* Simon and Schuster. New York, NY, 1986.

(Mondo-tronics 91) Mondo-tronics Inc. Biometal Robot DH-101. *1014 Morse Avenue, Suite 11.* Sunnyvale, CA 94089.

(Moravec) Hans P. Moravec. *Robot Rover Visual Navigation.* UMI Research Press. Ann Arbor, MI, 1981.

(Moroney, White and Howe 89) R.M. Moroney, R.M. White and R.T. Howe. Ultrasonic Micromotors. *IEEE Ultrasonics Symposium, Montreal, Canada.* October.

(Moroney, White and Howe 90) R.M. Moroney, R.M. White and R.T. Howe. Fluid Motion Produced By Ultrasonic Lamb Waves. *IEEE Ultrasonics Symposium.* Honolulu, Hawaii, Dec. 4–7.

(Motorola 88) Volumes I and II. *Microprocessor, Microcontroller and Peripheral Data.* Motorola Inc., Microprocessor Product Group, Microcontroller Division. Oak Hill, Texas 78735, 1988.

(Motorola 91) Motorola Inc., Microprocessor Product Group. *Motorola M68HC11 Reference Manual.* Microcontroller Division. Oak Hill, Texas 78735, 1991.

(Nilsson 84) Nils Nilsson. Shakey the Robot. *Artificial Intelligence Center, SRI International Technical Note 323.* Menlo Park, CA, April.

(Ohnishi, Myohga, Uchikawa, Inoue, Takahashi and Tomikawa 89) Osamu Ohnishi, Osamu Myohga, Tadao Uchikawa, Takeshi Inoue, Sadayuki Takahashi and Yoshiro Tomikawa. Paper Transport Device Using a Flat Plate Piezoelectric Vibrator. *Japanese Journal of Applied Physics.* Vol. 28., Suppl. 28-1, pp. 167-169.

(Pratt 92) Gill Andrews Pratt. EVs: On the Road Again. *Technology Review.* pp. 51–59, August/September.

(Ragulskis, Bansevicius, Barauskas and Kulvietis) K. Ragulskis, R. Bansevicius, R. Barauskas and G. Kulvietis. *Vibromotors for Precision Microrobots.* Hemisphere Publishing Co. New York, 1988.

(Riezenman 92) Michael J. Riezenman. Electric Vehicles. *Spectrum.* Vol. 29, No. 11, pp. 18–21.

(SAB) From Animals to Animats. International Conference on Simulation of Adaptive Behavior. *MIT Press.* Cambridge, MA.

(Sashida) T. Sashida. *Trial Construction and Operation of an Ultrasonic Vibration Drive Motor.* Oyo Butsuri. Vol. 51, No. 6, pp. 713-720, in Japanese, 1982.

(Seippel) Robert G. Seippel. *Transducers, Sensors, & Detectors.* Reston Publishing Company, Inc.. Reston, VA, 1983.

(Tanaka 81) T. Tanaka. Gels. *Scientific American,* pp. 124–138. January.

(Udaykumar, Chen, Brooks, Cross, Flynn and Ehrlich 91) K.R. Udayakumar, J. Chen, K.G. Brooks, L.E. Cross, A.M. Flynn and D.J. Ehrlich. Piezoelectric Thin Film Ultrasonic Micromotors. *1991 MRS Fall Symposium on Ferroelectric Thin Films.* Boston, MA, Dec. 1–4.

(Walter 50) W. Grey Walter. An Imitation of Life. *Scientific American.* 182(5), pp. 42–45, May.

(Walter 51) W. Grey Walter. A Machine That Learns. *Scientific American.* 185(2), pp. 60–63, August.

(Ward and Halstead) Stephen A. Ward and Robert H. Halstead, Jr. *Computation Structures.* MIT Press. Cambridge, MA, 1990.

(Wehner 87) Rüdiger Wehner. 'Matched Filters' - Neural Models of the External World. *J. Comp. Physiol. A 161.* pp. 511–531.

(Weiner 48) Norbert Weiner. *Cybernetics.* John Wiley and Sons. New York, NY, 1948.

(Weiner 61) Norbert Weiner. *Cybernetics.* Second Edition, MIT Press. New York, NY, 1961.

(Woodson and Melcher) Herbert H. Woodson and James R. Melcher. *Electromechanical Dynamics, Part II Fields, Forces and Motion.* Krieger Publishing Co. Malabar, FL, 1985.

(Zaks) Rodney A. Zaks. *From Chips to Systems: An Introduction to Microprocessors.* Sybex, Inc. Berkeley, CA, 1986.

Index